THE
WORLD MODEL

GOVERNED AI
FOR
HYPER-PERSONALIZED VENUES

THE
WORLD MODEL

GOVERNED AI
FOR
HYPER-PERSONALIZED VENUES

.

A Practitioner's Guide

Maris J. Ensing

ABBREVIATED CONTENTS

For a more detailed TOC, please refer to Appendix L

NOTICES AND DISCLAIMERS

Not legal advice.
This book is provided for general informational purposes and reflects engineering judgment. It is not legal advice, regulatory advice, medical advice, or a substitute for consultation with qualified professionals. Laws, regulations, standards, and enforcement practices vary by jurisdiction, venue type, and use case. You are responsible for determining and meeting all applicable requirements.

Privacy, identity, and consent.
This book discusses analytics, recognition, identity, and personalization. Any implementation involving personal data, biometrics, location data, recordings, or profiling should be designed for consent, minimization, proportionality, security, retention limits, and auditability, and should be reviewed by qualified legal, privacy, and security professionals before deployment.

Safety and engineering.
Descriptions of architectures, power practices, networking practices, and operational procedures are conceptual and may not be sufficient for a specific installation. Use qualified, licensed professionals for design, installation, testing, commissioning, and ongoing maintenance. Follow all applicable codes, manufacturer requirements, and safety standards.

Trademarks and terminology.
This book is written as a vendor-neutral operational reference for governed hyper-personalization in physical venues. Examples are expressed in capability-first language and can be implemented with multiple architectures, vendors, and system designs. Where the book uses named patterns or layers, they are used to make requirements, tradeoffs, governance constraints, and acceptance criteria unambiguous, not to imply a preferred product or supplier.

Versioning.

Some referenced technologies, standards, and policies evolve. Where practical, stable concepts are placed in the main chapters, and more time-sensitive material is placed in appendices or companion materials.

Terminology note.

This book uses descriptive layer names: Portable Identity, Consented Context, Agent Coordination, and AI Negotiation as primary terminology. The labels Web3, Web4, Web5, and Web6 appear parenthetically because they are already in use in industry RFPs and vendor discussions. These labels do not claim to represent an official standards progression or to predict formal Web versioning. Where these terms appear in RFPs or industry discourse, this book maps them to the patterns described here.

Where relevant, this book anchors identity and credential concepts to established specifications, including W3C Decentralized Identifiers (DIDs) and the W3C Verifiable Credentials Data Model, and to established digital identity guidance, including NIST SP 800-63.

Confabulation terminology.

Generative systems can produce fluent, plausible statements that are not grounded in verified sources. In industry this is commonly called hallucination. In this book, the formal term is confabulation (hallucination), and it is treated as a system-level risk that must be contained through constraints, evidence requirements, verification, and safe failure modes[i]. After first mention in a chapter, this book uses confabulation as the primary term[ii].

Standards, frameworks, and precedent informing this work.

This book is written as an operational reference for public venues. Where it relies on external standards or public frameworks, it anchors factual claims to primary sources. For generative AI risks, this book aligns terminology to NIST AI 600-1 and uses confabulation as the formal label. The terms used in this book are defined within the book, but the assurance posture aligns with established practice in accessibility, privacy, identity, and AI risk management.

Core standards and frameworks referenced include[iii]:

1. AI risk management and trustworthy AI: NIST AI RMF 1.0; NIST AI 600-1 (Generative AI Profile).
2. AI management systems: ISO/IEC 42001.
3. Security and abuse-resistance for AI systems: OWASP Top 10 for Large Language Model Applications.
4. Values-based AI principles: OECD Recommendation on AI.
5. Accessibility and inclusive design: WCAG 2.2; relevant ADA/Section 508 requirements where applicable.
6. Decentralized identity and minimal disclosure: W3C DID Core; W3C Verifiable Credentials Data Model.
7. Digital identity assurance: NIST SP 800-63.
8. Biometrics performance and demographic risk: NIST FRVT demographic analysis.
9. Hearing accessibility context: FDA OTC hearing aid category (effective October 17, 2022).

These references are used as anchors for terminology, risk framing, and assurance posture. This book does not claim that implementing any single pattern guarantees compliance in any jurisdiction. Where this book makes factual assertions about standards, regulations, or published research, citations are provided as endnotes. Where it proposes architectural requirements, it does so as engineering judgment expressed in verifiable requirements and acceptance criteria.

Capability-to-implementation mapping.
This book is written in capability-first terms to remain vendor-neutral. A separate capability-to-implementation mapping pack may be provided by the author for readers who need procurement-ready requirement translation, example architectures, and verification criteria. The mapping pack is not part of this publication and may change over time.

Intellectual property notice.

Some techniques and system behaviors discussed in this book may be subject to intellectual property rights held by one or more parties. This book is a public-safe architectural reference and does not grant any license to practice any patented method. Implementers should conduct appropriate diligence and obtain any required permissions or licenses before deployment. For permissions, diligence, or deeper implementation discussion under NDA, contact the author.

FRONT MATTER

The WorldModel™ Manifesto:

Ten Laws for the Living Venue

1. **The venue must work beautifully without requiring opt-in.**
   Personalization is offered, not extracted. A visitor who declines should still experience clarity, dignity, and full access. The default state is hospitality.

   Evidence: Core experience available without identity, clear opt-in paths, and acceptance tests showing full functionality in "no-identity" mode.

2. **Identity is earned, not demanded.**
   Real-world identity should be requested only when it provides a clear visitor benefit that cannot be achieved otherwise. Anonymous session continuity and preference persistence can often achieve the goal.

   Evidence: Documented identity decision rules, explicit consent flows, and logs proving identity use is limited to defined benefit cases.

3. **Collect only what is necessary, retain only what is justified, delete by design**
   Data minimization is not a compliance afterthought, but a design principle. The venue should explain, in plain language, why each data category is collected, how long it is retained, and what triggers deletion.

   Evidence: Data inventory with purpose and retention per category, automated deletion triggers, and auditable access, storage, and retrieval logs.

4. **Recognize the place first, the person second.**
   Most operational improvements stem from understanding the environment, occupancy, flow, congestion, acoustics, and dwell, not from identifying individuals. Place analytics should be the default; person analytics the exception.

   Evidence: A reconciled venue state model with timestamped updates, decision logs linking actions to state, and load tests proving state consistency under crowd conditions.

5. **Inclusion is the starting point, not the retrofit.**
   Multilingual support, accessibility modes, and comfort options belong in the
   initial architecture. At scale, language also becomes a governance artifact.

   A venue should maintain an approved terminology set for each supported
   language, including proper nouns, exhibit-specific terms, and any safety-
   adjacent or accessibility-adjacent phrases. The goal is consistency and
   dignity, not bureaucratic control. If a term is important enough to appear
   repeatedly, it is important enough to be standardized.

   This is how you prevent subtle drift, inconsistent translations across
   surfaces, and confabulation-driven errors that sound fluent but misrepresent
   the institution's meaning.

   Language dignity means no hierarchy of languages. Language dignity also
   means epistemic dignity: do not present confident guesses as facts.

   In multilingual systems, confabulation is often experienced as "translation
   that sounds right but is wrong." It can distort meaning subtly, especially
   where technical terms, names, dates, and proper nouns are involved.

   The correct posture is mode-aware delivery: for factual claims and
   interpretive assertions that matter, the system must remain inside approved
   sources, be able to point to them when required, and refuse to invent when
   grounding is insufficient. Multilingual at scale must not become
   misinformation at scale. Accessibility means participation is normal.

   Evidence: On-demand language rendering across surfaces and personal
   channels, QA evidence per language package, offline fallback behavior,
   defined accessibility and comfort modes, and measured intelligibility targets.

6. **Personalization must never become manipulation.**
   A helpful system offers choices. A hungry system steers behavior without disclosure. The visitor must always understand what the system is doing, and be able to decline without penalty.

   Evidence: Transparent explanations at decision points, a documented "no penalty" policy, and periodic audits for steering patterns, dark patterns, and unequal outcomes.

7. **Commands are not outcomes. Verify what actually happened.**
   A venue that sends commands and assumes success will eventually fail in public. Operational truth requires verification: confirming that displays show content, audio is present, and drift is detected before failure occurs.

   Evidence: Outcome verification checks with logged proofs, automated failure detection, drift instrumentation with thresholds, monitoring routed to tickets, and drift-to-incident reviews.

8. **Latency is part of hospitality, because timing is emotion.**
   If audio, visuals, translation, and interaction cues fall out of sync, the magic collapses. A living venue must treat timing as a first-class design constraint, including in degraded modes.

   Evidence: Instrumented end-to-end latency budgets across modalities, sync tolerances, and acceptance tests that pass in normal and degraded operation.

9. **The architecture must welcome change.**
   A venue system that requires demolition to upgrade will stagnate. Modular
   compute, clean interfaces, content versioning, and provisioning strategies
   keep a venue modern without constant rebuilding.

   Evidence: Node-based compute baseline using commercially available
   hardware, documented interfaces, substitution plans, and restore procedures
   proving nodes can be added, replaced, and recovered without rewiring or
   reprogramming.

10. **Governance is operational capability, not a policy document.**
    Decisions pass through explicit constraints, safety, accessibility, privacy, and
    fairness, before execution. Governance produces three outcomes: approval,
    modification, or rejection. Values that cannot be enforced at runtime are
    aspirations. Confabulation (hallucination) is one of the reasons this law
    exists.

    A system that speaks fluently can still be wrong. In governed environments,
    ungrounded output is treated as a known failure mode, not as a rare surprise.
    That is why the system must be able to constrain what it is allowed to say,
    require evidence when claims matter, and refuse to invent when grounding
    is unavailable. Governance is the layer that turns those requirements into
    enforceable behavior.[iv]

    Evidence: A documented governance boundary, operator authority and
    override, risk and impact assessments, decision audit trails, and third-party
    assurance readiness.

Proof Pack: What to Ask For, and What to Verify.

Check each item only when you have the named artifact, test result, or log.

1) Works beautifully without opt-in

1. "No-identity mode" documented (what works, what changes, what does not).
2. Acceptance test run: core experience passes with no app, no login, no identity.
3. Opt-in is optional and non-punitive (UI, signage, and staff SOPs).

2) Identity is earned, not demanded

4. Identity request register: each use case, benefit, and non-identity alternative considered.
5. Consent and revoke path tested (revocation returns to full baseline experience).
6. Logs show identity actions occur only in approved scenarios.

3) Minimize, retain only what is justified, delete by design

7. Data inventory: category, purpose, retention, access roles, deletion trigger.
8. Automated deletion evidenced (job logs, reports, or test run outputs).
9. Data access audit logs enabled and review cadence defined.

4) Recognize the place first, the person second

10. Venue state model exists (occupancy, flow, congestion, equipment state), timestamped.
11. Decision logs link actions to venue state inputs.
12. Peak load test confirms state consistency, including degraded modes.

5) Inclusion is the starting point

13. Supported language set defined, language-pack QA complete, offline fallback verified.
14. Accessibility modes defined and tested (participation is normal, not special-case).
15. Comfort and sensory modes defined and measured during peak occupancy.

6) Personalization never becomes manipulation

16. Transparency standard: what must be disclosed when adapting content, routing, prompts.
17. "Decline without penalty" tests pass (refusal does not degrade service).
18. Review evidence: steering or dark-pattern checks, with remediation path.

7) Commands are not outcomes, verify what happened

19. Outcome verification in place for display, audio, and critical actuations.
20. Failure detection and rollback behavior documented and tested.
21. Commissioning report includes verified outcomes, not just issued commands.

8) Latency is emotion, timing is a first-class constraint

22. End-to-end timing budgets defined across modalities.
23. Sync acceptance tests pass in normal and degraded operation within tolerances.
24. Timing drift monitoring enabled, with a correction runbook.

9) Architecture welcomes change

25. Node-based compute baseline documented using commercially available hardware.
26. Interfaces documented, and substitution plan approved (no vendor-only dependency).
27. Restore test proves nodes can be replaced without rewiring or rebuilding rooms.

10) Governance is enforceable at runtime

28. Governance boundary defined (allowed, disallowed, human-approval required).
29. Operator override tested, and override events are logged.
30. Risk and impact assessment completed, and decision audit trail enabled for assurance review.
31. Confabulation controls documented for any system component that generates or summarizes text at runtime, including: approved source boundaries, evidence pointers for factual claims, abstain behavior when grounding is insufficient, and monitoring of confabulation incidents over time.[v]

0

BOOK COVENANT

This book is intended for owners, designers, builders, and operators.

Owners and program leaders. Designers of venues. Builders of systems. Technical directors and systems engineers. Operators who have to keep the place running. Procurement, IT, privacy, and security stakeholders who must be able to defend the choices. The intent is the same: to help a venue become coherent, and to do so in a way that is practical, inclusive, and responsible.

Read this book before you choose vendors, before you approve sensing or data collection, and before you define KPIs. It gives you language for requirements such as minimization, consent, separability, and auditability, which are the difference between a system that scales and a system that becomes a liability.

To keep that intention honest, the book follows a covenant. It is not a slogan. It is a set of constraints that shaped every chapter.

Positive.

We do not frame progress as a denunciation of the past. The history of venues is not a story of failure. It is a foundation story. The purpose here is not to look back, but to show how quickly the ground has shifted, and how confidently we can build on what already works.

Helpful.

This is not a book of abstractions. It aims to be usable. When a concept matters, we explain it. When a trade-off is real, we name it. When a decision point is likely to arise in a project meeting, we give you a way to resolve it. The goal is that you can read a chapter and immediately improve a plan, a design, or an operational posture.

Idea-forward.

Venues win by offering experiences people cannot get at home. That requires ideas and permission to think beyond the default. This book is meant to serve as a generator. It introduces patterns, narratives, and architectures that can be recombined across museums, theme parks, cruise ships, retail, brand centers, and smart cities. The emphasis is not on a single right answer, but on a library of approaches that make change easier.

Privacy-forward.

The most capable systems in the world will fail in public space if they are not trusted. Trust begins with privacy. Throughout this book, privacy is treated as a design requirement, not a legal appendix. We assume anonymous operation by default. We assume opt-in identity only when it clearly benefits the visitor. We assume strict minimization, short retention, clear consent, and transparent purpose.

IDEA:
Inclusivity, Diversity, Equity, and Accessibility by design

In this book, IDEA is a practical design and operating posture: build environments that are inclusive, diverse in who they serve, equitable in how they work for different people, and accessible by default. IDEA is infrastructure, not decoration. If we do this properly, it is a first-class part of the experience, specified early, designed into the system boundaries, used as acceptance criteria, monitored in operation, and preserved through upgrades.

In practice, that posture has a concrete, testable baseline. It starts with multimodal access, continues through language dignity, and extends to sensory safety, acoustic comfort, and operational resilience:

1. Hearing access: captions for prerecorded content, live captioning where applicable, assistive listening support, and clear signaling that these aids are available.
2. Vision access: audio description for relevant video, tactile and Braille options where physical labeling is used, and digital channels that are compatible with screen readers and accessible navigation.
3. Language dignity: multilingual content is not a translation afterthought. That includes captions, and when sign-language interpretation is provided, it must be the appropriate sign language for the audience, not assumed to be universal.
4. Sensory safety and overstimulation, including neurodiversity: provide comfort modes, quieter mixes, predictable interaction pacing, and choices that reduce cognitive load and sensory intensity without degrading the experience for others.
5. Acoustic comfort and intelligibility: treat the acoustic layer as part of the accessibility stack. Manage noise, chaos, and sound bleed so that speech remains intelligible, spaces remain enjoyable, and people who rely on clear audio or who are easily overstimulated are not excluded in practice.
6. Equity by design: the accessible path is the primary path, not a separate workflow, and it remains functional during degraded and incident modes.

ESG:
Environmental, Social, and Governance outcomes by design

In this book, ESG is treated as an engineering constraint, not a marketing label. The environments we build consume energy, materials, labor, attention, and public trust. If hyper-personalization is done well, it can reduce waste, improve flow, and increase inclusion. If done poorly, it can increase compute load, shorten hardware lifecycles, amplify inequity, and create governance risk. ESG is how we keep the work defensible, measurable, and operable at scale.

In practice, this posture has a concrete, testable baseline. It aligns with the same four pillars used by major sustainability disclosure frameworks: governance, strategy, risk management, and metrics and targets.

For builders, that translates into requirements you can specify, commission, and audit:

- Environmental performance: treat energy, thermal load, and network load as first-class design variables and measure and manage them. Include power and heat budgets per zone, per show state, and per peak day, and verify them during commissioning.
- Compute efficiency: provision the right compute and prove it. Prefer architectures that reduce unnecessary data movement, idle draw, and scale gracefully under peak attendance. Track compute utilization, storage growth, and retention as operational metrics, not surprises.
- Longevity and circularity: design for repair, replacement, and modular upgrades. Favor components with clear serviceability, spare availability, and documented end-of-life pathways to prevent the venue from becoming an e-waste generator with every refresh cycle.
- Indoor environmental quality: treat comfort as environmental performance. Thermal stability, air quality, and acoustic control reduce churn, ease operational strain, and improve outcomes for visitors and staff.
- Social value in operation: protect staff and guests from operational overload. Define staffing assumptions, training requirements, and degraded-mode behaviors to ensure the venue remains safe, inclusive, and humane during peak load and incident mode.

- Responsible supply chain posture: specify what must be documented. Require vendor transparency regarding support life, update policy, security posture, and sustainability claims that affect procurement decisions.
- Governance and auditability: assign accountable owners for privacy, accessibility, safety, and sustainability outcomes. Establish a review cadence, incident response, change control, and logging that make the system explainable after upgrades, outages, and vendor changes.
- Metrics that matter: select a small set of venue-relevant ESG metrics, set target thresholds, and keep them visible. If a metric cannot be measured or audited, it is not yet a requirement.

This is how ESG becomes real: it is written into the specification, verified during commissioning and acceptance, monitored during day-to-day operations, and protected through upgrades, rather than rewritten after the fact as a glossy report.

If you need external reference points, refer to major sustainability disclosure frameworks including the GRI (Global Reporting Initiative) for broad impact reporting, ISSB/IFRS Standards (International Sustainability Standards Board) for global baseline financial materiality, TCFD (Task Force on Climate-related Financial Disclosures) for climate risk, and the EU's CSRD/ESRS for comprehensive European reporting, alongside voluntary tools like CDP, SASB, and emerging ones like TNFD, all guiding corporate transparency on ESG issues for investors and stakeholders. These frameworks are helpful, but the core discipline in this book is simpler: define what matters, measure it, assign accountability, and keep it under governance.

Patent-safe.

This is a reference book, and it is written with professional discipline. It explains what to do and why to do it, and it describes design patterns at a level that helps designers and operators make better decisions. It does not disclose implementation shortcuts, proprietary integration sequences, configuration specifics, source code, schematics, or other enabling details that would compromise issued patents, pending applications, or protectable know-how. The goal is to inform and inspire while keeping protected methods protected. The book stays practical by focusing on acceptance criteria, evidence requirements, and procurement-ready templates.

Where generative systems are discussed, this book treats confabulation (hallucination) as an expected risk, and it defines public-safe requirements for constraining outputs, requiring evidence where claims matter, and verifying behavior in operation. It does not publish prompt recipes, tool bindings, or implementation sequences that would compromise protected methods.

Public-safe reference vs NDA implementation appendix

This book is intentionally public-safe.

It explains the architecture, the operational patterns, the system boundaries, and the commissioning discipline required to build governed, inclusive, privacy-forward environments. It is written so owners, designers, operators, fabricators, and city stakeholders can make correct decisions, write correct specifications, and avoid fragile deployments.

It does not include implementation shortcuts, proprietary integration sequences, or detailed technical methods that would reduce the integrity of patented or pending work.

For qualified parties, a separate NDA Implementation Appendix is available. That appendix expands the material into deeper technical specifics, reference configurations, and implementation-level guidance suitable for serious deployment planning, while remaining aligned with the same privacy and governance posture.

IP posture and licensing discussions

Some techniques and system behaviors discussed in this book may be covered by issued and pending patents. This book is a public-safe architectural reference and does not grant any license to practice patented methods. For licensing, technical diligence, or deeper implementation discussion under NDA, contact the author.

If you hold this covenant in your mind as you read, the book becomes easier to use. You will see why some chapters are deliberately architectural instead of procedural. You will see why privacy is repeated so often. You will see why acoustics and comfort are treated as design requirements rather than "audio-visual (AV) details." You will see why we speak about systems as living things that must be maintained, audited, and upgraded, rather than installed and forgotten.

That is the posture. Now, here is how to use it.

0.1

HOW TO USE THIS BOOK

How to Read This Book by Role

Most books assume one kind of reader. This one assumes several, because modern venues are built by coalitions. You will get the most value if you read the chapters that match your role first, then circle back to the rest once you have the shared language.

This is a quick table, with some more options below.

READING PATHS BY ROLE

Your Role	Start Here	Core Chapters	Reference as Needed
Owner / Investor / Board	Ten Laws (p.15), Proof Pack (p.19)	1–4, 24, 26	Appendix J (RFP specs)
CEO / Program Director	Preface, Ten Laws	1–4, 22–24, 42	Appendix K (Assurance)
Experience Designer	Chapter 3 (Ladder)	6, 11, 13–16, 31–36	Part III (Personal Channel)
Technical Director / Systems Engineer	Chapter 5 (Compute)	5, 8–10, 23–25, 37–39	Appendices B–D, G–H
Operator / Venue Manager	Chapter 37 (Commissioning)	10, 37–41	Appendices B, C, H
Privacy / Legal / Compliance	Chapter 18 (Privacy)	4, 18–20, 26–27	Appendices E–K
Fabricator / Integrator	Part II (Substrate)	5–10, 37–38, 41	Appendix J (RFP specs)
Procurement / IT	Appendix J	4, 18, 24, 37	Appendices E, F, K
AI Governance / Academic	Chapter 4 (Governance Threshold)	3–4, 23–26, 42	Appendices G, K

By Immediate Goal

If You Need To...	Read This Path
Write an RFP or scope of work	Appendix J → Chapter 24 → Appendix K
Understand the privacy posture	Chapters 18–20 → Chapter 27
Evaluate an acquisition or investment	Ten Laws → Chapters 3–4 → Chapter 42
Commission a new venue	Chapters 37–38 → Appendices B, C, H
Deploy governed AI safely	Chapters 3–4 → 24–26 → Appendices G, K
Retrofit an existing venue	Chapter 41 → Part II → Appendix A
Prepare for regulatory review	Appendix K → Chapters 26, 18 → Appendix I

The Fastest Path to Governed AI

If your immediate goal is governed AI that is operable in public space:

Ten Laws and Proof Pack → Chapter 3 → Chapter 4 → Chapter 24 → Chapter 26 → Appendices G, H, K

This route keeps confabulation risk visible, testable, and containable - without turning the venue into surveillance.

If your immediate goal is governed AI that is operable in public space, take this path first: the Ten Laws and Proof Pack, Chapter 3 (delivery modes and bounded generation), Chapter 4 (governance threshold), Chapter 24 (implementation loop and evidence hierarchy), Chapter 26 (Constitution, constraints, and override), and Appendices G, H, and K (logging, monitoring, and assurance). This route is designed to keep confabulation (hallucination) risk visible, testable, and containable without turning the venue into surveillance.

For owners and investors

Read to clarify what you are buying, what risks you are assuming, and what you can demand from a team without forcing them into fragile solutions.

Recommended path:

- Chapter 1, the phone changed the room
- Chapter 2, the new psychology of going out
- Chapter 3, the personalization ladder
- The economics, operations, and governance chapters later in the book

What to listen for:

- Where personalization becomes operational, not cosmetic
- The difference between micro deployments and venue-scale systems
- What should be measured, and what should not be optimized blindly
- What privacy posture protects your brand and your visitors

The next pages summarize which parts will be most resonant with different readers of this book.

For designers and experience teams

Read to expand the palette, and to gain a language for proposing systems that remain beautiful, inclusive, and maintainable.

Recommended path:

- Chapter 3, the personalization ladder
- The chapters on Programmable Canvases and the Personal Channel
- The vertical focused pattern chapters for your venue type

What to listen for:

- How to treat language as runtime, not layout
- Designing for multiple rhythms, highlights and depth
- Creating peaks worth sharing without gimmicks
- Inclusion as posture, not retrofit

For operators and technical directors

Read to understand what makes systems stable over years, not just impressive on opening day.

Recommended path:

- The ladder chapter, to classify what you are operating
- The chapters on power management, verification, analytics, and preventive maintenance
- Governance and runbooks

What to listen for:

- What breaks first, and how to keep it from breaking
- How to use trend analytics to prevent failures
- How to design graceful degradation
- How to specify acceptance tests that staff can validate

For fabricators and integrators

Read to translate experience goals into robust physical systems, and to anticipate the operational expectations of the next decade.

Recommended path:

- The enabling substrate chapters
- Networking and reliability chapters
- Operations automation and verification chapters

What to listen for:

- How modular compute changes replacement and upgrade economics
- Why power sequencing, zoning, and sound bleed control matter to experience quality
- What the right edge capacity means in practical terms
- How to commission for maintainability, not just for completion

For cities and public-sector stakeholders

Read to understand the difference between "smart" and "governed," and how to build public trust in large-scale adaptive systems.

Recommended path:

- The ladder chapter
- Privacy and identity chapters
- Governance chapters
- Smart city vertical chapter

What to listen for:

- Remaining privacy-forward at scale
- How to use "yes or no" proofs rather than document collection
- How to make decisions auditable and policy-aligned
- How to design for equity and inclusion across demographics

This book is not intended to be read once, then shelved. It is intended to be used in meetings. If a debate becomes circular, return to the ladder. If a feature request starts to drift into surveillance, return to the privacy covenant. If a plan feels fragile, return to the operations chapters. The book is designed to be a shared language for teams that must collaborate across disciplines.

A note on overlap: This book can be read cover-to-cover, but it is designed as a reference for real meetings. Some core concepts appear in more than one place so individual sections can be read stand-alone. If you are reading linearly, treat repeated framing as a fast re-orientation, and feel free to skim to the new material using the cross-references and index.

Where the doctrine lives (quick map)

Use this map to jump directly to the doctrine you need.

Core architecture primitives

- The Personalization Ladder (Stage 0–2): Ch 3
- Governance threshold and externalities test: Ch 4
- Compute at the edge, node behavior, and acceptance criteria: Ch 5
- Programmable canvases and software-defined surfaces: Ch 6
- Non-visual navigation as first-class infrastructure: Ch 7
- The acoustic layer as operational state: Ch 8
- Networking, timing, resilience, and local-first posture: Ch 9
- Show control vs orchestration,
  and power as experience quality: Ch 10

Inclusive delivery

- Personal Channel (language, captions, personal audio): Ch 11
- Offline-first personalization (OnBoard posture): Ch 12
- Synchronized audio, captions, sign languages, and Braille: Ch 13
- Multilingual at scale without bias: Ch 14
- Visitor models that actually help: Ch 15
- Virtual Docents and embodied guides: Ch 16

Privacy, recognition, analytics

- Privacy covenant and consent posture: Ch 18
- Anonymous recognition explained: Ch 19
- Measuring a place vs measuring people: Ch 20
- Recognition across verticals
  (vehicles, POS, service-first phygital): Ch 21

Macro venues (WorldModel)

- Venue Concierge Interface at macro scale: Chapter 22
- WorldModel as operational truth and what it is not: Chapter 23
- Implementation loop (truth bus, state store, governance gate): Ch 24
- Evidence Hierarchy (what wins when signals conflict): Ch 24
- Outcome Verification is a First-Class Layer: Ch 24
- Why WorldModel cannot be piloted safely: Ch 24
- Decision architecture (ICL, EDE, MAOL, CGL): Ch 25
- Governance in practice (Constitution, constraints, override): Ch 26

Operations and lifecycle

- Commissioning and support posture: Ch 37
- Lifecycle automation, verified restoration, outcome verification: Ch 38
- Power and lifecycle loop: Ch 39
- Content operations and fast change without rebuild: Ch 40
- Upgrade path and future that behaves like an upgrade: Ch 41–42

Executive reading map

A separate high-level companion is available for board members and CEOs, titled "Hyper-Personalized Venues".

That book is the overview. This reference provides the detailed requirements, acceptance criteria, and evidence posture.

An optional, informative reference implementation mapping note is available on request. It is non-normative and separate from this reference standard.

New venues: Start with "Provisioning the Right Compute at the Edge" and Appendix J, Section 0, "Compute Node Architecture and Future Resistance."

Owners and program leaders:
Start with *The Personalization Ladder*, then *WorldModel as Operational Truth*, *Implementing WorldModel Venues*, *Governance in Practice*, and the operations chapters on verification and lifecycle.

Design leads:
Read *A New Psychology of Going Out*, *The Personalization Ladder*, *Programmable Canvases*, *The Acoustic Layer*, *Non-Visual Navigation and Guidance*, and the vertical chapter relevant to your project.

Technical directors and systems engineers:
Read *Provisioning the Right Compute at the Edge*, *Networking, Timing, and Resilience*, *Implementing WorldModel Venues*, *Decision Architecture*, and the verification, drift, and support chapters.

Operators:
Read *Commissioning, Monitoring, and Support*, *Lifecycle Automation Layer, Outcome Verification Layer, and Verified Restoration*, and *Power and Lifecycle Loop*.

Cities and public authorities:
Read the privacy and identity chapters, the Smart Cities chapter, *Governance in Practice*, and *Implementing WorldModel Venues*.

What makes this hard to replicate

- A WorldModel Venue is not a content system. It is a closed loop: operational truth, governed decisions, verified action, and drift-driven prevention.
- Most systems can propose. Few can prove. Verified restoration and outcome checks are what keep automation from becoming public embarrassment.
- Governance is not a slogan. It is an enforceable gate that constrains actions to ensure safety, accessibility, privacy, fairness, and integrity.
- The multi-plane delivery model, room, surfaces, Personal Channel, and spatial AR enable the venue to scale language and accessibility without turning the building into a cluttered space.
- Operability is the moat. Commissioning, monitoring, drift detection, and rollback discipline are what keep systems running for years rather than running demos for weeks.
- The upgrade path is architectural. When capability scales by provisioning compute rather than rebuild cycles, the system remains future-capable under real procurement constraints.

This book assumes familiarity with complex systems delivery and is written for practitioners responsible for implementation and governance, not as an introductory overview.

Independent research on immersive entertainment and culture also converges on the same conclusion: destinations that win are those that operate reliably, inclusively, and coherently at scale.[vi]

Owner acceptance test index

Use the index below to translate the reference into procurement language, acceptance tests, and commissioning evidence. It maps major chapters to Appendix J clause groups, and to the Evidence Pack artifacts in Appendices G, H, and I.

Where to start (chapters)	Procurement clause group (Appendix J)	Evidence artifacts (Appendices)
Ch. 3-4, 26 (ladder, governance threshold)	J4 Integrated Recognition & Governance	G1.1 Logging minimums; G3 Evidence Pack; I Assurance packet
Ch. 5 (node-based compute baseline)	J0 Compute Node Architecture & Future Resistance	G1.1 Update and rollback events; H Monitoring plan
Ch. 8 (acoustic layer)	J6 Digital Audio, Spatial Audio, and Adaptive Soundscapes	H Monitoring plan; G1.1 Verification and fault events
Ch. 9-10 (networking, timing, show control)	J5 Infrastructure & Maintainability	H Monitoring plan; G1.1 Incident events; I Third-party assurance notes
Ch. 11-14 (Personal Channel, multilingual delivery)	J2 The Personal Channel & Inclusive Delivery; J3 Shared Media Planes	G1.1 Session preference events; H Monitoring plan
Ch. 18-20 (privacy, anonymous continuity, analytics)	J4 Integrated Recognition & Governance	G1.1 Data access events; G1.2 Privacy controls; I Assurance packet
Ch. 23-24 (WorldModel operational truth and implementation)	J1 Operational Automation & Verified Restoration; J4 Governance	G3 Evidence Pack; H Monitoring plan; I Assurance packet
Ch. 37-38 (commissioning, verified restoration)	J1 Operational Automation & Verified Restoration	H Monitoring plan; G3 Evidence Pack; I Assurance packet

From Three Pillars to Six Layers of cognitive experience

One of the starting points for this book was an Ernst & Young (EY) article on theme park technology transformation. It served as a catalyst because it distills a complex operational reality into a practical baseline that any destination can recognize: connect, analyze, and act.

Ernst & Young (EY) describes a connectivity posture in which "Parks unify data from guests, rides and operations, creating a single source of truth that enhances decision-making and operational efficiency."

It then frames analytics as the layer that turns signals into operational advantage: "Advanced analytics reveals patterns, predicts outcomes and optimizes resources, enabling parks to enhance guest experiences and streamline operations."

Finally, it frames execution as the layer that moves the organization from insight to outcomes: "This technology enables automated decisions and balances traffic, facilitating safety and unlocking measurable value in guest satisfaction, operational efficiency, and revenue growth."

Those three pillars are necessary, but in public environments they are not a complete system. Public space imposes three additional requirements that do not disappear under load: explicit governance that constrains what the system is allowed to do, anticipatory planning that reduces friction before it forms, and continuity that lets the visitor carry preferences across spaces and sessions on the visitor's terms.

Once I started treating personalization as a complete system rather than an oversimplified one, those missing requirements became structural. That is why this book adds three layers to the baseline: Ethical Intelligence (governance), Anticipatory Intelligence (bounded prediction and planning), and Phygital Continuity (permissioned persistence across rooms, surfaces, and what we will call the "Personal Channel"). Each one of these layers has its place, as will become clear in this book.

The deeper consequence was unexpected, and it is the real reason the WorldModel framework exists. When you combine connectivity, analytics, and action with governance, anticipation, and continuity, you no longer have a set of features. You have an environment that must share operational truth, propose actions, constrain them by policy, execute them, and verify outcomes in the physical world. The WorldModel framework is the operating architecture that makes that coherence possible at venue scale.

That observation was the bridge from an operational baseline to the fully governed, continuity-aware, prediction-bound architecture the rest of this book develops.

The glossary that follows defines the vocabulary used throughout that journey.

Identity and Agent Layers: A Framework

This book organizes identity, consent, and coordination patterns into four layers. The industry often uses Web3, Web4, Web5, and Web6 as labels for related concepts - you will encounter these terms in RFPs and vendor discussions. Other authors may define them differently, which is fine. This book uses descriptive layer names as primary terminology, with Web labels shown parenthetically for industry reference.

For the purposes of this book, the following precise definitions apply:

Layer	Industry Label	Definition	Venue Application
Portable Identity	Web3	Identity you control	Verifiable credentials, decentralized identity, and smart contracts that delegate authority without surrendering it.
Consented Context	Web4	Context you consent to	Emotion, prosody, comfort preferences, and sensory context - shared only with explicit, revocable consent.
Agent Coordination	Web5	Agents that coordinate	Software agents acting on behalf of visitors and venues, negotiating preferences and managing interactions.
AI Negotiation	Web6	AI-to-AI communication	Visitor AI (on phone or wearable) communicating with venue AI to determine settings without human micromanagement

Portable Identity (Web3): Identity You Control

Web3 in the venue context means the visitor holds their own credentials. They can prove things about themselves, such as age, membership, professional certification, and accessibility needs, without handing over identity documents or creating accounts in venue databases.

The enabling technology is verifiable credentials: cryptographically signed attestations that a visitor presents from their own wallet. The venue verifies the signature and the claim without ever seeing the underlying identity. A visitor proves they are over 21 without revealing their birthdate. A member proves they hold an annual pass without the venue looking them up in a database.

Smart contracts extend this pattern by enabling programmatic delegation. A credential holder can grant limited, time-bound authority to another party without surrendering the underlying credential.

Example: A parent visiting a science center with their child's school group can temporarily delegate control of certain experience parameters to the teacher. The smart contract specifies: the teacher may adjust the child's interests to align with the curriculum and adjust the audio volume and pacing preferences, but only during school hours, only within the venue's education wing, and only until 3:00 PM. The parent retains the underlying credential. The delegation expires automatically. No one at the venue needs to manage permissions manually.

This is what "identity you control" means in practice: holding credentials and delegating limited authority under explicit, programmatic constraints.

Consented Context (Web4): Context You Consent To

Web4 adds emotional and sensory context to the identity layer. Where Web3 answers "who are you entitled to be," Web4 answers "how do you want to experience this moment?"

This includes explicit preferences: the comfort dial that lets visitors signal sensitivity to loud sounds, flashing lights, crowd density, or pacing. It also includes implicit signals that the visitor consents to share.

Prosody and emotion feedback: A visitor's device can detect stress indicators from voice patterns, hesitation in movement, or physiological signals (with appropriate sensors and explicit consent). If the visitor has opted in, this context can inform the venue's response. A guide system might slow its pace when it detects confusion in the visitor's voice. An attraction might offer a calm-down moment when it detects elevated stress. A retail assistant might recognize frustration and offer human help.

The critical word is consent. Web4 context is never extracted. It is provided by the visitor, with granular control over what is shared, with whom, and for how long. The visitor can revoke consent at any time. The venue cannot store Web4 data beyond the immediate interaction unless the visitor explicitly permits it.

This is the "comfort dial" made comprehensive: not just stated preferences, but also real-time emotional context that the visitor chooses to share.

Agent Coordination (Web5): Agents That Coordinate

Web5 introduces software agents that act on behalf of both visitors and venues. Rather than the visitor manually configuring every preference at every interaction point, an agent carries their preferences and negotiates on their behalf.

The visitor's agent knows their language, accessibility requirements, comfort preferences, and interests. When the visitor approaches an exhibit, the visitor's agent coordinates with the exhibit's agent to configure the experience. The visitor doesn't fill out forms or tap through settings. The agents handle it.

Example: A visitor with low-vision accessibility needs enters a museum. Their agent has been configured (once, at home) with their preferences: high-contrast visual modes, audio description enabled, and larger text on any interactive displays. As they move through the museum, their agent coordinates with each exhibit's agent. The exhibit reconfigures automatically. The visitor experiences a museum that seems to understand them without ever having to explain.

Agents also coordinate with one another. A family group's agents can share relevant context (with consent) so that experiences reflect the whole group - finding seating that works for everyone, pacing that accommodates varying mobility levels, and content that bridges different language preferences.

AI Negotiation (Web6): AI-to-AI Communication

Web6 is the emerging frontier: AI systems representing visitors communicating directly with AI systems representing venues. This moves beyond scripted agent negotiation to genuine inference and adaptation.

The visitor's AI, running on their phone or increasingly on wearables like AR glasses or other AI interface devices, builds a model of what the visitor wants. It learns from patterns, feedback, and context. It doesn't just carry static preferences; it infers dynamic ones.

Example: A visitor wearing AR glasses enters a history museum. The visitor's AI, with consent, has observed that they tend to linger at interactive exhibits, skip text-heavy panels, and engage more deeply with social-history content than with military-history content. The visitor's AI communicates with the museum's AI: "This visitor prefers interactive experiences, responds well to narrative over data, has shown interest in daily-life themes, and is currently 90 minutes into their visit, with energy levels suggesting another 45 minutes of engagement." The museum's AI responds by highlighting a path through the social-history wing, queuing up interactive stations, and preparing a culminating experience that fits the remaining time window.

Neither AI shares the visitor's identity. Coordination occurs through inference and intent.

Web6 is speculative today but directionally clear. As AI capabilities increase on personal devices, the negotiation between visitor intent and venue capability will increasingly happen at the AI layer.

The Progression

The four layers build on each other:

Portable Identity (Web3) establishes that identity belongs to the visitor. Without this foundation, every subsequent layer requires the venue to hold visitor data, which reintroduces the privacy and control problems the architecture is meant to solve.

Consented Context (Web4) adds context that identity alone cannot capture. Knowing that someone holds a membership tells you nothing about how they want to experience today's visit. This layer fills that gap with consented emotional and sensory context.

Agent Coordination (Web5) operationalizes both layers through agents that coordinate without human micromanagement. The visitor sets preferences once, and the agents then handle ongoing negotiation.

AI-AI Negotiation (Web6) extends coordination into inference. Rather than explicit preference-setting, AI systems on both sides develop models of intent and capability, coordinating at a level of nuance that explicit configuration cannot match.

If the Web3/4/5/6 labels conflict with definitions you have encountered elsewhere, that is expected. Web3, in particular, carries baggage from cryptocurrency discourse that may not align with the usage here. This book uses these labels parenthetically because they appear in industry RFPs and vendor discussions, but the architectural patterns are what matter. Focus on Portable Identity, Consented Context, Agent Coordination, and AI Negotiation. The layer names describe what each pattern does; the Web labels are simply industry shorthand.

Part VI of this book (Chapters 27–30) develops these patterns with implementation details, including worked examples and acceptance tests. Readers who prefer to focus on the architecture rather than the terminology can proceed directly to those chapters.

0.2

GLOSSARY OF KEY TERMS

This glossary is deliberately brief. Each term will be expanded with context later in the book. The purpose here is to establish a shared vocabulary early, so that technical and non-technical readers can follow the same argument. For updates to the Glossary of Key Terms, refer to the accompanying web site, worldmodel.global .

Accessibility

The measurable ability of an environment, interface, or service to be perceived, understood, navigated, and used by people with a wide range of physical, sensory, cognitive, and situational constraints. Accessibility in this book includes both digital and physical layers: content, interfaces, wayfinding, acoustics, visibility, mobility, and assistive compatibility.

Acoustic layer

The set of acoustic conditions and controllable behaviors that determine how sound is experienced in a space, including intelligibility, noise floor, reverberation characteristics, spatial zoning, directional sound, and the interaction between adjacent zones. The acoustic layer matters because it directly affects comprehension, comfort, inclusion, and perceived quality, and it can be actively managed, for example by reducing sound bleed in empty galleries, or by offering quieter narration-only mixes for visitors who are easily overstimulated.

Acoustic state

A snapshot of the current sound environment in a space, for example measured or inferred values such as ambient noise level, program level, intelligibility conditions, or crowd-generated sound. Acoustic state is part of operational reality, and can influence decisions about routing, content delivery modes, and power-saving behaviors.

Age-Appropriate Experience

The practice of tailoring content, language, pacing, interaction patterns, and responsibility expectations to the visitor's developmental stage and typical capabilities for their age group. Age-appropriate experience ensures that children are not exposed to adult framing, complexity, or implied responsibilities they cannot reasonably process, and that adults are not treated as if they require simplified, childlike explanations. It supports safety and comprehension while preserving dignity.

Age Dignity

The condition where a person is treated with respect that fits their age and role, without infantilization, stereotyping, or inappropriate expectations. Age dignity is maintained when the experience neither talks down to adults nor burdens children with adult-level complexity, and when it offers choices that let people self-select the level of depth without embarrassment.

Agent Coordination Layer (Web5 in industry usage)

Shorthand for an identity and data layer using DIDs, verifiable credentials, and decentralized web nodes, as publicly described.

AI-to-AI Negotiation Layer (Web6 in industry usage)

Author-defined forward label for multi-agent, cross-system coordination requiring enforceable governance and verification.

Anonymous recognition

Recognition that preserves privacy by avoiding identification. The system can recognize a context or a returning session or match a pattern that is useful for experience continuity or analytics, without learning a person's real-world identity. Anonymous recognition must still be treated as sensitive, because linkability can exist, and it must be governed by minimization, short retention, and clear purpose.

Audience Appropriateness

The principle that content and interaction depth should match the visitor's context, including age, literacy, cognitive development, cultural expectations, time available, and domain expertise. Audience appropriateness avoids both under-serving and over-serving a person by delivering information that is too advanced, too simplistic, irrelevant, or patronizing. In this book, audience appropriateness is a dignity requirement: people should not be forced into experiences that misjudge who they are.

Board Oversight (ESG)

Formal accountability of the board or a delegated governance body for ESG strategy, risk appetite, performance indicators, and incident response. For personalization systems, board oversight covers privacy boundaries, consent integrity, accessibility commitments, language equity targets, vendor compliance, and safety-related governance.

Body of Knowledge (BoK)

The authoritative, structured collection of verified information, interpretive frameworks, and contextual data that informs all AI-generated content within a deployment. The Body of Knowledge serves as the source material from which the system draws when responding to visitor inquiries, generating narratives, or constructing personalized experiences. It can contain text, media, and anything else that defines the background from which content should be derived.

BoK

See "Body of Knowledge"

CGL™

Cognitive Governance Layer™. A governance layer that evaluates proposed actions against explicit objectives and hard constraints, ensuring that systems behave safely, inclusively, and in alignment with venue policy. It is the layer that makes powerful automation trustworthy in public space. CGL™ is a runtime decision mechanism. It may be implemented using rules, AI, or hybrid methods, but its defining property is enforceable constraint evaluation with auditable outcomes.

Channel-less Experience

An experience goal where the visitor does not feel channel boundaries at all; the system behaves as one continuous service across touchpoints. This is frequently discussed as a phygital outcome

Confabulation

Confidently stated but erroneous or false output produced by generative AI systems, especially when the system lacks sufficient grounding in approved sources or state. In this book, confabulation is treated as a system-level risk managed through delivery modes, constraints, evidence requirements, verification, and safe failure behavior[vii].

Consent

A visitor's explicit, informed permission for specific data use and experience behavior, granted for a defined purpose, scope, and duration, and revocable at any time. Consent is not a one-time checkbox. In this book, consent is treated as an ongoing operational condition that gates what the system may infer, remember, personalize, or share.

Consented Context Layer (Web4 in industry usage)

Author-defined label for consent-driven, device-mediated context negotiation in physical venues.

Consent Integrity

The property that consent is captured, enforced, and honored correctly in runtime behavior. Consent integrity requires that consent has clear scope, can be proven, can be withdrawn, is respected across all subsystems, and cannot be silently expanded by convenience, vendor defaults, or later integration changes.

Consent Receipt

A visitor-visible record of what was consented to, including purpose, scope, retention expectations, and how to revoke. Consent receipts support dignity and trust by making consent legible, and they support governance by making consent auditable without forcing visitors to become privacy experts.

Constitution

The foundational rule set governing system behavior, output boundaries, and ethical constraints. The Constitution defines what the AI may and may not say, do, or generate - functioning as the normative layer that shapes all downstream decisions made by the CGL and other governance components.

Constrained generation (also called bounded generation)

The process by which AI-produced content is bounded by predefined rules, approved source material, and governance parameters. Constrained generation ensures outputs remain factually grounded, institutionally appropriate, and aligned with the Constitution. Where constrained generation is permitted, the system must be able to point to its approved sources when factual claims matter, and must refuse to invent when grounding is insufficient.

Curated delivery

A content distribution model in which human experts pre-select, sequence, and contextualize material before it reaches the visitor. Curated delivery prioritizes institutional voice and narrative coherence over real-time personalization.

Data Minimization

The discipline of collecting, generating, and retaining the least data and the weakest identifiers needed to deliver the intended outcome. Minimization applies to raw inputs, derived attributes, logs, embeddings, and linkable tokens, not just obvious personal data.

Decision record

A minimal, auditable record of a consequential decision, including the operational state used, the applicable constraints or rules checked, the chosen delivery mode, the outcome (approve, modify, or reject), and any operator override. Decision records support accountability without turning logs into personal dossiers.

DID

Decentralized Identifier. A standards-based identifier designed to be controlled by the subject rather than issued as a permanent identity by a single centralized authority. DIDs are used to support modern credential and permission models that allow a person to prove something without disclosing unnecessary personal data - so instead of asking someone's age, we verify that they are over the required age to participate, without accumulating unnecessary information such as their birthdate.

Digital twin

In general usage, a dynamic virtual replica of a physical asset, system, or environment that maintains synchronization with its real-world counterpart through sensor data, operational telemetry, or periodic updates. Digital twins are widely deployed in manufacturing, infrastructure management, and smart city applications to enable monitoring, simulation, and predictive maintenance.

Digital twin (in the Agentic Museum framework)

Within this framework, digital twin functionality is distributed across multiple components rather than residing in a single mirrored replica. The WorldModel framework maintains the representational state of the venue and its objects. The Lifecycle Automation Layer tracks equipment condition and maintenance history. The Outcome Verification Layer confirms alignment between intended and actual states. Together, these layers achieve digital twin objectives - real-time awareness, predictive capability, and operational coherence - but through a governance-integrated architecture rather than a passive mirroring system.

Usage note: This framework does not employ a standalone "digital twin" component. Readers familiar with conventional digital twin implementations should understand that equivalent functionality emerges from the interoperation of the WorldModel framework, the Lifecycle Automation Layer, and the Outcome Verification Layer - with the Constitution ensuring that the system's self-model remains aligned with institutional intent, not merely physical reality.

Dignity
The condition where a person can engage without humiliation, dependency, avoidable confusion, or loss of agency. In this book, dignity is tightly coupled to comprehension and autonomy. If the system fails someone's language or accessibility needs, the person is forced into reliance on companions, staff, or guesswork, which is a dignity failure.

Diversity
The presence of different identities, backgrounds, languages, abilities, and perspectives among visitors, staff, and stakeholders. In this book, diversity is a baseline reality of modern venues, and systems must be designed to operate correctly across that real-world variety.

Drift (equipment)
The gradual deviation of calibrated equipment - such as sensors, displays, or projection systems - from specified operational parameters. Drift degrades experience quality and triggers maintenance workflows within the Lifecycle Automation Layer.

EDE™
Environmental Dynamics Engine™. A layer that models the evolving dynamics of a venue, such as flow, congestion, audio levels, temperature and solar ingress, and state changes, so the system can anticipate and respond rather than react late.

Embodied Docent Interface
A physical or robotic instantiation of the Virtual Docent Layer, capable of gesture, locomotion, and spatially situated interaction. The Embodied Docent Interface extends AI-guided experiences into three-dimensional space.

Emergency Stop (E-Stop)

A physical or logically equivalent control that immediately halts a defined set of system behaviors to reduce imminent risk. An E-Stop is for urgent conditions, not normal shutdown. In this book's context, an E-Stop must map to specific behaviors it stops (for example motion, actuators, automated doors, dynamic lighting sequences, audio above defined limits, or any autonomous agent action that could affect guests), and it must force a transition to a defined Safe State. Recovery must require deliberate, authorized action, not automatic resumption.

Equity

The practice of designing for fair outcomes, not equal inputs. Equity acknowledges that different people need different supports to achieve the same level of access, safety, comprehension, and autonomy. In venue systems, equity is achieved by intentionally removing barriers that otherwise create predictable disadvantage.

ESG (Environmental, Social, and Governance)

A governance framework for how organizations define, manage, measure, and report their impacts and risks across environmental, social, and ethical domains. In this book's context, ESG applies to how venues oversee personalization, identity, consent, privacy, accessibility, and safety as board-level responsibilities with measurable controls, auditability, and accountability.

ESG Governance

The operational system that assigns responsibility, incentives, policies, controls, and reporting to ensure ESG commitments are executed rather than advertised. For personalized venues, ESG governance includes how decisions are made about what the system may infer, store, remember, recommend, and disclose, and how harm, bias, and privacy failures are prevented and remediated.

ESG Risk Management

Integration of ESG factors into enterprise risk management, including prevention, detection, response, and learning loops. In this book's context, ESG risk management covers reputational, regulatory, operational, and safety risks arising from personalization failures, privacy breaches, bias, accessibility gaps, and vendor non-compliance.

Ethical Behavior (ESG)

Policies and practices that prevent corruption, manipulation, discrimination, and misuse of personal data or influence. In personalized environments, ethical behavior includes guardrails against discriminatory targeting, coercive interface patterns, opaque profiling, and undocumented data sharing.

Executive Alignment (ESG)

The linkage between leadership objectives, performance evaluation, and measurable ESG outcomes. In this book's subject matter, executive alignment means leadership is accountable for results such as accessibility conformance, language parity, consent capture quality, privacy incident rates, and the effectiveness of remediation processes.

Expertise-Appropriate Experience

The practice of matching content depth, vocabulary, and assumptions to a visitor's professional or domain knowledge when that expertise is explicitly indicated by the visitor or is strongly implied by the setting. When a visitor identifies themselves as an expert, the system should default to expert-level content, including technical detail, constraints, tradeoffs, and operational realities, rather than introductory explanations. In executive briefing centers, industry visitor centers, and professional demonstrations, this means moving quickly to substantive material while still offering optional pathways to definitions or refreshers. Expertise-appropriate experience preserves dignity by respecting stated competence, intent, and time.

Extended Reality (XR)

An umbrella term for AR and VR. XR is frequently used to deliver phygital effects, such as AR overlays that augment physical space with digital interpretation.

Governance

The practice of constraining system decisions under explicit policy, including objectives, hard constraints, auditability, and operator override. Governance is not bureaucracy. It is the mechanism that turns capability into responsibility.

Governance gate

A decision checkpoint within the CGL at which proposed system actions or outputs are evaluated against Constitutional rules before being permitted to proceed. Governance gates enforce compliance at runtime.

Governance (Operational)

The runtime enforcement of constraints that determine what actions are permitted, prohibited, or required within a system. Operational governance differs from policy by being executable, auditable, and enforceable under live conditions, including degraded and incident states.

Graceful Degradation

A deliberate reduction in system capability that preserves safety, accessibility, and trust when full operation cannot be maintained. Graceful degradation prioritizes conservative behavior, refusal when truth is insufficient, and operator escalation over speculative automation.

Grounding

The requirement that factual outputs and consequential decisions are anchored in approved sources, current operational state, or verifiable evidence, rather than invented detail. Grounding is the primary countermeasure to confabulation in venue contexts, where false certainty can create safety, reputational, or legal harm.

Hallucination

Common industry label for confabulation. See: Confabulation.

Hybrid Commerce

A retail-focused label for seamless blending of physical and digital journeys; some sources treat "phygital" as a synonym or market-specific term for this concept.

Hybrid delivery

A content distribution model that blends curated delivery with adaptive, AI-driven personalization. Hybrid delivery allows institutions to maintain editorial control over core narratives while enabling the system to tailor pacing, emphasis, and supplementary content to individual visitors.

ICL™

Identity Continuity Layer™. A layer that maintains continuity of visitor preferences and interactions over time, using privacy-forward principles and opt-in identity mechanisms when appropriate.

IDEA (Inclusion, Diversity, Equity, and Accessibility)

This book treats IDEA as a delivery requirement, not a slogan. Inclusive experiences only work when every visitor can understand, navigate, and participate with dignity, regardless of language, ability, or familiarity with the venue. That includes equity of languages, meaning the venue does not treat non-dominant languages as a reduced-feature fallback. Language access is a dignity issue because comprehension controls autonomy: if a guest cannot understand instructions, safety cues, consent prompts, or context, they lose agency and are forced into dependence.

Incident Mode

An explicit operating state in which automation, messaging, and guidance are constrained to a conservative, auditable posture in response to authoritative safety or hazard signals. Incident Mode suppresses optimization behavior, restricts public messaging to pre-approved content, and requires verified restoration before return to normal operation.

Inclusion

The property of an experience where people are not merely admitted, but can participate meaningfully without being singled out, marginalized, or forced into workarounds. Inclusion requires that core experiences, instructions, and services remain available across different needs and contexts.

Internet of Things (IoT)

Networked sensors and actuators embedded in physical environments that provide state awareness and enable responsive behaviors. IoT is a common substrate for phygital implementations because it connects physical conditions to digital logic.

Language Equity

A specific form of equity where languages are supported with parity of quality, completeness, and timeliness. Language equity means non-dominant languages are not treated as reduced-function translations, partial summaries, or out-of-date variants, especially for safety cues, consent prompts, navigation, and critical context - but also for content delivery.

Linkability

The ability to correlate a person, device, or session across time, locations, or systems, even if the person is not identified by name. Linkability is the central risk in "anonymous recognition," because it can still reconstruct a behavioral identity unless minimization, retention limits, and governance gates are enforced.

MAOL™

Multi-Agent Orchestration Layer™. A coordination layer that allows specialized agents to propose actions in parallel, then resolves those proposals into coherent actions under governance constraints.

NDEF (NFC Data Exchange Format)

A standard message encapsulation format defined by the NFC Forum for storing and exchanging NFC application data. NDEF supports one or more records per message, where each record includes a type and length, plus optional identifiers, enabling consistent interpretation of common payload types such as URLs and other structured records.

NFC (Near Field Communication)

A standards-based, short-range wireless communication method designed for simple, quick data exchange when two devices are within a few centimeters of each other, typically using a "tap" interaction. NFC operates at 13.56 MHz and is commonly used to exchange small payloads, trigger actions, or perform credential-like interactions.

NFC Code

Colloquial shorthand for the data payload carried by NFC, usually stored on an NFC tag or exchanged between an NFC-capable device and a reader. In practice, the "code" is typically an NDEF message that may contain a URL, an identifier, a small structured record, or an application-defined payload. NFC codes are phygital bridge mechanisms because they link a physical touchpoint (the tag location) to a digital action (open, launch, redeem, pair, continue, or log an interaction).

NFC Tag

A small, usually passive (battery-less) NFC device embedded in a sticker, card, wristband, token, placard, or exhibit element, which stores a small payload for an NFC reader to retrieve. NFC tags commonly hold NDEF-formatted messages and can be read-only or writable depending on the tag type and configuration. In venue contexts, NFC tags are used for tap-to-continue content, exhibit handoffs, staff workflows, access control patterns, and consented continuity, but they should be treated as potentially linkable touchpoints if the payload encodes stable identifiers.

Node-based compute

A distributed processing architecture in which computational tasks are handled by localized Edge Compute Nodes rather than centralized servers. Node-based compute reduces latency and supports real-time interactions at the point of experience.

Omnichannel

A design and operations approach that coordinates multiple channels and touchpoints so the visitor's journey remains coherent across in-person, web, mobile, and staff-assisted interactions. Phygital is often discussed as a practical expression of omnichannel in physical environments.

Online-to-Offline (O2O)

A journey pattern where a digital step intentionally triggers a physical step (or the reverse), for example scanning a code in-gallery to continue content, redeem access, or personalize a path. This is a common phygital mechanic in retail and can map cleanly into venues.

Operator Override

A controlled mechanism that allows authorized staff to approve, block, or modify system behavior in real time, with a decision record. Operator override is essential for public-space safety, incident response, and edge-case handling, and it must be constrained to prevent ad hoc policy drift.

Personalization Ladder

A practical classification system for the scope of personalization: exhibit-scale localization, micro hyper-personalization across multiple exhibits, and macro hyper-personalization at venue scale. The ladder determines what architecture is required, and when governance becomes mandatory.

Phygital

A portmanteau of "physical" and "digital," describing experiences, environments, or retail moments where physical space and digital interaction merge seamlessly. In venue contexts, phygital refers to touchpoints where a visitor's in-person presence and their digital identity, preferences, or device interactions converge - such as a retail display that recognizes a returning customer's preferences, or a museum exhibit that continues a narrative begun on a visitor's phone. The term originated in retail and marketing but now applies broadly to any environment where the boundary between physical presence and digital engagement becomes operationally irrelevant. Phygital design assumes the visitor arrives with a device and may have context the venue can use - if offered, consented, and handled with appropriate restraint.

Portable Identity Layer (Web3 in industry usage)

Shorthand for portable, cryptographically verifiable identity and claims, anchored to DIDs and verifiable credentials.

Purpose Limitation

A privacy and governance principle that data, inferences, and identifiers may only be used for the explicit purposes communicated at collection time. Purpose limitation prevents function creep, where data gathered for experience continuity is later repurposed for marketing, profiling, enforcement, or unrelated analytics.

Prosody

The rhythm, stress, intonation, and tempo of speech: the musical layer of how something is said, distinct from the words themselves. Prosody carries emotional and contextual signal: a rising tone may indicate uncertainty, a rushed cadence may signal frustration, a softened pace may suggest fatigue or reflection. In venue contexts, prosodic analysis allows systems to infer visitor state (confused, delighted, overwhelmed, hurried) from voice interactions without requiring explicit statements. This capability enables adaptive responses, slowing down delivery for a visitor who sounds lost, or abbreviating guidance for one who sounds impatient. Because prosody reveals emotional state, it falls squarely within consented context: systems should not analyze prosody without visitor awareness, and the analysis should inform service quality rather than build persistent profiles. Prosodic signals are inherently ephemeral and should be treated as momentary context, not stored attributes.

Provisioned compute

Dedicated processing capacity allocated to a specific venue, experience, or operational function. Provisioned compute ensures predictable performance by reserving resources in advance rather than relying on shared or on-demand infrastructure.

Proximity Interaction

A class of interactions triggered by location or nearness (beacons, UWB, Wi-Fi RTT, vision zones, geofencing), enabling content to respond to where the visitor is without requiring explicit input each time. This is a common enabling pattern in phygital environments.

QR Code Interaction

A lightweight phygital bridge that links physical objects, signage, or artifacts to digital content, transactions, or continuation states. Often used because it avoids specialized hardware and is visitor-controlled.

Reporting and Transparency (ESG)

The practice of producing accurate, auditable disclosures about ESG performance, controls, and incidents. For venue personalization, reporting and transparency includes what data is used, why decisions are made, how consent is obtained, what is retained, how accessibility is measured, and how language equity is maintained.

Retention Policy

A defined lifecycle for data and derived context, including retention duration, deletion triggers, and exception handling. In this book, retention is treated as a control surface for risk. Short retention reduces linkability, breach impact, and misuse opportunity.

Runbook

A governed operational playbook that defines permitted actions, required verification, and escalation paths when a system enters degraded or incident conditions. In this architecture, runbooks are not troubleshooting guides or vendor manuals. They are the executable interface between governance and operations, ensuring consistent, auditable behavior under stress.

Safe Failure

A defined behavior for when the system lacks sufficient grounding, consent, policy clearance, or operational certainty. Safe failure prioritizes safety, dignity, and clarity, for example by refusing, switching to curated delivery, asking for minimal additional context, or routing to a human, rather than guessing.

Safe State

A predefined system condition that reduces risk to an acceptable minimum during an abnormal or emergency situation. A safe state is not simply "off." It is the state in which hazardous behaviors are stopped, outputs are made predictable, safety-critical information remains available as appropriate (for example static exit signage, baseline lighting, and clear paging capability), and the system remains stable until an authorized operator explicitly returns it to normal operation under defined procedures.

Safety Override

A controlled mechanism that preempts normal automation, personalization, and agent behavior when safety-relevant conditions exist. Safety override defines precedence: safety constraints supersede experience goals. In this book's context, a safety override forces the system into a Safe State, constrains outputs to safety-approved delivery modes, and creates a decision record or incident record indicating trigger condition, affected zones and subsystems, actions taken, and the identity of any operator who initiated, confirmed, or cleared the override.

Session continuity

The system's ability to maintain coherent context, memory, and personalization across multiple interactions, visits, or touchpoints within a single visitor relationship. Session continuity is enabled by the ICL.

Sound bleed

Unwanted transmission of audio from one space into another, which can reduce intelligibility, increase fatigue, and degrade perceived quality. Managing sound bleed is both an acoustic design problem and an operational problem, because zone occupancy and program content change throughout the day.

Spatial AR Interface

An augmented reality presentation layer that anchors digital content to physical locations, objects, or surfaces within a venue. The Spatial AR Interface enables visitors to encounter contextual information as they move through space.

Spatial Computing

A computing model where digital content is anchored to real-world locations, objects, or surfaces, enabling persistent, location-aware digital layers in physical space. Spatial computing is a common technical pathway to phygital experiences.

Tap Interaction (NFC Tap)

An interaction pattern where the visitor brings a device within a few centimeters of a tag or reader to trigger a data exchange or action. Tap interactions are favored in phygital design because they are explicit, visitor-initiated, and lower ambiguity than passive proximity triggers, but they still require governance if the tap creates continuity, logging, or linkability.

Touchpoint

Any moment where a visitor interacts with the venue, including exhibits, kiosks, staff, signage, mobile prompts, queues, entrances, and post-visit follow-up. Phygital design is primarily about making touchpoints behave as one joined-up experience.

URL (Uniform Resource Locator)

A text string that specifies the address of a resource on the internet, and how to retrieve it, most commonly via HTTP or HTTPS. In this book's venue context, URLs are a primary "phygital bridge" payload carried by QR codes, NFC tags, short links, and kiosks, because they can launch content, continue a story, trigger a workflow, or hand off context to a device. A URL is typically composed of a scheme (for example, https), an authority (host, and optional port), a path, and optional query and fragment components. URLs can create linkability if they embed stable identifiers or visitor-specific tokens, so they should be handled under minimization, purpose limitation, and retention controls, especially when used for continuity, analytics, or personalization.

Verifiable Credential

A cryptographically signed credential that can be presented in whole or in part, depending on the policy requirements of a transaction. In practical terms, it supports interactions where a venue can ask for a proof and receive a minimal answer, such as "yes" or "no," rather than collecting an entire document

Verified restoration

The controlled process of returning a system from degraded or incident operation to normal operation only after outcomes have been confirmed in the physical world. Verified restoration requires evidence that the triggering condition has cleared and that affected systems are behaving correctly, rather than relying on assumption or command completion.

World model

In general usage, a computational representation of an environment, its entities, and their relationships that enables a system to reason about states, predict outcomes, and plan actions. World models are foundational to robotics, simulation, and increasingly to large-scale AI systems seeking to understand context beyond immediate inputs.

WorldModel™

The implementation of World Model architecture within this book's governed, agent-mediated framework. The WorldModel™ framework integrates venue-specific spatial data, object registries, operational states, and visitor context into a unified representational layer. Unlike generic world models, WorldModel™ architecture is purpose-built for large venues, encoding not only physical reality but also interpretive relationships, narrative pathways, and governance constraints defined by the Constitution. The WorldModel architecture encompasses the full stack: the operational truth layer (world model), the four decision layers (CGL™, EDE™, ICL™, MAOL™), the substrate systems (edge compute, programmable surfaces, acoustic layer, Personal Channel, Avatars and more), the Web3/Web4/Web5/Web6 identity and consent patterns, and the operational disciplines (lifecycle automation, outcome verification, content operations). When this book refers to "a WorldModel venue" or "WorldModel architecture," it means a venue implementing this complete framework, not merely a venue that maintains state. WorldModel is discussed at the architectural level in this book.

PREFACE

A place is a set of promises made physical

Safety. Dignity. Clarity. Belonging. And what happens when a crowd arrives at the same moment a subsystem fails.

Designers understand those promises. Operators inherit them. Owners fund them. Visitors feel them instantly when they hold, and even faster when they break.

For most of the last century, we built public environments for the mythical median person. One narrative. One language. One pace. We asked everyone to adapt, and for a long time, they did.

That era ended in their pockets.

Stop and consider the word "phone." It understates what this device did to behavior, expectations, attention, and trust. Calling is now a minor function inside a handheld interface to the world, one that made instant translation, personalized guidance, and real-time answers feel normal. Physical venues are now judged against that standard, whether they planned for it or not. You might prefer visitors not to hold devices all the time, but it is hard to avoid, because the device has become the default tool for certainty and continuity. People arrive fluent in on-demand interfaces. A child and a professor can stand side by side in front of the same artifact and require different vocabulary to understand the same truth. A visitor who is easily overstimulated can love the story and struggle with the noise. A guest can want the shortest path forward, or a slow dive into detail.

The audience has not become "difficult." It has become visible.

Personalization is often treated as a feature. In physical space, it behaves more like infrastructure. Once visitors experience adaptive guidance, the absence becomes friction, showing up as confusion, crowding, accessibility failures, and staff burden.

But most "smart venue" efforts fail in one of two ways. They either drift into surveillance, or they ship as impressive demos that operations cannot trust on a normal Tuesday.

This book exists to prevent both failures.

It introduces the WorldModel™ framework: a staged approach for building venue intelligence that can scale without becoming invasive, fragile, or unmanageable. WorldModel™ is not a single product. It is not a mandate to "AI everything." It is a method for deciding what coherence actually requires, and what to demand in procurement so that outcomes remain measurable long after launch.

Personalization becomes dangerous when it is used to justify surveillance, data hoarding, or identity exposure. This is not about knowing more about people. It is about treating people well, with restraint. Many visitor needs can be honored with bounded consent, short retention, and proofs that confirm preference without creating dossiers. A venue can be hospitable without becoming invasive.

The second failure is equally lethal: systems that perform in controlled demonstrations, then degrade in the real world.

A system that works once is not a system. A system is something that can be trusted. Trust is earned through commissioning, maintained through verification, and protected by maintenance that catches drift before the public becomes your diagnostic tool. This book treats the venue as a living operational environment, not an installation frozen on opening day.

The argument is not that places need more technology. The argument is that places need coherence.

Coherence has a cost. Fragmentation costs more.

When systems remain isolated, every new capability increases complexity. Staff become translators between subsystems. Experience fractures across zones. Accessibility becomes an afterthought. Safety becomes a patchwork. The venue becomes harder to operate, not easier. Many of the most advanced destinations in the world are already encountering this ceiling, and they are asking for what this book describes: a venue-wide intelligence layer that helps the environment behave as one place.

This is not a manifesto. It is a reference.

It is written for owners who need outcomes, designers who need patterns, engineers who need structure, and operators who need proof. It does not argue that every venue must become a fully governed WorldModel™ environment immediately. It advocates a deployment ladder that starts with practical wins at the exhibit level, then scales when flow, scheduling, safety, and cross-zone orchestration become the real problem.

The ideas here did not emerge from theory alone. They were forged over decades of commissioning systems that had to work on opening day, and maintaining them through seasons of wear and change. Every pattern has been shaped by the physics of crowds, the behavior of hardware under load, and the quiet expectations of visitors who never read the manual but know immediately when something feels wrong.

The most sophisticated technology in a venue is not the one with the most features. It is the one that disappears into hospitality: the system that anticipates without intruding, the interface that speaks your language without extracting your identity, the environment that adapts without becoming fragile.

If you are an owner, I hope you find clarity about what to demand and what to question. If you are a designer, I hope you find patterns worth stealing. If you are an engineer, I hope you find structure that survives contact with reality. And if you are an operator - the person who inherits whatever gets built - I hope you find evidence that someone was thinking about your Tuesday morning, not just your ribbon-cutting ceremony.

The venues we build today will outlast the technology we install in them. The governance frameworks we establish will shape decisions we cannot yet imagine. The promises we make to visitors, about safety, dignity, and what we will and will not do with their presence, will define whether these places earn trust or merely demand it.

A destination that can change is impressive. A destination that can change while remaining truthful, safe, inclusive, privacy-forward, and operable is the future.

This book is an invitation to build that future, with discipline, imagination, and respect for the people these places exist to serve.

I hope it helps.

-

Maris J. Ensing, Orange, California, February 2026

Part I

THE WORLD CHANGED, AND PHYSICAL PLACES NEED A NEW PLAYBOOK

Physical places did not merely "go digital." They inherited the visitor's pocket, attention, expectations, and comparison set. This section names the change without nostalgia and without blaming audiences for behaving like modern humans.

Chapter 1 begins with the most consequential shift: the phone turned every venue into a context-switching environment.

Chapter 2 follows with the psychological knock-on effects: why visitors now seek agency, reassurance, translation, and continuity, not just information.

Chapter 3 introduces the Personalization Ladder, a practical way to distinguish "nice-to-have novelty" from "venue-scale outcomes."

Chapter 4 draws the bright line: when personalization stops being decoration and becomes governance.

Read this section if you need to explain the "why" to decision-makers, align a team on language, or set criteria for what is worth building. If you are already convinced that visitor experience is now algorithm-shaped, you can skim Chapters 1-2, go straight to the ladder and governance threshold, and return later for the narrative foundation.

1

THE PHONE
CHANGED THE ROOM

If you design, commission, operate, or fund public experiences, this chapter explains the single shift that changed visitor behavior and what it means for what you build next.

If you have watched a family split across languages, a school group drift because no one knows where to go next, or a donor ask why "the AV budget" quietly turned into networks and compute, you have already seen the effects of that shift. This chapter names it and turns it into a practical stance you can bring to design meetings.

There was a time when the most polite thing a museum could ask of a visitor was silence. Silence as focus, the condition that makes attention possible. A quiet gallery made a promise: if you entered it, you would be given the one thing modern life rarely offers, uninterrupted attention.

In that era, phones were mostly instruments of interruption. They rang at the wrong moment, shattered the room, and dragged a visitor out of the present and into the elsewhere. When institutions posted signs asking visitors to put phones away, they were not being prudish. They were protecting a fragile, valuable kind of experience.

> *Then the phone changed, and with it,*
> *the visitor's posture in the world changed as well.*

The modern phone is a universal interface

The modern phone has quietly become the most common piece of hardware in every public space, serving as a universal interface. In the United States, Pew Research Center reports that about nine-in-ten adults, 91%, own a smartphone.[viii][ix]

Treat that number as a design constraint. When something is that widespread, the room behaves as if it exists, regardless of policy preferences or personal taste.

The phone rewired expectations

The more profound shift is psychological. The phone trained people to expect that the world should answer them quickly, that translation should be immediate, and that context should be available on demand. Waiting, searching, and guessing now feel like friction, even when the content is good.

It also trained people to expect frictionless navigation rather than navigation that requires confidence, familiarity, or local knowledge. Finally, it trained people to treat a moment as something they can capture, share, and place in a personal archive.

Physical places still have power. What changed is convenience: screens now own the default path for quick answers. Venues remain special for reasons screens cannot replicate, so they must earn attention in a world where convenience is always in a pocket.

The day the device became a platform

A hinge moment is useful because expectations reorganized around it, even though the forces behind the shift were already in motion. Apple's newsroom press release states that the iPhone went on sale in the United States on June 29, 2007.[x]

You do not need to be an Apple enthusiast to recognize what followed. The phone ceased to be a device you used occasionally and became a platform you lived through. It absorbed the camera, the map, the browser, the ticket, the wallet, and the library, and it turned every person into a publisher and every outing into a potential story.

For venues, the implications were immediate and enduring. Visitors began translating labels themselves, searching for context in real time, and photographing text instead of reading it on the spot, turning "I will learn now" into "I will learn later." They began documenting rather than merely experiencing, often because the moment mattered and they wanted to keep it.

At first, many institutions viewed this as a loss. Phones seemed like a distraction and a barrier between a person and a place. In practice, they were often a symptom of something else. They were a symptom of friction.

What visitors do with phones, and what it means

Watch a visitor in a gallery long enough, and you will see a pattern. People reach for their phones when they do not feel confidently oriented: when they cannot tell what they are looking at, when they want the name pronounced correctly, when they need translation without struggling or guessing, and when they are uncertain about where to go next. They also reach for their phones to coordinate with their group and to carry the moment home.

At its core, this behavior is uncertainty management. This is why conversational systems must be governed.

A fluent assistant can reduce uncertainty, or it can amplify it by confabulating (hallucinating) plausible details that are wrong. In public space, that failure mode is unacceptable in any context where guidance, safety, accessibility instructions, hours, pricing, eligibility, or operational state are involved. The design implication is simple: treat model output as a proposal, require grounding when claims matter, and degrade safely when grounding is unavailable.

That is the first design insight the phone makes unavoidable: your visitor is constantly negotiating uncertainty, and if you do not help them, they will help themselves.

Once you accept that, the debate about phones shifts. It becomes less moral and more practical. Instead of asking whether phones belong in the room, you ask what uncertainties the room is producing and how you can reduce them without degrading the beauty of the experience.

The goal is reliable, inclusive identity and continuity, with phones as one optional carrier. Design so a phone can be used when it adds value, and design so the experience still works when a phone is unavailable. Treat this as an architectural principle. The experience must still work when a phone is unavailable.

In practice, this includes patterns such as Near Field Communication (NFC) tags for quick, consented recognition at an exhibit and Quick Response (QR) codes for onboarding flows. One example of an onboarding pattern is what some teams call the Virtual Docent OnBoard Layer, also referred to elsewhere in this book as an OnBoard Virtual Docent Layer.

The intent is the same: reduce friction and keep the experience inclusive without making the room carry all the complexity in public view. All these patterns must still support alternatives for guests who cannot or do not want to use a phone.

The phone as an opt-in personal layer

A museum has always carried a tension between two kinds of design. One kind focuses on the room as a shared experience: coherence, atmosphere, rhythm, and a narrative voice that belongs to a place. The other focuses on the visitor as an individual: language, accessibility, pacing, comfort, and interest.

Before the phone, these often collided. If you wanted to support multiple languages, you added more signage, and the room became visually cluttered. If you wanted captions, you added screens that competed with the exhibit itself. If you wanted accessibility modes, you relied on specialized devices that were expensive, hard to maintain, and often stigmatizing.

The phone is a way out of that collision. It allows a venue to keep the room coherent while offering individualized modes as a personal layer. It can deliver content in a visitor's preferred language without turning every wall into a compromise, provide captions without forcing them on everyone, and offer alternate audio mixes, including quieter, narration-forward options for visitors who are easily overstimulated. It can deliver sign-language video as an option rather than a permanent overlay and do so without asking the room to carry the entire burden of inclusion in public view.

Inclusive design works best when it is normal, dignified, and easy to adopt. A visitor should be able to participate with dignity, and the personal layer enables that. This is also why the phone becomes a bridge between what we call micro hyper-personalization and what becomes possible later at the macro scale. In the early stages, the phone is the simplest, most universal Personal Channel. In later stages, it increasingly becomes a privacy boundary and a trust boundary.

The phone as a privacy boundary

The more capable personalization becomes, the more critical it is to prioritize privacy. Privacy is what makes adoption possible, and it deserves first-class design attention.

A common failure mode in technology projects for public spaces is that people over-collect, under-govern, and then wonder why visitors do not trust what they built. A phone-based personal layer offers a different posture. It allows the visitor to receive value without the venue collecting a dossier.

Preferences can be stored locally. Sessions can be anonymous. Language can be selected without revealing identity, and accessibility modes can be activated without registration. Even cross-exhibit continuity can be managed with minimal, rotating identifiers, short retention, rather than permanent profiles. This is more ethical and operationally smarter. It reduces liability, fragile integrations, and the number of sensitive systems that must be defended, and increases trust, which increases uptake.

Later in this book, we will discuss decentralized identity and the direction in which personal technology is moving toward minimal disclosure. The critical point here is the attitude: the phone can become the place where privacy is enforced and where trust is expressed through constrained proofs rather than indiscriminate collection.

The same posture applies to content integrity: the personal layer can carry evidence, choices, and explicit modes, so the venue does not rely on confident, ungrounded output.

The selfie moment, reconsidered

If you want a quick test of whether a venue has embraced modern behavior, watch how it treats the camera. Many institutions still treat selfies as if they undermine seriousness. In practice, selfies are a modern way people mark meaning.

When someone takes a selfie, they are not necessarily saying, "Look at me." They are often saying, "I was here, and it mattered. I want to attach myself to the memory. I want to share it with my friends." That has practical implications.

A modern venue needs peaks: moments of beauty, awe, humor, surprise, intimacy, and scale that punctuate experience and become memory anchors. Peaks are a form of hospitality rather than an add-on. A peak is a moment that punctuates the experience. If you design peaks well, the phone becomes a memory anchor, an amplifier of your experience rather than a competitor because visitors create the artifacts that bring the next visitor, and they do it because the moment was real. This is what people talk about afterwards, and what they return for.

What the phone forces us to admit about change

There is another way the phone changed the room that is less visible but more consequential. It shortened the perceived half-life of experiences. When a visitor can step outside a gallery and sees a thousand other options on their screen, the implied question becomes: how current is this place, how alive is it, and does it respond to what is happening now, or is it frozen in the assumptions of its build year?

This is where the conversation shifts from interaction design to systems design. A modern venue cannot be built as a static installation expected to endure unchanged for a decade. That does not mean you rebuild every year. It means you build in a way that allows change without trauma.

This is where compute becomes the quiet lever of modern venues.

When you can provision the right compute in the right place, you can replace nodes without drama, upgrade capabilities without rewriting the entire system, and add modalities, languages, content formats, and operational intelligence as the venue evolves. Replacement becomes a provisioning problem, not a reinvention problem. Upgrades become capacity decisions, not construction crises.

This is not about chasing novelty. It is about building a venue that can remain relevant, inclusive, and maintainable as expectations continue to shift.

The phone did not merely introduce a competing screen. It introduced a culture of continuous update.

Venues that respond gracefully to the culture of continuous updates become magnets. Venues that resist it become brittle.

A practical stance to carry into every design meeting

The most productive outcome of the phone era is not a particular app or technology choice. It is a stance: design for phones without requiring phones.

At the micro scale, this can be deployed with minimal physical change. Start with decision points. When a visitor hesitates, offer a fast opt-in path that reduces uncertainty.

QR codes or NFC tags at decision points can connect visitors to the Personal Channel, and an offline-first module, such as an OnBoard Virtual Docent Layer, can cache and deliver language, age, and interest choices locally. The room stays coherent while the personal layer carries the adaptable content.

If you adopt that stance, several design problems become easier to address. You stop trying to police behavior and start shaping experience. You stop treating multilingual delivery as a graphic constraint and start treating language as a runtime capability. You stop treating accessibility as a special-case retrofit and start treating comfort and modality as normal. You stop building fixed experiences that ossify and start building systems that can evolve.

The good news is that physical places still have an advantage no screen can fully replicate. A room has scale, objects have texture, sound has presence, and a shared human reaction creates a kind of electricity. The phone does not diminish that advantage. It raises the standard for how gracefully a venue must guide someone into it.

Design for phones without requiring phones.

This chapter is the book's beginning for that reason. It names the ground truth on which everything else rests: the phone changed the room. In the next chapter, we turn to what that change reveals about the deeper question every venue now faces: why people still leave home and what kinds of experiences make the trip feel inevitable.

2

A NEW PSYCHOLOGY OF GOING OUT

This chapter identifies the psychological expectations visitors bring with them now and provides you with practical language for explaining those expectations to decision-makers, designers, and operators.

For most of modern history, physical places did not have to justify themselves. A museum, a visitor center, a themed environment, and even a shopping street benefited from a kind of default authority. They were where the day happened. They were where you went to see what could not be brought to you.

That assumption has quietly but completely dissolved.

Today, the alternative to going out is not boredom. The alternative is a near-limitless stream of content, community, and comfort, available at the tap of a thumb. A person can learn, laugh, browse, and socialize without leaving the couch, paying for parking, standing in line, or negotiating a crowd. The friction is low, and the rewards arrive quickly.

This does not mean the public has become shallow. It means the public has become selective. The decision to go out has become about time and emotional return, not just minutes. Was it worth it? Did it feel good? Did it feel easy? Did it feel meaningful? Did I feel seen? Did I feel safe? Did I feel like myself?

Once you accept that, exhibit design stops being an argument about content and becomes an argument about hospitality. Hospitality, in this context, means reducing friction, increasing agency, and creating moments that can exist only in physical reality. It also means designing for how modern people actually behave, including behaviors designers sometimes wish they could outlaw.

The goal of this chapter is simple. It provides a psychological foundation for the technology in the chapters that follow. If you understand why people choose to leave home now, you will build different things and build them in a way that remains positive, inclusive, and sustainable.

Standards for attention are higher

The popular story is that people have shorter attention spans. The more accurate story is that people have higher standards for what earns attention. The common complaint that audiences cannot focus is usually wrong. People focus intensely when they feel ownership of the focus.

They binge on series, learn skills for hours, research purchases until midnight, and fall into deep curiosity when an experience respects their agency.

> *People do not have shorter attention spans. They have higher standards for earning attention*

What has changed is the threshold for earning attention in a public place. In a venue, the old model assumed that the visitor arrived willing to follow the intended path, and the venue set the pace, the order, and the interpretation. Many visitors still enjoy that. A strong linear narrative can be extraordinary. What has changed is that a single mode is no longer sufficient.

Modern audiences expect choice. They want to decide how deep to go, when to linger, what to skip, and what to save. They want the experience to meet them, rather than forcing them to conform to a single ideal visitor. This is why micro personalization can feel powerful even when the underlying technology is modest. The effect is dignity, the sense that the venue did not design only for a hypothetical average person.

In practice, this means the experience must offer at least two rhythms. One is for the visitor who wants the highlights. They are not disrespectful. They are time-constrained or scanning for what matters most to them. If the venue punishes them with friction, they will not suddenly become contemplative scholars. They will disengage. The second rhythm is for the visitor who wants depth. They want nuance, context, and detail, along with the satisfaction of mastery. If the venue offers only shallow summaries, they will feel underfed.

A modern venue does not choose one rhythm. It offers both, without shaming either visitor.

Phones revealed how people behave

Phones did not ruin the room. They revealed how people behave. There was a period when many exhibit teams believed the correct goal was to keep phones away.

The hope was understandable. Phones can fragment attention, pull a visitor out of the room, and turn a shared space into a cluster of private screens. It was tempting to treat the phone as the enemy of presence.

Then reality intervened.

People will have their phones out, whether we like it or not. They will translate labels, search for names, check maps, coordinate with friends, take photos, record short clips, and capture the moment, often because the moment matters to them. Trying to fight this behavior is futile and also wastes an opportunity.

The phone is the most universal accessibility device humanity has ever deployed. It is a screen, a speaker, a microphone, a captioning surface, a personal audio channel, and a tool. Increasingly, it is also a secure identity wallet. A venue that designs with the phone gains a Personal Channel that can deliver individualized value without cluttering the shared space. It can provide language, captions, sign-language video, quieter mixes, and other accessibility modes as opt-in experiences, while keeping the public room coherent and beautiful.

The commitment is straightforward. The experience remains complete without a phone. For visitors who choose to use a phone, the experience becomes richer and more adaptable. The design stance is simple: design for phones without requiring them.

Social proof is participation

In older venue thinking, the camera was often treated as a nuisance. A visitor taking a photo was seen as distracted. A visitor taking a selfie was seen as vain. A visitor filming a moment was seen as rude. Those judgments miss what is actually happening.

Humans preserve meaning.

When a visitor captures a moment, they are not necessarily leaving the experience. They are trying to keep it. They are creating a memory artifact. They are also, often unconsciously, creating social proof. They are telling their world that the trip was worth it.

This has a direct design implication.

This is why a modern venue needs peaks. Peaks are not loudness, spectacle, or forced gimmicks. They are emotional punctuation: moments of wonder, beauty, humor, awe, intimacy, or scale. These peaks do not cheapen the venue. They increase its power. They become anchors in memory and invitations to the next visitor.

The point is to make authentic moments easy to capture when a visitor feels moved to do so. The best social proof is not fabricated. It is a byproduct of genuine experience quality.

Learning works best as discovery

Learning works best as discovery, not instruction. There is a word that tries to merge education and entertainment, and many people dislike it because it suggests a compromise. It suggests that learning must be sugar-coated or that entertainment must be morally justified. That framing is unhelpful.

People do not resist learning. People resist being managed. They resist being lectured. They resist experiences that feel like obligations. In physical venues, learning is most powerful when it feels like discovery. Discovery begins with curiosity, respects pacing, and rewards the visitor quickly enough to keep them engaged.

This is one of the quiet reasons personalization matters. Personalization does not exist primarily to impress. It exists to reduce the friction between curiosity and reward. If a visitor wants the quick story, they should get it. If they want the deeper story, it should be available without penalty. If they want an alternate modality, such as captions, sign-language video, audio description, Braille, or a quieter mix, the venue should not make them feel like they are requesting a special exception. It should feel normal.

When a venue gets this right, it removes the anxiety of doing it wrong. Many visitors feel a low-grade stress in traditional exhibits. Am I missing something? Am I going too fast? Am I going too slow? Am I taking up space? Is there a right way to be here? A good system dissolves that stress by giving visitors paths that fit them and permission to choose. This is modern cognition treated with respect, not dumbing down.

Comfort is access

Comfort is not a luxury. It is access. Accessibility is often framed as compliance. Compliance is important, but it is not the same as participation.

People avoid experiences for reasons that never appear on a checklist. Noise can overwhelm. Crowds can exhaust. Rapid sensory changes can disorient. A lack of language support can make a visitor feel invisible. A lack of control can make a person feel unsafe. Comfort is the condition that enables more people to participate.

This is why inclusive design belongs at the beginning, not at the end. If you treat inclusion as a retrofit, you will build a system that excludes by default and then scramble to patch. If you treat inclusion as a posture, you build different capabilities. You build alternative mixes. You build multiple modalities. You build language as a runtime choice. You build mechanisms for a visitor to control their experience, including intensity, pacing, and sensory load.

You also build trust. Trust in venues grows when a visitor senses that the system is there for them, not for the venue's data appetite. Trust also collapses when a system speaks with certainty it has not earned.

In practice, confabulation (hallucination) is not only a technical quality issue. It is a hospitality issue. If a visitor follows guidance that turns out to be wrong, they stop trusting the system, and they often stop trusting the venue. This is why the book treats confabulation as a risk category that must be contained by mode selection, constraints, evidence requirements, verification, and safe failure behavior, not by hoping the model will "get better."

The more powerful the technology becomes, the more important it is that the visitor feels the venue has boundaries. This book will press the issue of privacy repeatedly because privacy is ethical and practical. Privacy-forward systems are adopted more readily, used more confidently, and defended more easily in public conversation.

The new bargain is coherence

The new bargain is not content. It is coherence. It is tempting to think the path forward is a stack of features: add a few interactive stations, add projection, add a Virtual Docent, add personalization, add analytics, add an app, add a kiosk, add an AI voice, add clever signage. Features do not add up to a reason to go out unless they cohere.

Coherence is what a visitor feels when a place understands itself. It has a voice, a rhythm, and a consistent sense of hospitality. It respects time, attention, diversity, and privacy, and it remains calm under load. Coherence is also what operators feel when a venue is maintainable: the system can be monitored, failures are predictable, recovery is graceful, and change is possible without construction trauma.

This is why the rest of the book treats personalization as an architectural topic, not a feature. At a small scale, personalization can be about content relevance and accessibility. At a larger scale, it becomes part of operations. It begins to touch flow, queues, scheduling, staffing, and safety. That is when governance becomes mandatory, not as a brake but as the mechanism that makes powerful systems responsible in public space.

Even before you reach that scale, the psychology of going out tells you what you must deliver. You must deliver an experience that feels worth it and do so in a way that fits how people behave now.

A practical set of design commitments

This chapter is not a theoretical essay. It is a set of commitments you can bring to every planning meeting.

1. Design for phones without requiring a phone.
2. Offer at least two rhythms: highlights and depth.
3. Treat language as a runtime choice, not as graphical clutter.
4. Make accessibility and comfort the norm, not the exception.
5. Create a few authentic peaks that visitors will want to remember and share.
6. Treat opt-in as a benefit exchange, never as a data grab.
7. Aim for coherence, not a pile of features.

The next chapter introduces the Personalization Ladder. It is the discipline that keeps promises honest. It keeps systems aligned with the scale of the problem, the appropriate privacy posture, and the operational reality of physical places.

If you want to build something that makes people leave home, start here.

3

THE PERSONALIZATION LADDER

This chapter introduces the Personalization Ladder as a practical tool for distinguishing exhibit-level novelty from venue-scale outcomes and for deciding what to build first.

Personalization is one of those words that can mean almost anything, which is exactly why it often disappoints. In venues, it is used to describe a bilingual label, a recommendation engine, a camera that recognizes a returning visitor, and a fully orchestrated environment that seems to anticipate what people need before they ask. When a word stretches that far, it stops being a tool and becomes a fog. Teams talk past one another, budgets drift, and systems are built that look impressive in a demo and quietly unravel in the first month of operation.

The simplest way out of that fog is to stop arguing about labels and start describing scale. Here, scale means the scope of a decision, not the square footage of a building. A venue can personalize at the level of a single exhibit. It can personalize across multiple exhibits. It can also personalize at the venue level, treating the environment as a living system with flow, capacity, scheduling, and operational constraints.

These are not marketing tiers. They are distinct types of problems, with different risks and different architectures. The ladder helps you decide what you are building, what you must not promise, what you can responsibly measure, and what must be governed. It is also a permission slip. You do not need the most complex version of personalization to deliver value.

A well-executed micro deployment can transform a museum. A well-executed macro deployment can transform a small city. The ladder is how you know which one you are actually trying to do.

Hyper-Personalization as a Governed System

Hyper-personalization is frequently described as tailoring content to individuals. That framing is insufficient, and in practice, misleading.

At venue scale, hyper-personalization is not a content feature. It is a governed system that determines how approved material is selected, assembled, adapted, and, where explicitly permitted, generated at runtime. The system operates across media, space, time, and modality. It must remain coherent under load, under partial failure, and under changing operational priorities.

Confabulation is one of the reasons this must be a governed system.

Generative systems can produce fluent, plausible statements that are not grounded in approved sources or operational truth. In public environments, that failure mode cannot be treated as an edge case. It must be assumed, and designed around.

The design posture used throughout this book is simple: model output is treated as a proposal. Delivery becomes a decision only after the system applies constraints, checks evidence requirements where factual claims matter, and selects an appropriate delivery mode for the risk level of the moment.

This distinction matters. A feature can be added. A system has to be designed.

From personalization as output to personalization as process

Most legacy "personalization" approaches treat variation as an output choice. A visitor selects a language. A different clip plays. A different audio guide is launched. The underlying experience remains static, and the system simply chooses between pre-authored variants.

Hyper-personalization, as used in this book, refers to a different model. Variation is not an endpoint decision. It is the result of a controlled process that operates continuously during an interaction. The system is aware of context, state, and constraints, and it uses that awareness to determine how the same canon should be delivered at a given moment.

That process must be explicit, testable, and governable. Otherwise, personalization becomes brittle, opaque, or unsafe.

In practice, "governable" includes four non-negotiables:

- First, the system must be able to constrain what sources are allowed for factual claims.
- Second, the system must be able to point to those sources when claims matter.
- Third, the system must be able to refuse to invent when grounding is insufficient.
- Fourth, the system must degrade safely without punishing the visitor, including switching to curated or Body of Knowledge modes, or escalating to staff when necessary.

The governed media pipeline

A robust hyper-personalization system operates through a media pipeline with four distinct operations. These operations are architectural primitives. They should be named explicitly rather than implied.

1. Select.
    Approved assets are chosen based on context. Context may include language preference, accessibility mode, interest domain, pacing preference, proximity, time of day, crowd state, and operational mode. Selection never alters canon. It determines relevance.

2. Assemble.
    Selected assets are sequenced into a coherent presentation. Order, duration, emphasis, and transitions may change. Assembly allows the same material to support different depths, time budgets, and narrative paths without duplicating content.

3. Adapt.
    Delivery characteristics are modified without altering meaning. This includes pacing, verbosity, layout, typography, subtitles, audio balance, modality, and spatial routing. Adaptation is where inclusion, accessibility, and dignity are actually implemented.

4. Generate, bounded (also called constrained generation in this book).
    Where synthesis is permitted, new delivery instances may be produced under explicit constraints. Approved sources, allowed transformations, tone limits, length limits, and governance rules apply. Generation adapts delivery. It does not extend canon. When bounded generation is used, the system must be able to point to approved sources for factual claims, and it must refuse to produce confident claims when grounding is insufficient.

These operations may occur independently or in combination. The system must be able to express each one distinctly in order to be governed.

Delivery modes are policy choices, not technical accidents

From the same pipeline emerge multiple delivery modes. These are not marketing categories. They are policy choices that determine how the system constrains confabulation, how it expresses evidence, and what kind of verification is required.

- Curated delivery.
  Content is delivered exactly as authored. The system selects between pre-approved variants and does not synthesize new claims. This mode is appropriate where regulatory sensitivity, curatorial intent, or brand requirements demand strict control.

- Body-of-Knowledge driven delivery.
  Content is generated only from an institution-owned Body of Knowledge. The system can vary depth and phrasing, but it must remain inside approved sources. Where factual claims matter, the system must be able to point to the approved source material it used.

- Hybrid delivery.
  A curated narrative structure is combined with Body-of-Knowledge delivery and controlled adaptation. This mode balances authority with responsiveness. It is often appropriate for core exhibitions where the institution wants editorial control while still supporting different depths, languages, and accessibility modes.

- Bounded generation.
  Synthesis is permitted only within explicit governance constraints. This mode is used sparingly, deliberately, and auditable. Where claims matter, evidence pointers are required. When grounding is insufficient, the system must refuse to invent and either fall back to a stricter mode or escalate to a human.

A single venue may use all four modes across different spaces, audiences, and operational conditions. The critical point is that the mode is chosen intentionally, and it is visible to operators through decision evidence, not inherited accidentally from tooling.

Hyper-personalization requires runtime state

Hyper-personalization cannot be reduced to precomputed choices. It requires runtime state.

State may be derived from many sources:

1. Interaction history,
2. Spatial position,
3. Recognition signals,
4. Show-control cues,
5. Operational mode,
6. Crowd conditions,
7. Accessibility settings,
8. Consent boundaries.

The system must be able to accept state, reason about it, and act on it deterministically. This is why compute at the node, orchestration, and governance are inseparable from personalization. A system that cannot reason at runtime cannot personalize safely at scale.

Governance is what makes variation safe

The primary risk of hyper-personalization is not technical. It is loss of control.

Governance exists to prevent that loss. It constrains what the system may do, when it may do it, and why. Governance defines:

1. What sources are permitted,
2. What transformations are allowed,
3. What correlations require consent,
4. What behaviors are prohibited,
5. What must be auditable.

Without governance, personalization becomes arbitrary. With governance, it becomes reliable. This book treats governance as continuous. It is not a review step. It is a live constraint that shapes every decision in the pipeline.

Why this definition matters

The rest of this book assumes that hyper-personalization is a system property, not a feature flag.

Recognition, multilingual delivery, non-visual navigation, concierge behavior, programmable canvases, and adaptive audio all depend on the same underlying capability: the ability to vary delivery at runtime under governance.

If personalization is treated as an add-on, these capabilities fragment. If it is treated as a governed system, they compose.

This is the line that separates "personalized content" from adaptive environments. The remainder of the book is concerned with how to build, govern, and operate the latter.

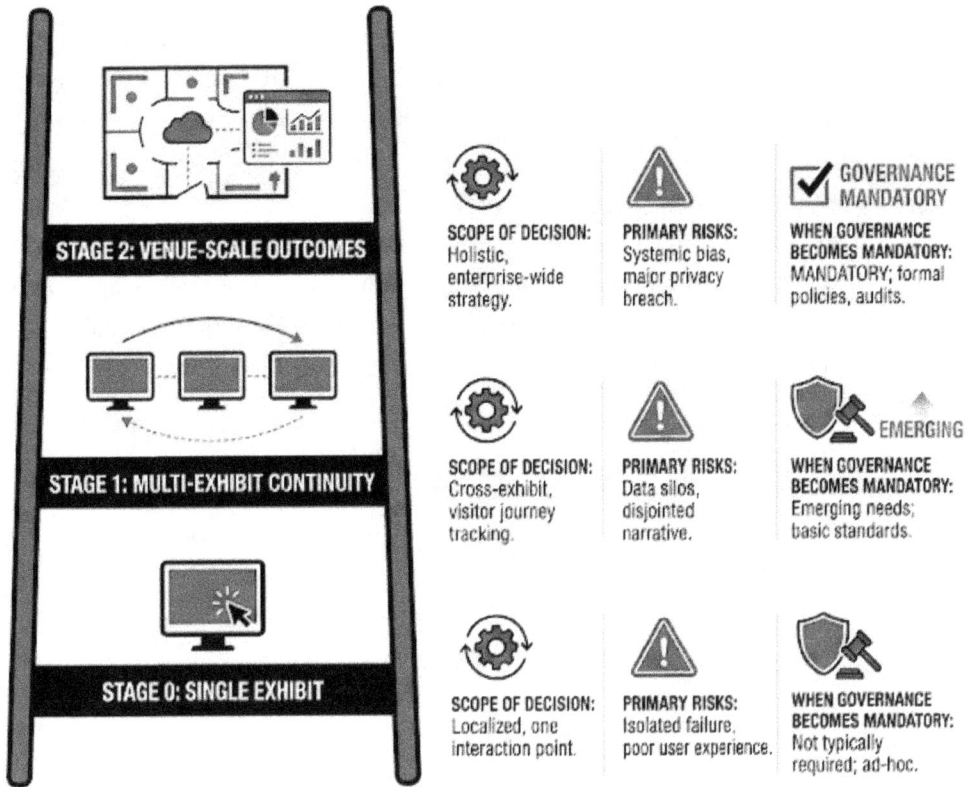

STAGE 2: VENUE-SCALE OUTCOMES

SCOPE OF DECISION: Holistic, enterprise-wide strategy.

PRIMARY RISKS: Systemic bias, major privacy breach.

GOVERNANCE MANDATORY

WHEN GOVERNANCE BECOMES MANDATORY: MANDATORY; formal policies, audits.

STAGE 1: MULTI-EXHIBIT CONTINUITY

SCOPE OF DECISION: Cross-exhibit, visitor journey tracking.

PRIMARY RISKS: Data silos, disjointed narrative.

EMERGING

WHEN GOVERNANCE BECOMES MANDATORY: Emerging needs; basic standards.

STAGE 0: SINGLE EXHIBIT

SCOPE OF DECISION: Localized, one interaction point.

PRIMARY RISKS: Isolated failure, poor user experience.

WHEN GOVERNANCE BECOMES MANDATORY: Not typically required; ad-hoc.

Stage 0: Exhibit-scale localization

Stage 0 is where most venues begin, and where many should remain for longer than they think.

At this stage, personalization means choice at the point of experience. A visitor approaches a single exhibit, and the exhibit adapts to their preference in the moment. The most common example is language selection. The interface asks, politely and plainly, "Which language would you like?" The visitor chooses, and the exhibit responds.

This is foundational, and it does not depend on identity. No profile is required. No long-term memory is required. The system does not need to know who the visitor is, where they live, or what else they have seen. It only needs to honor a preference for the next few minutes. That is not a trivial capability. It is the difference between access that feels normal, and access that feels like a workaround.

The Stage 0 design trap is the temptation to hard-bake language into graphics. You have seen it: large English, smaller Spanish, and perhaps one more language squeezed in as an afterthought. It looks inclusive, but it performs exclusion. It suggests which language is "default," and which languages are "accommodations." It also fails for tourist sites and multilingual cities, because it assumes two languages are enough.

Stage 0 done well treats language as a runtime parameter. Text is rendered, not printed into a layout. Audio is selected, not embedded as a single track. Captions are available as a mode, not as a permanent overlay that competes with the visuals for everyone. When the system is designed this way, language becomes flexible, equitable, and expandable.

Technically, Stage 0 is also where a venue begins to learn what "compute at the edge" really means. You can run an exhibit locally, with the right compute provisioned in a node near the experience. That node does not need to be exotic. It needs to be robust, maintainable, and replaceable. If the node fails, the exhibit should fail gracefully or fall back to a simpler behavior. In a real venue, resilience is part of experience quality.

Stage 0 has a clear boundary: adaptation stays inside the exhibit. It remains local, avoids venue-scale flow control, and does not allocate shared resources. The venue is not yet making venue-wide decisions. That boundary is a virtue. It lets you deliver real value while keeping privacy straightforward and risks local.

Stage 0 signals that you are doing it well

You can say "yes" to most of these:

- A visitor can choose language without creating an account.
- The exhibit can deliver multiple language tracks without redesigning graphics.
- Captions can be enabled without cluttering the experience for everyone.
- The exhibit works even when the network is degraded.
- Replacing the hardware is a provisioning exercise, not a redesign.

If that is your reality, you have done something rare. You have built something that respects the visitor, and respects operations.

Stage 1: Micro hyper-personalization across multiple exhibits

Stage 1 begins when the venue wants continuity. Not identity in the surveillance sense. Continuity in the experience sense.

At this stage, personalization is no longer confined to a single exhibit. The visitor moves through a sequence, and the venue begins to remember how that visitor wants to be treated. Language persists. Content depth persists. Interests persist. Accessibility modes persist. The venue becomes coherent, rather than episodic. Stage 1 is where you earn loyalty, because you stop making people repeat themselves.

The most important concept in Stage 1 is optionality. A visitor should be able to experience the venue without registering anything. If they choose to register, it should be because the value is obvious, immediate, and dignified. Registration is not the price of admission. It is a benefit exchange.

A Stage 1 profile is not a dossier. It is a set of declared preferences. Think of it as a personalization contract between the visitor and the venue.

A well-formed Stage 1 profile typically includes:

- Language preference, and possibly a ranked list, not a single choice.
- Interests, expressed as a small set of themes appropriate to the venue.
- Delivery style, such as concise versus deep, playful versus technical.
- Accessibility preferences, such as captions, sign language, quieter mixes, or simplified narration.
- Visiting archetype, best treated as a depth mode rather than a label.

Those depth modes can be simple, and surprisingly powerful. Some visitors are Streakers: they read headlines and want the core idea with minimal friction. Some are Strollers: they read headlines and a bit more, enough context to orient and connect. Some are Students: they read everything they can, and want the full narrative, the technical detail, and adjacent sources. These are modes, not identities, and a person can shift between them across a single day.

Stage 1 is also where the venue's content strategy matures. You stop writing a single narrative. You begin to write a narrative space. In a car exhibit, a visitor might care about design, aerodynamics, engines, and mechanical detail. In an art exhibit, the interests might be technique, history, symbolism, or the artist's life.

In a science exhibit, the interests might be fundamentals, applications, and contemporary relevance. Stage 1 gives you a way to honor those differences without building separate exhibits for every kind of person.

This is where platforms like Virtual Docent Layer and Personal Channel become practical in a way that is visible to visitors. The Virtual Docent Layer can behave like a Virtual Docent that adapts to declared interests and preferred depth. It can keep the venue coherent for someone who wants a consistent narrative voice. It can serve school groups differently from tourists and do so without forcing a choice that feels like a quiz.

The Personal Channel extends that coherence into the visitor's Personal Channel. When people use their own devices for synchronized audio, captions, or sign-language video, you gain a privacy-forward way to deliver individualized accessibility modes without turning the entire room into a compromise. The same show can be experienced in different languages and different sensory intensities without cluttering the shared space. Stage 1 is where inclusion stops being a retrofit and becomes a design posture.

Consider a theatre show in a venue where the dominant language in the room changes by the hour. If the main room audio is presented in Spanish for one showing, and in English for the next, you have already improved relevance for many visitors. But the world is more diverse than any two-language plan. A visitor who prefers Mandarin should not be pushed into silence.

A visitor who wants sign language should not be forced into a separate seating section with specialized hardware. A visitor who is easily overstimulated should not be excluded by sheer volume and layered sound design. Stage 1 can solve this with dignity: the room presents a main mix, the Personal Channel delivers alternatives, and the visitor opts in without spectacle.

Privacy remains central here. Most Stage 1 systems can be built to operate with minimal identity, because what you really need is preference continuity, not a real-world name. You can maintain continuity across multiple exhibits with anonymous session mechanisms, rotating tokens, and short retention. When a visitor wants cross-visit continuity, you can shift to stronger identity methods, but those should be opt-in and minimal, by design.

Stage 1 done well feels like the venue is listening. Stage 1 done poorly feels like the venue is collecting. The difference is not a marketing statement. It is architecture, and policy.

Stage 1 signals that you are doing it well

You can say "yes" to most of these:

- Visitors can get value without registering, and registration is clearly optional.
- Preferences persist across multiple exhibits, locations, or stations without forcing a data-heavy identity.
- Language and accessibility modes are supported without graphic clutter.
- Content has multiple depths, and the system can select them coherently.
- Operators can update content without rebuilding hardware.
- The experience remains stable when networks are imperfect.

Stage 1 is the point where hyper-personalization becomes real for most venues, and it is often enough to justify the next decade of investment. It is also the point where many teams become tempted to promise more than they can support. That is why the ladder has a third stage.

Stage 2: Macro hyper-personalization in WorldModel Venues

Stage 2 begins when the environment itself becomes part of the decision.

At this stage, personalization is no longer primarily about which story a visitor hears. It is about what the venue does. The system starts making decisions that have externalities. It affects the distribution of people across space, influences queues, shifts demand between galleries, lands, decks, or districts, and coordinates show timing and capacity. It balances experience quality against throughput, accessibility requirements, and safety constraints. It may even coordinate commerce and staffing, because those are not separate systems in a real venue. They are part of the lived experience.

This is the stage where venues begin to resemble small cities because they contain the same operational complexity: transportation, scheduling, crowd dynamics, staffing constraints, and public safety requirements. Human diversity is present at scale, so fairness and inclusion must be expressed as venue-wide policy rather than handled exhibit by exhibit.

Stage 2 is where a layer like Venue Concierge Interface becomes essential, because it provides an interface between the visitor's intent and the venue's operational state. A request for guidance becomes a negotiation with reality: what paths are open, what zones are congested, what shows have capacity, what accessibility routes are available, what noise levels are tolerable, what the weather is doing, what the staff load is, and what the venue is trying to accomplish in the next thirty minutes.

Stage 2 cannot function without a representation of state that is more than a map. It needs a WorldModel framework: a living representation of the venue as it is now, including constraints, resources, and the consequences of actions.

This is also where systems must care about layers that are sometimes treated as infrastructure footnotes, because at Stage 2 those layers become decision variables. The acoustic layer matters because sound bleed affects comfort, comprehension, and perceived quality. The power layer matters because energy consumption, thermal constraints, and equipment life shape what the venue can sustain. The reliability layer matters because the cost of failure is no longer a broken exhibit. It is cascading disruption.

Stage 2 makes governance mandatory because decisions become consequential. Routing can misroute. Queue balancing can produce unfair outcomes. Scheduling can exclude visitors who need accessibility accommodations unless those constraints are explicit. Throughput optimization can damage meaning, and engagement optimization can create congestion and fatigue.

Governance is the mechanism that turns capability into responsibility. It is how the system ensures it is aligned with the venue's values, constraints, and obligations. In practice, governance means decisions are evaluated against explicit policy. The venue can specify objectives and hard constraints. The system can be audited. Operators can override. Privacy and inclusion are enforced rules, not aspirational slogans.

Stage 2 is also where privacy must be pressed hardest, because the temptation to over-collect grows as operational stakes rise. At macro scale, the question is not "can we recognize people?" The question is "what do we actually need to operate responsibly?" Many macro decisions do not require identity. They require flow metrics, occupancy, queue times, and state. Anonymous analytics can often do most of the work.

When identity becomes necessary, it should be opt-in and minimal, with well-defined purposes, retention rules, and clear benefits to the visitor. This is where decentralized identity and credential models belong. The book will cover those later, because they enable a critical shift: a venue can request a proof rather than a document. A system can ask a question and receive a yes or no, rather than acquiring a packet of sensitive personal data that it cannot justify retaining. That is how trust scales without turning venues into liability factories.

Stage 2 done well feels like the venue is alive in a practical sense. It responds, adapts, stays calm under stress, respects privacy, and includes more people by default. Stage 2 done poorly feels like an optimization engine applied to humans. The ladder is how you keep it on the right side of that line.

Stage 2 signals that you are doing it well

You can say "yes" to most of these:

- The system can guide and balance flow without requiring personal identity by default.
- Objectives and constraints are explicit, written, and enforceable.
- Operators can see what the system is doing, and why.
- Safety and accessibility constraints are treated as hard constraints, not recommendations.
- The venue can degrade gracefully under load, rather than failing abruptly.
- Decisions can be logged and audited without turning logs into personal dossiers.

The ladder as a design discipline

The ladder is not a maturity model that shames small deployments. It is a discipline that prevents category errors.

If you are building Stage 0, do not promise Stage 2. Promise excellence at the point of experience: multilingual delivery that feels equitable, and accessibility modes that are immediate and dignified.

If you are building Stage 1, do not let registration become a data grab. Make it a value exchange. Give visitors control, and make continuity feel helpful rather than invasive.

If you are building Stage 2, do not treat governance as bureaucracy. Treat it as the enabling layer that allows powerful systems to operate safely, fairly, and transparently in public spaces.

Across all stages, treat compute as the underlying lever. When you can provision the right compute in the right place, you can keep the system modular. You can expand capabilities without rewriting the venue. You can incorporate future hardware advances without rebuilding the architecture. That is how systems become sustainable.

The point of this book is not to sell a ladder. It is to give people a language for building the right thing. In the chapters that follow, we will walk down the ladder in detail. We will start with the substrate, because the substrate determines what is possible. Then we will build up into multi-exhibit coherence. Then we will step into venue-scale orchestration and governance, where systems stop being about content delivery alone and begin to shape the lived behavior of a place.

Hold one idea now, because it will keep returning in different forms: personalization is not a feature. It is an agreement between a human and an environment. The ladder is how you decide what kind of agreement you are making.

A short, practical self-assessment

You are in Stage 0 if:

- The experience is mostly self-contained at a single exhibit.
- Language and accessibility are chosen locally.
- No continuity across exhibits is required.

You are in Stage 1 if:

- Preferences need to persist across multiple exhibits.
- Visitors can opt in to a lightweight profile for better continuity.
- The system is optimizing the individual or group experience, not the venue's operational state.

You are in Stage 2 if:

- You are managing flow, queues, scheduling, or capacity across zones.
- Decisions have externalities that affect other guests.
- You need explicit constraints, auditability, and operator control, because outcomes matter at venue scale.

If you classify a project honestly at the beginning, you will build faster, and you will build better. If you classify it incorrectly, you will either overbuild an exhibit that never needed it or underbuild a venue that cannot safely operate what it is trying to become.

That is why the ladder is Chapter 3. It is the hinge the entire book turns on.

Decision checkpoint: know which stage you are building

If you cannot state the stage, you cannot commission it.

- Stage 0: the exhibit behaves the same for everyone and still works beautifully.
- Stage 1: the venue remembers a session, with consent, without creating a permanent identity by default.
- Stage 2+: decisions create externalities, so governance, evidence, and operator authority become non-optional.

If any component generates visitor-facing claims, guidance, or instructions at runtime, confabulation controls become non-optional, including approved source boundaries, safe refusal behavior, and operational monitoring.

For why this is hard to replicate, see the front matter section titled "What makes this hard to replicate."

4

WHEN PERSONALIZATION BECOMES GOVERNANCE

Personalization begins as a courtesy.

A visitor chooses a language at an exhibit, and the exhibit responds. A family selects a shorter narrative because the children are restless, and the system respects their pace. Someone enables captions, or chooses a quieter mix, and the room becomes available to them in a way it was not before. These are not small things. They are the difference between a venue that feels indifferent and a venue that feels hospitable.

Then personalization crosses a threshold. It stops being only a courtesy and becomes consequential.

As personalization crosses the governance threshold, intent, authority, and accountability become system properties, not feature attributes. The appearance of agentic governance frameworks in 2026 is a lagging indicator of this shift. They formalize what operational deployments reveal quickly: once autonomy becomes consequential, governance must be enforced structurally, not documented aspirationally.

That moment arrives when the system's decisions begin to shape the experience of other people, not only the person receiving the personalization. It arrives when the system starts allocating shared resources such as time, space, attention, seats, pathways, and staff. It arrives when the system begins to move people through an environment, not merely deliver content to them.

At that point, personalization is no longer just an experience feature. It becomes public infrastructure, and public infrastructure demands governance.

Governance, in this book, means responsibility expressed as design. It means making the system safe to operate at scale, in a place where choices have consequences, trade-offs are unavoidable, and the public has every right to expect fairness, transparency, and restraint. The aim of this chapter is to make that shift crisp, so you can recognize when the rules change, and build with confidence instead of improvising under pressure.

The externalities test

The cleanest way to identify the governance threshold is to stop talking about technology and start talking about externalities. An externality, in this context, is any effect of a decision that lands on someone other than the person who triggered it.

When a visitor chooses their preferred language at a single exhibit, the externalities are near zero. Other visitors are not harmed. Capacity is not reallocated. Safety is not affected. If something goes wrong, the failure is local. The visitor can try again, choose another mode, or move on.

When a system starts directing visitors away from a crowded gallery, the externalities appear immediately. That redirection changes who sees what, when they see it, and how long it takes to get there. It changes congestion elsewhere and it changes staffing needs. It can change accessibility outcomes. It can also change mood, because a congested space feels different from a calm one, and visitors carry that feeling into the rest of their day.

The externalities test is simple to state: if the system's decisions can alter other people's experience, safety, access, or opportunity, the system requires governance.

Confabulation makes this threshold sharper.

When decisions have externalities, a system cannot be allowed to invent. Any visitor-facing claim that affects routing, safety, accessibility, eligibility, timing, closures, or instructions must be treated as safety-adjacent. In those contexts, model output is treated as a proposal, and the system must either ground it in approved sources and operational truth, or refuse and escalate.

That is a practical claim, not a moral one. Without governance, a system at this scale will behave like a well-intentioned amateur. It will solve the problem it sees and accidentally create a new problem somewhere else. It will optimize for what it can measure easily, and it will degrade what is harder to quantify, such as fairness, comfort, or meaning.

A venue can survive small mistakes at exhibit scale. A venue cannot survive systemic drift at venue scale.

Why "optimization" is a trap word

When a system begins to shape flow, it is tempting to speak the language of optimization. It sounds modern, scientific, and reassuring. In public environments, optimization is rarely the real goal.

A venue is not a factory line. It is a cultural, emotional, and operational ecology. It has competing goals that must be held in balance. A museum cares about meaning, learning, and emotional resonance, and it also cares about throughput, safety, and inclusivity. A theme park cares about awe, pace, and spectacle, and it also cares about queues, crowd behavior, staff load, and incident response. A cruise ship cares about delight and convenience, and it also cares about drills, capacity constraints, and the reality that you cannot simply leave.

As soon as a system is asked to do venue-scale work, it is asked to balance competing objectives. That is not a technical detail. It is the nature of the job. Governance is how those balances become explicit rather than accidental. It is how a venue decides, ahead of time, what it will prioritize when trade-offs appear, which they always do.

What changes at venue scale

At exhibit scale, personalization is mostly about content and comfort. It is about relevance and mode. At venue scale, personalization touches the operating functions of a place:

- Dynamic wayfinding, including accessibility-aware routing.
- Load balancing across galleries, lands, decks, or districts.
- Queue management, including predicted waits and routing choices.
- Show management, scheduling, and capacity allocation.
- Staff deployment and operational interventions.
- Food and beverage (F&B) distribution, timing, and availability.
- Safety posture and incident response coordination.
- Acoustic comfort at the venue level, including sound bleed, and sensory load management.
- Power behavior, equipment life, and maintenance posture, because what you run, and when you run it, has real operational costs.

Notice what all these share. They affect more than one person at a time. They also blend guest experience and operations. In a real venue, those are not separate systems. A beautifully written narrative means little if a family spends the day stuck in bottlenecks. An efficient flow plan means little if it destroys the sense of discovery that makes a museum feel human.

Venue-scale systems must therefore be designed to behave like good operators. They must be calm, constraint-aware, and accountable.

That is governance.

Governance as a positive capability

Some teams fear governance because they imagine it as a brake. Governance is what makes ambitious systems deployable. Without governance, the venue cannot trust the system to act responsibly under stress, so the system must remain modest. With governance, the system can be allowed to do more, because its actions are constrained by explicit policy.

A governed system is not timid. It is capable, and dependable. A governed system can make suggestions without becoming manipulative, route without becoming exclusionary, adapt without becoming unpredictable, measure without becoming invasive, and learn without becoming unaccountable. The goal is trustworthy autonomy.

The three ingredients of governance

Governance can be discussed in philosophical terms, but in a venue it must be operational. A workable governance layer has three ingredients.

1. **Explicit objectives**
   The venue must state what it is trying to accomplish, and it must do so in language that is concrete.

   For example:
   - Reduce congestion without reducing discovery.
   - Improve accessibility outcomes without segregating the experience.
   - Increase throughput without lowering experience quality.
   - Increase dwell in certain zones without creating fatigue or bottlenecks.
   - Improve clarity without turning the venue into a signage farm.

   Objectives can conflict. That is not a failure. That is reality. Writing them down is the beginning of coherence.

2. **Hard constraints**
   Objectives can be negotiated. Constraints cannot. Constraints are the lines the system must not cross, even if it could improve a metric by doing so.

   Examples:
   - Do not route a visitor into a path that is not accessible for their declared needs.
   - Do not recommend an experience that is closed, unsafe, or beyond capacity.
   - Do not increase volume beyond a defined comfort threshold in certain zones.
   - Do not collect or retain personal data beyond defined purpose, consent, and retention policies.
   - Do not make identity a requirement for basic participation.
   - Do not produce outcomes that systematically disadvantage certain groups.

   Do not present ungrounded output as fact in any context where it can affect safety, accessibility, eligibility, timing, or the visitor's choices.

Where a claim matters, require approved sources or operational truth, and refuse to invent when grounding is insufficient.

Constraints are where values become enforceable behavior. They are also where a venue's obligations become technical.

3. **Auditability and override**
   If a system influences public behavior, it must be accountable. That does not require turning every action into a bureaucratic log.

4. It requires that the venue can answer three questions when it matters:
   - what did the system do?,
   - why did it do it?, and
   - what were the conditions at the time?

   For any action or output that is consequential, the evidence should include the decision record: the operational state used, the constraints checked, the chosen delivery mode, the outcome (approve, modify, or reject), and any operator override.

   It also requires that staff can intervene as a normal part of operations, not as a panic button. Good venues already rely on operator judgment, because reality is messy. A governed system respects that. Auditability and override are signs of professional maturity.

Privacy pressure increases at venue scale

At venue scale, the temptation to over-collect grows. That is a predictable risk, and it must be addressed with clarity. The privacy principle remains simple: anonymous by default and opt in only when the visitor receives clear benefit. Most venue-scale decisions do not require identity. They require state.

State includes counts, dwell, congestion, closures, schedules, and capacity. It includes acoustic and operational conditions. It includes knowledge of what is open, what is full, what is quiet, what is loud, what is broken, and what is about to start.

A WorldModel™ Venue, in the sense used in this book, is built to reason about state, not to accumulate dossiers. Identity becomes relevant when the visitor asks for continuity, entitlements, accessibility persistence, or membership benefits. Even then, the goal is minimal disclosure, and consent is explicit.

Later chapters will treat decentralized identity, credentialing, and "yes or no" proofs as a natural way to scale trust without scaling risk. The point here is posture: venue-scale systems must be privacy-forward, or they will not be adopted, defended, or sustainable.

The turning point: from displays to decisions

It is worth naming the turning point in a way that designers and operators can recognize. The turning point is when the venue stops treating technology as display infrastructure and starts treating it as decision infrastructure.

Display infrastructure shows things. Decision infrastructure chooses. Once a system chooses, it needs a disciplined relationship with policy, constraints, and accountability. That is governance.

This is also where a layer like Venue Concierge Interface becomes meaningful as an interface concept. At exhibit scale, the visitor interacts mostly with exhibits. At venue scale, the visitor begins to interact with the venue as a whole. They ask the place for outcomes: guide me, help me, reduce my friction, find what fits me, and keep my day coherent.

The venue, in response, must negotiate between what the visitor wants and what the venue can responsibly do under current conditions. That negotiation is a governed decision process grounded in operational reality.

The WorldModel solution is the idea that the venue's operational reality can be represented as state. The Cognitive Governance Layer™ (CGL™) is the idea that proposed actions can be evaluated against explicit objectives and hard constraints. Those are architectural ideas. They remain patent-safe when described at this level, and they remain crucial.

What governance feels like to the visitor

Done correctly, governance is largely invisible. A governed venue feels calmer and more coherent. It feels as if the place is paying attention, in a hospitable way, not in a creepy way. Bottlenecks ease rather than intensify. Guidance feels timely rather than nagging. Accessibility feels normal rather than exceptional.

Most importantly, a governed venue respects the visitor's agency. It offers, suggests, and guides. It does not coerce through obscurity. It can be clear about why it is making a suggestion when that clarity matters. It allows the visitor to ignore the suggestion without punishment. This is how sophisticated systems remain human.

What governance feels like to the operator

To operators, governance is the difference between a clever system and an operable system. Operators do not want novelty. They want reliability, visibility, and predictable behavior under stress.

A governed system supports operators by making its objectives explicit, making its constraints enforceable, making its behavior visible when needed, providing intervention points staff can use, and degrading gracefully rather than failing dramatically.

This is also where analytics and verification layers become central later in the book. A venue cannot operate a macro system on hope. It needs observability, and it needs trust that what it thinks is happening is actually happening.

Those are part of governance in practice.

A practical governance threshold checklist

If you answer yes to any of them, you are in governance territory.

1. Does the system route visitors, or influence where they go next?
2. Does it allocate shared capacity, such as seats, queue positions, time slots, or limited experiences?
3. Does it balance load across zones, galleries, lands, or decks?
4. Does it affect staff deployment, scheduling, or operational interventions?
5. Does it change environmental conditions, such as audio levels, lighting states, or power modes, based on occupancy or predicted demand?
6. Does it create recommendations that could systematically advantage some visitors over others?
7. Does it use recognition, entitlements, or identity continuity that, if mishandled, would erode trust?
8. Would a mistake affect more than one person at a time?
9. Does any component generate or summarize visitor-facing information at runtime in a way that could be acted upon, including routing guidance, safety instructions, accessibility instructions, eligibility, or hours, where confabulation risk must be actively contained?

Governance is not a label you add to a project at the end. It is the posture you adopt at the beginning when these conditions are present.

The chapter's conclusion

The ladder in Chapter 3 is not a climb toward complexity. It is a way to match architecture to responsibility.

Stage 0 and Stage 1 can be transformative, and they can remain largely local, largely anonymous, and largely independent of venue-wide policy. They can be built to be elegant, inclusive, and practical.

Stage 2 is different. Stage 2 is where the venue itself becomes a system that acts. When a system acts at venue scale, it must do so under explicit objectives, hard constraints, and accountability. That is how we respect the operators who must keep them running, and the visitors whose day is shaped by every decision the system makes.

Governance is not the opposite of imagination. Governance is what allows imagination to be deployed.

In the next part of the book, we move from psychology into substrate. We will talk about the enabling structures that make modern venues flexible and maintainable, and why provisioning sufficient compute in the right nodes is not a technical preference, but a strategic choice.

Part II

THE ENABLING SUBSTRATE,
PROVISIONING
THE RIGHT COMPUTE

This part is the engineering spine. If Part I explains why the old playbook broke, Part II explains what has to exist in the venue for personalization to be reliable, accessible, and operable.

Chapter 5 defines what "provisioning sufficient compute" means in a physical place, and why edge capability is not a luxury.

Chapter 6 introduces programmable canvases and software-defined surfaces, the shift from fixed endpoints to reconfigurable experience layers.

Chapter 7 covers non-visual navigation and guidance as first-class infrastructure, not an accessibility afterthought.

Chapter 8 puts audio where it belongs: as a controllable layer with measurable outcomes, not as an uncontrolled byproduct.

Chapter 9 addresses networking, timing, resilience, and observability, the difference between "it worked in the lab" and "it behaves in the field."

Chapter 10 ties orchestration, show control, and power quality to experience quality.

If you are an owner, this part tells you what to require. If you are a designer, it clarifies what constraints are real. If you are an operator, it shows what must be instrumented so the venue can be supported without heroics.

Without this substrate, personalization becomes brittle, expensive to support, and easy to overpromise.

5

PROVISIONING THE RIGHT COMPUTE AT THE EDGE

The origin story is on-site reality

The Compute Nodes concept did not come from a theoretical desire to modernize AV. It came from the lived experience of building and supporting real venues, where the difference between a clever system and a usable system is the moment it must be fixed quickly by the people who actually work there.

One mature implementation of this approach was engineered by seasoned venue professionals and built on robust, non-proprietary hardware designed for serviceability. The phrase "non-proprietary" matters because it is not a branding preference. It is a survival strategy. When you rely on proprietary black boxes, you inherit product cycles, unfixable unwanted behaviors, spare parts availability, integration limits, and failure modes. When a device goes obsolete, the system becomes a hostage negotiation. When you build on non-proprietary, commercial-grade compute, you can substitute, upgrade, and scale without re-architecting the entire experience.

The shift: from a pile of devices to a small number of capable nodes

Traditional AV systems often grow by accumulation. You add a device for each function: a player, a scaler, an extender, an audio unit, a controller, a router, a touch panel, plus the driver layer and glue logic that tries to keep everything coordinated. It works, but it is difficult to maintain, and it is rarely designed for change.

A compute-node approach takes the opposite path. It consolidates many of those functions into a smaller number of solid-state nodes running software services. Installation becomes simpler, infrastructure becomes less brittle, and long-term operational burden drops. The practical consequence is not merely fewer boxes. The practical consequence is a venue system that can be understood, supported, and evolved.

Because modern compute is predominantly solid state, these nodes can be deployed closer to where work is needed, making edge execution practical. Fewer discrete devices reduces cabling, reduces integration complexity, and reduces failure modes. It also makes resilience easier to design: replacement becomes a planned logistics event rather than a redesign event.

This model is intentionally familiar to IT operations.

Nodes can be monitored, patched, and managed using standard practices, with secure remote access for proactive support. In field terms, "provisioning sufficient compute" means the system is sized, structured, and documented so it can be maintained and upgraded incrementally, without turning routine changes into capital rebuilds.

Provisioning the right compute

Compute-First Venue Substrate Architecture

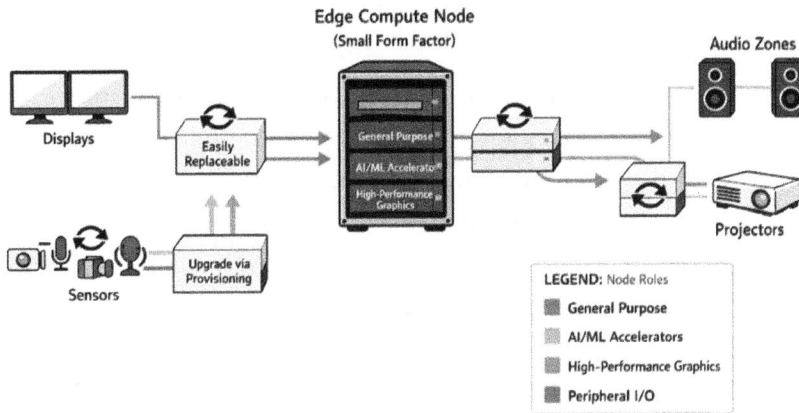

Compute has two virtues in this context: it is scalable, and it is substitutable.

When you size a node, you are sizing capacity rather than committing to a fixed hardware destiny. If you need more capability later, you provision more compute at the relevant nodes. If a node fails, you replace it with one that meets the same functional envelope, then restore configuration and content.

Crucially, an edge compute node is not a proprietary black box. It is a standard computing envelope. The goal is simple: when hardware inevitably ages, it should be replaceable with standard market components without rewriting the venue's logic.

That is the difference between an installation that survives product cycles and one that becomes brittle. The value of this architecture lives in the software and configuration that define the experience, not in a proprietary chassis.

Once you have a modular compute foundation, ambition can expand without turning every new capability into a new rack. This is the structure that makes a venue capability ecosystem practical: change becomes routine, and upgrades become cumulative rather than disruptive.

What a node can do when it is designed as compute

Compute becomes real when you stop describing a node as a player and start describing it as a participant.

A conventional endpoint plays a file. A compute node renders an experience. Rendering means the node makes governed decisions at runtime, and it expresses those decisions through synchronized audio, video, text, overlays, interaction state, and operational mode switching. It also means the node can operate coherently when the network is degraded, when the venue is busy, and when the system must continue to serve people safely and consistently.

This is the core shift that enables the rest of this book. The moment the node becomes compute, the venue stops being a static broadcast system and becomes a governed adaptive environment.

Topology is part of the design

A practical venue architecture cannot assume a single deployment model. Modern systems must support:

1. **Wired nodes.** Fixed installations, predictable bandwidth, and hard reliability.
2. **Wireless nodes.** Retrofits, temporary exhibits, pop-ups, and expansions that cannot tolerate construction cycles.
3. **Hybrid topologies.** The default for real venues. Core components are wired, while edge elements and temporary surfaces can be wireless. Hybrid designs are also how systems degrade gracefully instead of failing abruptly.

The topology matters because it determines whether personalization and governance are theoretical or operational. If adaptation depends on perfect connectivity, it will not survive a crowded venue day.

The node is a media computer, not a media deck

At a minimum, a compute node behaves like a compact media server. It can store assets locally, deliver high-quality playback, and maintain timing without brittle chains of extenders and conversions, although there are always systems and situations where extenders will continue to be used, and that is fine. The value is not that it can play media. The value is that it can vary delivery without multiplying assets.

Variation is where static systems traditionally collapse. If you need a separate export for every language, every subtitle style, every accessibility mode, and every pacing profile, you quickly end up with an unmaintainable library. Sometimes, specifically in the case where curated content is a requirement, this library approach has to be implemented. Compute nodes eliminate that failure mode by rendering presentations dynamically under governance.

Subtitles and captions become rendered layers, not burned-in pixels

In older pipelines, subtitles were burned into video because that was the easiest way to ensure timing. The cost was that every language and every subtitle style became a new video master.

In a compute-based system, subtitles and captions are delivered as data. A timed text file specifies release times and text. It can also specify presentation parameters such as font, placement, size, color, background treatment, line breaks, and any other rules the venue chooses to standardize. The node renders the subtitles at runtime.

This is not cosmetic. It changes what "multilingual" means operationally. It becomes feasible to support many languages without creating many parallel video masters. It also becomes feasible to tune accessibility presentation without reauthoring core media.

Sign-language inserts become composited streams, not embedded edits

A sign-language inset is not a special edition. It is an additional stream. The node can composite the sign-language video over the primary video, synchronized at runtime, under the same governance and timing rules as every other layer.

Once you accept multi-stream composition, you also accept that a presentation can be built from multiple synchronized video and audio elements, layered and mixed deterministically. That is the technical foundation for accessibility overlays, contextual inserts, and controlled variation.

Audio becomes adaptive, because the node can output digital audio

Audio is where most venue systems still behave like the 1990s. They treat audio as a single fixed output that is mixed elsewhere and then "played."

A compute node can output multi-channel, polyphonic digital audio directly, with built-in equalization and serious DSP capabilities. That allows a fully digital signal path, tight synchronization with visuals, and runtime control of audio behavior. It also enables a class of experiences that static playback cannot deliver:

- Language-specific audio delivery by exhibit or by zone, without duplicating the entire exhibit footprint.
- Controlled mixing so multiple languages can coexist in the same space without turning into noise.
- Spatialized or steered audio behaviors where the venue design supports them.
- Consistent loudness and intelligibility management at the node, as part of the experience definition.

This is not an "audio feature." It is part of the governing substrate. If the system can adapt visuals but not audio, it cannot serve multilingual or accessibility needs reliably.

Non-repeating and procedural sound becomes practical, not exotic

Compute nodes also make sound design operational.

A node can generate non-repeating, randomized soundscapes that do not fall into obvious loops. It can vary layers, densities, transitions, and timing. It can respond to operational state, crowding state, or the presence of visitors. It can trigger additional sounds or layers of sound, and do transitions between day and night mode, creating the right environment for the right time of day. It can be governed so the result remains within approved aesthetic and safety boundaries.

This matters because a venue that repeats the same short loop is perceived as mechanical. A venue that evolves its sound field feels alive. This is one of the simplest ways to raise perceived quality without adding content volume, and it becomes straightforward when the node is compute.

Sensors are context inputs, and show control is part of the same system

Sensors in this model are not mere triggers. They are context inputs. A compute node can accept signals from buttons, presence detectors, distance or height sensors, and touchless interfaces. It can accept zone occupancy and proximity signals from devices that deliver media. It can also accept triggers from the show-control layer, including scheduled cues, state-machine cues, and cues generated by operational mode switching.

This is where the book's thesis becomes concrete. Show control is not just about timelines. It is about state.

When show control is integrated with recognition, the exhibit stops being a sequence that plays regardless of who is present. It becomes an adaptive system that selects, assembles, and adapts delivery based on who is there, what they need, what has already happened, and what the venue is trying to achieve at that moment.

Recognition-driven triggers are the bridge from AV to governed environments

Recognition does not have to mean identity. Recognition can be anonymous continuity, a consented membership, a staff role, a group classification, a language preference, or a proximity condition. The point is that recognition produces state, and state drives behavior.

A compute node can consume recognition state and express it through governed delivery. That includes:

- Selecting the language and modality that best fits the visitor.
- Switching pacing between streaker, stroller, and student styles.
- Choosing between curated, Body-of-Knowledge selection, hybrid assembly, or bounded generation, depending on context and risk posture.
- Triggering concierge workflows where, and only where, consent and operational policy allow it.
- Adjusting routing, guidance, and non-visual communications for accessibility needs.

This is the practical definition of "hyper-personalization" that survives contact with reality. It is not a marketing claim. It is a controlled system behavior enabled by compute at the node.

Simultaneous multilingual delivery becomes a spatial design tool

Once the system can deliver multiple languages simultaneously, the venue can stop treating language as a menu and start treating language as spatial behavior.

There are multiple correct patterns:

- Static language zones, where different areas in front of a surface are assigned to different languages.
- Dynamic language zones, where zones adapt based on who is present, where they stand, and what is needed operationally.
- Mixed modality zones, where audio is steered and visual subtitles are rendered to reduce cross-talk.

The venue can make these zones legible. Projection mapping, lighting, or on-surface cues can indicate which language is active in which area, so visitors can step into the right region without friction. This is not a gimmick. It is a dignity feature. It reduces contention between audiences who have different language needs, and it does so without multiplying the number of exhibits.

As display technologies evolve, including displays that allow multiple viewers to perceive different images simultaneously, position-aware delivery becomes more powerful. The compute node is what makes that power usable, because it can ingest position signals and render the correct layers deterministically.

Compute nodes are the substrate for programmable canvases

A programmable canvas is not limited to walls. It can be ceilings, floors, an entire building, a ship, a vehicle, or a temporary structure. It can be realized through projection mapping or by cladding surfaces with image-emitting hardware such as LED systems.

The enabling condition is not the surface. It is the ability to render, synchronize, and govern what the surface does at runtime. Compute nodes provide that condition. They allow the same governed delivery system to drive the environment itself, not just the screens mounted on it.

Why this matters

The purpose of this section is not to list node features. It is to establish the necessary substrate for everything that follows.

If you do not have compute at the node, you do not have reliable governance at runtime. If you do not have reliable governance at runtime, you do not have safe bounded generation, safe multi-domain correlation, safe multilingual scaling, or safe adaptive accessibility. You have fragile demos and brittle deployments.

Confabulation is one of the most practical reasons to care about this.

A system that generates language without sufficient local grounding will eventually produce fluent, plausible statements that are wrong. In a venue, those errors land in the real world: routing guidance, accessibility instructions, eligibility, hours, closures, or safety-related prompts.

Provisioning the right compute at the edge is what allows you to contain that risk operationally. It allows the venue to enforce delivery modes, apply constraints before output is delivered, require evidence where claims matter, and degrade safely when grounding is insufficient, without relying on an upstream service that may be unavailable, slow, or inconsistent.

If you do have compute at the node, the rest of the book becomes implementable. The venue becomes capable of governed variation, where truth remains stable, but delivery adapts to people, context, and operational reality.

Retrofit is a strategy, not a compromise

One of the clearest expressions of the compute-first discipline is retrofit. Most venues cannot justify tearing out monitors, projectors, and speakers that still have usable life. They also cannot afford long closures, and they cannot accept a future in which every upgrade requires construction disruption. A system that insists on replacement as the default upgrade path is not advanced. It is not deployable.

A compute-first retrofit posture treats existing audiovisual endpoints as reusable surfaces. You modernize incrementally by attaching maintainable nodes at the edge, then evolving software, content, and control over time. If you want the deeper significance of retrofit, read it as a statement about time. A venue that can modernize continuously can keep evolving. A venue that must rebuild wholesale to change becomes brittle, and eventually stagnant.

This is where the ladder in Chapter 3 becomes practical. When the substrate is modular, you can deploy in stages. You can begin with exhibit-scale localization, expand into micro hyper-personalization across multiple exhibits, and then decide, with clarity and governance, whether the venue is ready for macro behavior. The ladder becomes possible because the base is designed for evolution.

Acceptance Criteria for Node Compute in Real Venues

If node compute is treated as a foundational capability rather than a convenience, it must be specified, tested, and accepted as such. The following criteria are not vendor features. They are operational requirements. A venue that cannot meet them does not have a governed adaptive system. It has a demonstration.

In practice, acceptance is not only a technical checklist. It is an assurance posture.

If a venue claims governed behavior, it should be able to show evidence that the system held its mode boundaries, did not collapse under degraded conditions, and did not deliver ungrounded visitor-facing instructions as fact. This is why the acceptance criteria in this chapter should be paired with the evidence artifacts defined later in the book, including logging minimums and monitoring plans.

Deployment and topology

A compute-node architecture must support wired, wireless, and hybrid deployments as first-class configurations. Hybrid operation must be normal, not exceptional. Nodes must continue to function coherently under partial network failure, degraded bandwidth, and temporary disconnection. Loss of upstream connectivity must not result in loss of basic experience integrity.

Acceptance requires that:

- Core behaviors continue locally when connectivity is degraded.
- Nodes can rejoin orchestration cleanly after reconnection.
- Operational modes do not collapse into undefined states.
- Governance posture remains intact during degraded connectivity, including mode enforcement and safe failure behavior for any visitor-facing guidance.

Latency and determinism

Compute nodes must support deterministic timing for synchronized media delivery. Latency budgets must be explicit and testable.

Acceptance requires that:

- Audio and visual streams remain synchronized within defined tolerances.
- Triggered behaviors respond within predictable bounds.
- Recognition-driven state changes do not introduce perceptible lag.
- Any shift in delivery mode or safe-failure fallback occurs within predictable bounds and is observable to operators.

Determinism matters more than raw throughput. An unpredictable system cannot be governed.

Media rendering and composition

Nodes must render experiences dynamically rather than relying on pre-baked media variants.

Acceptance requires that:

- Subtitles and captions are rendered from timed text data, not burned into video.
- Presentation parameters such as font, placement, size, and color are controllable at runtime.
- Sign-language inserts are supported as independent composited video streams.
- Multiple synchronized video and audio streams can be layered deterministically.
- Where factual claims are presented in visitor-facing text or speech, the system can be configured to require approved-source boundaries and to refuse to invent when grounding is insufficient.

If every variation requires a new media export, the system does not scale.

Audio capability

Audio must be treated as an adaptive delivery channel, not as an external dependency.

Acceptance requires that:

- Nodes output multi-channel, polyphonic digital audio directly.
- Built-in equalization and DSP are available to support intelligibility and consistency.
- Audio routing and steering can be controlled per zone and per language.
- Audio behavior can change at runtime based on operational mode and context.

If audio cannot adapt, multilingual and accessibility strategies will fail under real crowd conditions.

Sensor and interaction inputs

Nodes must accept and interpret a range of input signals as context, not as simple triggers.

Acceptance requires that:

- Physical controls, presence detectors, distance or height sensors, and touchless interfaces are supported.
- Signals from show-control systems are accepted as state inputs.
- Recognition-derived signals can be consumed as context under governance.
- Input handling supports both instantaneous triggers and continuous state.

A trigger-only system cannot express adaptive behavior.

Show control and orchestration integration

Compute nodes must integrate with show control as peers, not as slaves.

Acceptance requires that:

- Nodes can receive scheduled cues and state-machine cues.
- Nodes can react to operational mode changes without reauthoring.
- Orchestration can coordinate behavior across multiple nodes deterministically.
- Failures in one node do not cascade unpredictably.

Show control without state awareness produces sequences. Show control with compute produces systems.

Recognition and consent boundaries

Nodes must support recognition-driven behavior without assuming identity.

Acceptance requires that:

- Anonymous continuity is supported by default.
- Identity binding is possible only under explicit opt-in.
- Recognition-derived state is constrained by purpose and policy.
- Nodes do not require persistent identifiers to deliver adaptive behavior.

This boundary must be testable, not aspirational.

Multilingual and multi-audience operation

Nodes must support simultaneous delivery to multiple audiences.

Acceptance requires that:

- Multiple languages can be active concurrently.
- Language selection can be spatial, device-based, or contextual.
- Audio and visual outputs can be zoned to reduce contention.
- Language behavior can be changed operationally without rebuilding media.

If a system can only serve one language at a time, it is not suitable for real public space.

Non-visual delivery

Nodes must support non-visual communication as a first-class output.
Acceptance requires that:

- Guidance and navigation can be delivered without reliance on screens.
- Verbosity and pacing can be adapted.
- Non-visual outputs respect the same governance constraints as visual media.

Accessibility cannot depend on best-effort workarounds.

Upgradability and longevity

Nodes must be designed to evolve.

Acceptance requires that:

- Software updates do not require physical replacement.
- New modalities and sensors can be integrated without architectural breakage.
- The node ecosystem supports incremental expansion.

A system that cannot evolve will be obsolete before the venue is finished.

Why acceptance criteria matter

These criteria exist to prevent category errors.

Without them, it is easy to confuse a playable prototype with a durable system. With them, procurement, design, and operations align around the same reality: governed adaptation at scale requires compute at the node, treated as infrastructure, not decoration.

This is the line between experiments and environments.

Boundary: what must run local

Decide up front which behaviors must work when the WAN is down or the venue is saturated.

- Any safety-adjacent behavior, access behavior, or core show behavior must be local-first.
- Cloud assistance can improve quality, but it must not be a single point of failure.
- The local node must be replaceable without rewriting the experience.

For any new venue build, treat compute nodes as baseline infrastructure. A node-based architecture is the simplest way to keep the experience future-resilient across product cycles, supply chains, and new capabilities.

Lock-in failure modes

If the baseline substrate is proprietary, the venue inherits failure modes that are structural, not incidental.

- Vendor end-of-life forces rebuild cycles instead of incremental upgrades.
- Closed toolchains limit auditability, independent support, and substitution.
- Supply chain shocks become redesign events, not like-for-like replacements.
- Opaque licensing and per-seat models turn into a recurring operational tax.

- Tightly coupled integrations make changes risky and increase time-to-restore.
- Upgrades require broad retesting because components cannot be swapped cleanly.

A node-based architecture built on non-proprietary hardware, standard interfaces, and documented substitution paths is the simplest way to keep a new venue future-resistant. Treat it as base infrastructure, then let capability grow by adding or upgrading nodes, rather than rebuilding the venue.

Most venues inherit their audiovisual systems the way cities inherit streets: one decision at a time, one project at a time, one vendor at a time, until the result feels less like an architecture and more like an archaeological record.

A rack becomes a stack, the stack becomes a dependency graph, and the graph becomes a fragile truce among black boxes, adapters, extenders, firmware revisions, and the occasional miracle. It works until it does not, and then someone is in a back-of-house closet at 7 a.m. with a flashlight, trying to persuade an ecosystem to behave.

For decades, that fragility was treated as normal, the cost of doing business. We built exhibits by assembling devices, then adding show control to glue them together, hoping the glue would hold across time, supply chains, product cycles, and staff turnover. The approach could produce beautiful results, but it also produced a predictable kind of operational debt: a system that was hard to understand, hard to repair, and hard to evolve.

The concept of Edge Compute Nodes begins with a refusal to accept that debt as inevitable. The present has changed. Visitors now expect adaptability, multilingual support, and a level of inclusion that cannot be bolted on gracefully. Operators need systems that can be monitored and maintained without relying on a specialist who remembers a custom wiring diagram from eight years ago. Budgets, especially in cultural institutions, demand reuse, retrofit, and staged upgrades rather than periodic demolition.

The Edge Compute Nodes idea is simple enough to state, and deep enough to reshape everything that follows: treat compute as the atomic unit of capability. If you can provision the right compute in the right place, you can replace and upgrade without rewriting the venue. You can distribute intelligence to the edge, consolidate single-purpose hardware into a smaller number of maintainable nodes, and reduce obsolescence risk by refusing to bind the venue's future to a single vendor's black box.

Why this belongs early in the book

It can seem premature to talk about compute before content, narrative, or visitor psychology. In practice, the substrate determines what you can promise. If your system cannot be upgraded without trauma, every new language, modality, and accessibility improvement becomes expensive.
If the system cannot be monitored and maintained by the people who own it, reliability degrades over time. If the system depends on proprietary black boxes, it eventually pays an obsolescence tax.

The Edge Compute Node approach is a design discipline that preempts those predictable failures: provision the right compute, keep it non-proprietary, distribute it where it is needed, and design the system so replacement and upgrade are routine. The rest of the book builds on that discipline. Programmable Canvases depend on it because programmable surfaces require reliable local capability. The Personal Channel depends on it because synchronized personal delivery requires stable, low-latency nodes. Venue-scale orchestration depends on it because you cannot govern what you cannot observe, and you cannot observe what you cannot maintain.

Compute is not an implementation detail. It is the lever that allows a venue to stay modern without living in permanent rebuild mode.

That is why the Edge Compute Node solution came about, and why it matters. In the next chapter, we will take the compute substrate and make it visible to visitors. We will move from nodes to surfaces, and from displays to canvases, and show how Programmable Canvases turns a wall into something more than a screen: a programmable layer that can change as quickly as the world around it.

6

PROGRAMMABLE CANVASES AND SOFTWARE-DEFINED SURFACES

A Programmable Canvas is an experience surface whose layout and behavior can be changed by software without rebuilding the physical installation. A canvas can be a video-mapped wall, a projection field, a lighting zone, an audio zone, an interactive tabletop, a kiosk, a room surface, or a personal device layer. The defining feature is that the surface is treated as a runtime target rather than a fixed endpoint.

In a governed environment, "runtime target" must also mean "governed target."

If a surface can change at runtime, the venue must be able to constrain what changes are permitted, when they are permitted, and under what authority. Otherwise, the canvas becomes an uncontrolled actuator. A programmable canvas is only an asset when it can be operated safely: mode-aware, evidence-aware where claims matter, and capable of safe fallback when context is ambiguous or signals conflict.

Programmable Canvases Are Not Just Walls

When programmable environments are discussed, the conversation often collapses to screens.

That framing is too narrow.

A programmable canvas is any surface whose behavior can be addressed, governed, and changed at runtime. This may be a wall, but it may also be a ceiling, a floor, an entire building, a ship, a vehicle, or a temporary structure. It may be realized through projection mapping, through emissive surfaces such as LED systems, or through future display technologies that have not yet stabilized.

The defining characteristic is not the surface. It is addressability.

Addressability changes what must be governed.

Once surfaces are addressable, the venue can change what is shown, said, or signaled without construction. That is power. In a public environment, power must be bounded. The same governance threshold in Chapter 4 applies here: if a change can affect safety, accessibility, eligibility, or the visitor's choices, it must be constrained by policy and auditable through decision evidence.

Once an environment is addressable, the same governed delivery system can drive it.

Content can be layered, multilingual, modality-aware, and adaptive. Behavior can change operationally within defined modes and constraints, without requiring construction or wholesale rebuild. The environment itself becomes part of the narrative system.

This is where media stops being decoration and becomes infrastructure.

Infrastructure must be testable and supportable.

A programmable surface that cannot be monitored, verified, and restored is not infrastructure. It is a fragile art piece. Treat canvases like systems: define modes, define safe defaults, define rollback posture, and require evidence that the surface behaved as intended under normal load, partial failure, and operational stress.

Programmable canvases are the natural endpoint of the ideas in this book. They are not a special category. They are what happens when governed personalization is allowed to extend beyond devices into space itself.

In a programmable canvas, content, language, modality, pacing, and interaction rules are rendered dynamically based on context and chosen modes, rather than hard-baked into graphics, wiring, or one-off device configurations. A software-defined surface is the engineering expression of the same idea. The surface is driven by addressable compute, timing, and control, so capabilities can be reconfigured through software: mapping can change, layouts can change, audio routing can change, captioning can turn on and off, and accessibility modes can be enabled without adding permanent clutter to the room.

In practical terms, programmable canvases turn a venue from a collection of fixed outputs into a set of adaptable experience layers. That shift matters because it makes inclusion, iteration, and staged upgrades possible without construction trauma.

If you want to understand why you design with programmable canvases, begin with a familiar failure. You walk into a beautiful space, carefully lit, carefully paced, and carefully designed, and then you see the language compromise. English is large, confident, and visually integrated. Spanish is smaller, squeezed in beneath it, sometimes literally treated as a footnote. Other languages are absent, even when the city outside the doors is clearly multilingual.

The problem is not only aesthetic, although it is that. It also becomes political the moment you acknowledge it. The deeper problem is structural. Fixed signage turns language into architecture. It hardens a choice that should remain dynamic. It makes inclusion expensive, slow, and brittle, because each new language becomes a fabrication project rather than a configuration choice. It forces a venue to guess, in advance, what its audience will need over the next decade, in a world where demographics, tourism patterns, and social expectations change faster than construction cycles.

Programmable canvases are a refusal to accept that brittleness as inevitable. They are a design principle, stated plainly: deliver the interpretive layer through pixels rather than fixed signs. When content lives in pixels, it can be updated, reshaped, restyled, and relocalized without tearing out walls, printing new panels, or reopening the same debate every time language support expands. In other words, programmable canvases turn the wall back into a living surface.

Programmable Venue Canvas

NARRATIVE MODE
Storytelling, content & events

GUIDANCE MODE
Real-time directions & wayfinding

MULTIMODAL RENDERING
Welcome
Bienvenido
欢迎
Multilingual, inclusive formats

SPECIALIZED CANVASES
Custom modules & interactive zones

The constraint we inherited

For much of the modern exhibit era, we treated surfaces as containers. A wall held text. A monitor played a loop. A projector told a story. Each surface had a job, and that job was largely fixed. The venue's creative ambition lived inside the boundaries of those jobs.

There is a quiet economic logic behind that inherited model. A printed panel is predictable. A printed panel does not crash. A printed panel does not require a software update. A printed panel does not need compute.

But that predictability comes with a cost that is easy to underestimate until you are running a venue. Fixed signage does not merely display information. It locks the venue into a slow refresh cycle. It makes every change expensive. It makes seasonal updates painful. It makes response to current events awkward. It makes language expansion politically charged because it requires visible negotiation over space and hierarchy. It also makes a museum's promise of inclusion dependent on fabrication budgets rather than on intent.

Programmable canvases change the cost structure, and they raise the creative ceiling. They do not argue that everything must become a screen. They argue that the parts of an experience that benefit from adaptability should not be frozen into materials that resist change.

The idea of the canvas

A Programmable Canvas asks you to treat a surface as a canvas, not as a single-purpose display.

A canvas can hold many things. It can be divided, layered, rethemed, and scheduled. It can present different compositions at different times of day, or for different audiences, without rebuilding the space. In its more powerful form, multiple surfaces participate together rather than behaving as isolated displays.

This is the point where a venue stops thinking in screens, and starts thinking in states. In the simplest form, a wall becomes a multilingual surface that can switch cleanly. In the more powerful form, a single physical surface can host multiple independent story zones, each with its own rhythm, purpose, and audience.

A single projection can behave less like a loop and more like a platform, supporting multiple concurrent narratives and visual shapes through mapping and composition.

This matters because most exhibits do not need one giant story occupying an entire wall for an entire day. They need several smaller stories delivered in a way that can adapt. A family with young children needs a different entry point than a graduate student. A tourist needs a different pace than a local repeat visitor. A visitor who wants technical depth needs a different density of information than a visitor who wants the high-level arc. Programmable canvases give you a physical place to express that plurality without building a new wall every time the audience shifts.

Language as a runtime choice

The most obvious benefit of programmable canvases is language dignity. When language is printed, it becomes a spatial negotiation. When language is rendered, it becomes a configuration.

Rendered language is not only more languages. It is better language. It allows typography that fits the script, line lengths that fit the translation, and layouts that respect readability rather than forcing every language into the same box. It allows you to treat language as a first-class part of the experience rather than as a secondary caption tucked beneath the "real" label.

This is a modern public-space issue. Museums, visitor centers, and corporate briefing centers deal with multilingual audiences. Tourist sites face diverse audiences by definition. Large cities are multilingual by reality. Corporate experience centers face global visitors. Cruise ships operate as bounded cities with constantly shifting language mixes. Smart-city environments increasingly use digital signage and guidance as public infrastructure.

A Programmable Canvas gives you a way to keep the physical environment elegant while keeping the interpretive layer adaptable. This chapter is not asking you to replace every label with a display. It is asking you to stop building language into the walls when the walls can instead host language as a living layer.

The enabling substrate is compute

A canvas is only as useful as the engine that can drive it. A programmable canvas is not magic paint. It is an operational capability that depends on provisioning sufficient compute near the point of experience.

This is where Edge Compute Nodes re-enter the story. Edge Compute Nodes matter for programmable canvases in three ways.

First, they make the canvas reliable. A programmable surface cannot be a science-fair project. It must be stable, serviceable, and replaceable. When the surface is driven by a node with appropriately provisioned compute, replacement becomes a provisioning exercise rather than a redesign.

Second, they make the canvas upgradeable. As display technology evolves, rendering demands increase, and experience ambitions expand, you can scale capability by provisioning more compute where it is needed, rather than tearing out the conceptual foundation of the system.

Third, they can make the canvas more affordable. If one node and one projector can render multiple simultaneous, separate media streams, costs can come down, often drastically.

The programmable canvas is therefore not only a visual concept. It is a commitment to modularity, and modularity is impossible without a compute-first substrate.

Programmable canvases without identity

It is important to say this clearly, because the temptation to overreach is real. Programmable canvases do not require identity to be valuable.

A surface can adapt to time of day, occupancy, programming schedules, and exhibit state without knowing who any individual is. It can become clearer when crowds increase. It can switch to guidance when a route changes. It can reduce sensory load when a space is already loud. It can support language selection at the point of interaction without storing anything about the visitor.

When identity is introduced, it should be introduced as a benefit exchange, and it should remain consent-bound and minimized. In practice, this means a visitor can opt into a preference profile, or use a personal device as a privacy boundary, and then the environment can respond more specifically, including switching programmable canvas surfaces to a preferred language or delivery style. The technology becomes more helpful, but the venue does not become a collector of dossiers.

This is the line held throughout this book: adaptive environments should behave like good hosts, not like hungry databases.

A canvas is also an operations tool

There is a second, less celebrated consequence of turning surfaces into pixels. It changes how a venue operates.

Fixed signs assume a stable world. A Programmable Canvas assumes a world that changes, and it is built to respond without drama. That response can be mundane, and still transformative. A gallery closes. A queue forms. A show time shifts. A school group arrives. A special event begins. A fire alarm triggers an evacuation path. Even in the calmest museum, reality is not static.

When a venue has programmable canvases, it can express operational intelligence through the environment itself. Guidance can replace promotion when guidance is needed. Clarity can replace spectacle when clarity is required. A space can retheme for a donor event and then return to its everyday interpretive voice without reprinting a single thing.

This is the point where programmable canvases stop being content delivery and become a layer of venue behavior. That is why, later in the book, they belong naturally inside a WorldModel Venue. At macro scale, you are not merely delivering stories. You are managing flow, comfort, fairness, and operational coherence. Programmable surfaces become one of the venue's most humane levers, because they can reduce confusion before confusion becomes anxiety.

The design discipline

Programmable canvases do not absolve you from design. They increase the importance of design. When you can change anything, you must decide what should change and what should remain stable.

A venue still needs a voice, a rhythm, and restraint. The goal is not to flood walls with motion. The goal is to let the environment speak when it should and remain quiet when it should not.

That discipline is where the best teams distinguish themselves. They use pixels to remove compromises that never served the visitor, such as awkward bilingual hierarchies, stale messaging that should have been updated years ago, and interpretive layouts that could never accommodate the audience that actually arrived. They use pixels to add capabilities that increase inclusion, such as language dignity, optional depth, and personal comfort modes. They also use pixels to reduce operational friction, such as replacing brittle printed guidance with adaptive guidance that matches reality.

Programmable canvases are therefore not gadgets. They are a way to make the built environment more honest, because it can reflect the world as it is.

Key takeaways

By now, the central shift should feel clear. Edge Compute Nodes are the discipline of provisioning sufficient compute, so systems remain flexible and maintainable. Programmable canvases are what that discipline looks like when it becomes visible to the visitor: surfaces that can change, speak in the right language, and carry more than one story without forcing the room into a permanent compromise.

This matters because it changes what a venue can promise. A venue can promise that language will not be treated as a footnote. It can promise that exhibits can be refreshed without construction trauma. It can promise that the environment can adapt to programming, demographics, and the lived rhythms of the day. It can promise that inclusion is structural, because the medium itself is adaptable.

In the next chapter we turn from surfaces to guidance and non-visual navigation, because a venue is not only what it displays. It is what visitors can do, and how confidently they can move. Buttons, sensors, readers, triggers, and small physical affordances are points where the environment meets the hand, the wheelchair, the stroller, the school group, and the person who simply wants the experience to respond without friction.

That interaction layer is where many venues quietly become brittle, not through grand failures, but through a thousand bespoke exceptions.

7

NON-VISUAL NAVIGATION AND GUIDANCE

A venue that assumes vision for navigation cannot call itself inclusive.

Most wayfinding systems, even well designed ones, still treat sight as the primary interface. Signs, maps, icons, arrows, and screens work for many visitors. They also leave visually impaired guests dependent on staff, companions, or luck. That is not accessibility. That is vulnerability.

Non-visual navigation is the corrective posture. It treats independent movement as a first-class outcome, and it treats wayfinding as a service the environment provides, not a puzzle the visitor must solve.

One implementation of this pattern is the Cane-based guidance Layer. Cane-based guidance extends the mobility tools cane users already rely on, adding clear directional cues, tactile feedback, and optional audio, subject to safety and policy constraints.

Why this layer matters more than most people admit

Wayfinding is not a minor convenience. It is the first condition of participation. A visitor who cannot navigate cannot choose depth, cannot choose pacing, and cannot choose comfort. They are forced into dependence, and that dependence changes the emotional character of a visit.

In large venues, the problem becomes more severe. Congestion shifts routes, closures appear, and queues form. The route that was calm ten minutes ago may no longer be calm. The route that was safe may no longer be safe. Non-visual navigation must therefore be dynamic, not static.

Non-visual navigation as an interface to operational truth

A Cane-based guidance Layer is not a separate feature floating beside the venue. It is an interface to what the venue knows right now: which routes are open, which routes are crowded, which routes are accessible, which routes are acoustically overwhelming, and which routes are safe under the current operating mode.

This is why non-visual navigation belongs inside the same governed architecture as everything else. Guidance must be derived from operational truth and constrained by explicit policy. A wrong cue is worse than no cue, because it creates false confidence.

Confabulation risk, and why guidance must be conservative

In generative systems, the most dangerous failure mode is not silence. It is fluent error. In AI risk terminology, this is confabulation (often called "hallucination"): confident output that is wrong. In navigation, confabulation shows up as an authoritative instruction that sounds plausible but does not match the venue's current reality.

For non-visual guidance, confabulation is unacceptable. The system must not "fill in the blanks" when operational truth is missing, stale, or ambiguous. When the system cannot establish correctness, it must adopt a conservative posture: guide to a safe default, or escalate to human assistance through an explicit, opt-in mechanism.

Design principles for non-visual navigation

Start with explicit priorities, then constrain everything else under them.

1. Independence first. The default experience should enable a visually impaired guest to move confidently without staff intervention, unless the guest requests help.
2. Safety as a hard constraint. Guidance must never route a guest into unsafe conditions, restricted zones, or crowd hazards. When uncertainty rises, the system must become conservative.
3. Predictability over cleverness. Cues should be consistent, stable, and easy to learn. A visitor should not need training to trust the interface.
4. Uncertainty must be representable. Guidance must carry an explicit confidence posture. When confidence drops below a venue-defined threshold, the system must abstain from precise routing and switch to safe defaults, or request human assistance. "I am not sure" is a valid output state in safety-adjacent guidance.
5. Privacy by default. Non-visual navigation does not require identity. It should work anonymously, and it should not rely on biometric identification.
6. Multi-plane inclusion. Some visitors are fully blind, some have low vision, and some have fluctuating ability. Non-visual navigation should coexist with other planes, such as the Personal Channel, programmable surfaces, and spatial overlays, without forcing any one plane to become mandatory.

How the cues should behave

The cues can be tactile, auditory, or both, but the behavior must follow a clear contract.

- Directional cues must be unambiguous and must reflect current route availability.
- Graceful degradation is mandatory. If the system cannot guarantee correctness, it must not guess. It should guide toward safe defaults, reduce specificity, and, when necessary, request human assistance through an explicit, opt-in mechanism.
- Mode awareness is required. In busy mode, guidance should bias toward routes that preserve safety and reduce stress. In incident mode, guidance must follow emergency procedure constraints.

Emergency-aware guidance, and the inviolable boundary

In incident mode, guidance is not "helpful." It is safety-adjacent infrastructure.

The system shall treat the following as Level 1 authoritative constraints, not as advisory signals:

- Life-safety state (for example, fire alarm or emergency communications activation).
- Emergency stop state (E-stop, show-stop, ride-stop, or equivalent hard-stop chains).
- Operator-declared hazard, closure, responder-only zone, or hold state.

Routing rule (non-negotiable): the system must never recommend, hint, or guide a visitor toward an area that is under emergency response, flagged as hazardous, or declared closed. If the destination lies inside a restricted area, the system must refuse routing, offer a safe alternative, or escalate to staff.

Messaging rule (non-negotiable): during incident mode, any safety-related message shown on canvases, spoken by agents, or delivered via Personal Channel must be pre-authored and owner-approved. Freeform generation is not permitted for evacuation, hazard, or life-safety instructions. See Chapter 26, "Emergency overrides and the Emergency Constraint Set," and Chapter 24, "Evidence Hierarchy," for how incident posture is enforced venue-wide.

How it integrates with other layers

Non-visual navigation becomes strongest when it is treated as part of the same multi-layer inclusion posture described throughout this book.

- The Personal Channel can provide additional context, such as step-by-step instructions, language-specific narration, or confirmation prompts, without requiring the room to change.
- The acoustic layer matters because sound is both information and fatigue. Routes that are technically accessible can still be oppressive if they are loud and chaotic.
- Programmable surfaces can support companions and staff with clear operational truth so assistance, when requested, is fast and consistent.
- Spatial overlays, when used, can support low-vision guests with opt-in, high-contrast guidance, while Cane-based guidance supports fully non-visual navigation.

Commissioning and acceptance tests

A non-visual navigation layer should not be treated as a demo. It should be commissioned like any other safety-adjacent system.

At minimum, a venue should prove:

- Route correctness under changing conditions, including closures, congestion, and schedule pulses.
- Safety constraint enforcement, with no guidance into restricted or unsafe zones, ever.
- Mode behavior, where normal, busy, and incident modes produce predictably different guidance posture.
- Latency envelope, where cues update quickly enough that guidance feels trustworthy in motion.
- Degraded network behavior, where the system fails conservatively rather than guessing.
- Operator override, where staff can place areas into hold or restriction states immediately, with clear expiry rules.
- Confabulation control: when closures, congestion inputs, or route availability are uncertain or stale, the system demonstrably abstains from precise instructions, switches to safe defaults, and logs the uncertainty state.

Operator controls

Operators need simple, bounded controls. The goal is trust.

At minimum, operators must be able to:

- Mark temporary closures and hazards that affect routing.
- Set restricted zones and time windows.
- Switch operating modes: normal, busy, and incident.
- View current guidance posture by zone.
- View confidence posture by zone (high, medium, low), including why confidence is reduced.
- Force conservative mode for a zone or route class for a time-bounded window, with clear expiry rules.
- Override automation when needed, with auditability.

The chapter's conclusion

Non-visual navigation is not a niche feature. It is a declaration that the venue was designed for the public as it actually is.

Cane-based guidance becomes powerful when it is treated as an interface to governed operational truth, rather than as a gadget. That is how non-visual navigation stays safe, stays calm, and stays trustworthy under the real conditions of public space.

In the next chapter, we turn to the acoustic layer, because sound is the most underestimated constraint in public environments. Intelligibility, sound bleed, and comfort are not aesthetics. They are access.

8

THE ACOUSTIC LAYER

Sound is the most underestimated design material in public space.

We lavish attention on sightlines, lighting, typography, and finishes because those are visible, and because they photograph well. Sound rarely photographs well, but it does something more consequential. It governs comprehension, comfort, pace, and fatigue, often without a visitor being able to articulate what is wrong.

When the acoustic layer is right, a room feels calm and intelligible. When it is wrong, the room feels chaotic, even if every exhibit works perfectly. That is why acoustics belongs in this book as a first-class layer. It is an operating condition of the experience itself.

When sound becomes a decision input, it also becomes a governance concern. Acoustic adaptation must be evidence-led, bounded by explicit comfort and intelligibility constraints, and auditable in operation. If the venue uses acoustic state to influence guidance, routing, or mode switching, it must be able to show what it believed about the sound environment at that moment, how confident it was, and what action it took as a result.

In museums, the problem is especially acute because modern exhibition design often favors openness: large volumes, hard surfaces, coupled rooms, and long sightlines. Those same characteristics can undermine speech intelligibility and create an environment where sound spills into adjacent spaces. The Acoustical Society of America has described this museum pattern directly, including open-plan exhibition halls, coupled rooms, high volumes, and reflective surfaces, and the persistent tension these create between intelligibility and privacy[xi].

In themed entertainment, the stakes change, but the problem remains. Sound is part of spectacle, which makes it tempting to push levels upward, layering effects until every zone competes with every other zone. In cruise ships, sound becomes both experience and logistics because guests move through a tightly constrained footprint with simultaneous programming. In retail, sound is often the difference between "pleasant" and "leave quickly," which is an economic difference. In restaurants, sound directly shapes dwell time and perceived service quality. Research on restaurant noise has linked higher background noise to communication disturbance, and reduced willingness to spend time and money on a meal[xii]. I have restaurants I will not visit because of their acoustic properties.

Across these verticals, the acoustic layer often determines whether a place invites participation or quietly repels it.

Why sound is an inclusion layer

A venue can be accessible on paper and inaccessible in practice if it is not intelligible. Part of this is obvious: hearing ability varies, and it varies more than most teams assume. The National Institute on Deafness and Other Communication Disorders reports that approximately 15% of American adults ages 18 and over report some trouble hearing.[xiii] This is not a niche population, and the fraction generally increases with age.

But the acoustic layer is not only about hearing loss. It is also about language comprehension, cognitive load, and sensory comfort. A visitor can have "normal" hearing and still struggle in a space where reverberation, noise floor, and competing sources make speech hard to parse.

A visitor listening in a second language often needs better signal-to-noise, not louder sound.

A visitor who is easily overstimulated can be excluded by a soundscape that never gives their nervous system a place to rest.

This is the first principle of the acoustic layer:

Intelligibility is not the same thing as loudness.

Intelligibility means the listener can correctly understand the words, not merely detect that speech is present. In practice, intelligibility is driven by how much of the speech signal arrives unmasked and unblurred, so the brain can separate consonants, syllables, and meaning without excessive effort. Speech can be loud and still unintelligible if the room, the noise floor, or competing sources overwhelm the information that carries clarity.

Intelligibility is shaped by both the origin and the environment. The origin includes the talker or recorded voice, microphone technique, processing, loudspeaker directivity, placement, and whether the system delivers strong direct sound where people actually stand.

The environment includes reverberation, reflections, background noise, and competing soundscapes. Together, these determine signal-to-noise and the direct-to-reverberant balance, which is why better sound is often achieved by distribution and control rather than by adding volume.

For teams that want an objective measure, intelligibility can be quantified. One widely used approach is the Speech Transmission Index (STI), defined in IEC 60268-16, which produces a value on a 0 to 1 scale for speech transmission quality through a given system and space[xiv].

> *Loudness is easy. Intelligibility is* design.

Sound bleed is architecture

Sound bleed, the unwanted spill of audio from one zone into another, behaves like an invisible demolition of narrative boundaries. A visitor reading about a delicate historical moment should not be hearing the triumphant soundtrack of a nearby interactive. A contemplative gallery should not be punctured by a loop designed to be energetic in a different context. When this happens, it is not only annoying. It rewrites meaning and makes a carefully shaped venue feel accidental.

In museums, the challenge is structural. Open-plan halls, hard floors, glass, and large volumes can amplify bleed. In interactive environments, the challenge becomes exponential because every additional sound source is also a competing story, a masking signal, and a contributor to fatigue.

The practical implication is simple. A venue cannot treat sound as local. Sound is always environmental. Once you accept that, you start designing sound the way you already design light: zoned, controlled, and intentional.

The room mix, the personal mix, and the dignity of choice

In modern venues, the best acoustic posture is not to force a single mix on everybody. It is to provide a coherent room mix and then offer optional personal mixes.

This is where the phone, earbuds, and hearing aids stop being distractions and become part of inclusion. A Personal Channel can deliver a visitor's preferred language, a narration-only option, captions, or alternate mixes without turning the room into a compromise. The room remains coherent for the group, and individuals can still receive what they need.

This also solves a problem that is otherwise difficult to solve gracefully: the mixed-language room. If a theatre experience is running a main narrative in one language for the majority of that audience, there will still be guests who prefer another language. Without a Personal Channel, you either ignore them, or you clutter the room with fixed compromises. With a Personal Channel, you can remain elegant and inclusive: the room plays one coherent mix, and the visitor receives the language they prefer privately.[xv]

The deeper value here is not convenience. It is dignity. A visitor should not have to choose between comprehension and belonging.

Programmable sound needs programmable compute

When sound becomes a managed layer, it becomes a compute problem. You need the ability to route multiple audio feeds, switch languages cleanly, run alternate mixes, manage zones, and do so reliably in real time. This is where a compute-first substrate becomes enabling rather than merely convenient. A mature approach emphasizes multi-channel capability and stable output routing, so audio and visual experiences remain coherent across zones and modes.

This is not about a brochure channel count. It is about that once you treat audio as programmable, you stop being trapped by the "one exhibit, one loop, one speaker system" model. You can build a venue where sound behaves like a designed layer, and where inclusion is delivered by configuration rather than special-case hardware.

Programmable sound also needs a stable transport layer. In practice, audio is often the first system to suffer when "show control" becomes a catch-all network, because audio transport is continuous, latency-sensitive, and timing-disciplined. That does not mix well with general management traffic.

A serious venue treats the audio network as its own first-class system layer, separate from guest traffic, separate from general operations, and separate from show control, with its own quality-of-service and timing posture. When audio has a protected path, intelligibility improves, bleed policies hold, synchronization remains reliable, and the rest of the experience stack becomes calmer because the venue is no longer fighting invisible instability.

Acoustic State Model

If the acoustic layer is first-class, it must be representable.

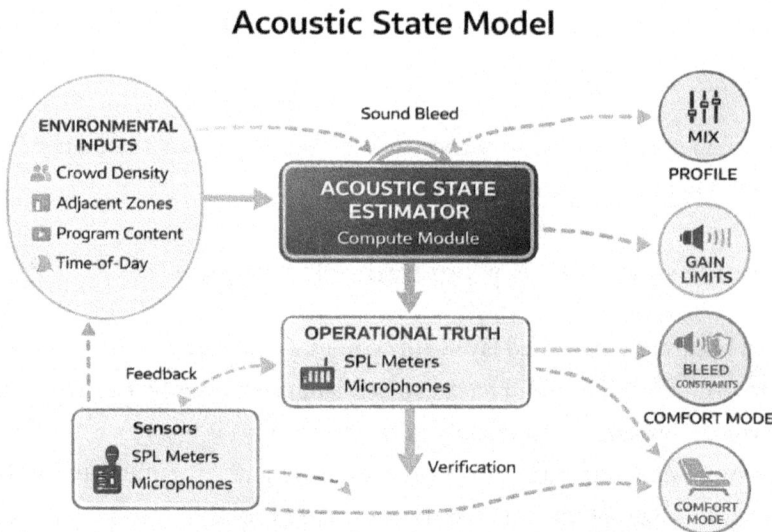

Acoustic State Model

A venue cannot govern what it cannot express as state. It cannot route comfort-aware paths if it does not know which zones are loud. It cannot reduce bleed deliberately if it cannot describe bleed risk between adjacent spaces. It cannot offer quiet mixes intelligently if it cannot distinguish between a calm room and a room that has become acoustically busy.

The goal is not to create an acoustics research project. The goal is to give the venue a practical, operational model of sound conditions that can be used for decisions and can be verified.

The model must also represent uncertainty. If acoustic inputs are missing, contradictory, or stale, the system should not invent precision. It should lower confidence, adopt conservative defaults, and avoid making aggressive comfort or routing decisions until the state is re-established.

Acoustic state fields

At a minimum, each zone should carry an acoustic state object with fields shaped like this:

- noise_floor_db: estimate of ambient noise level in the zone
- program_level_db: estimate of the current program level being played into the zone
- intelligibility_flag: categorical estimate such as good, marginal, poor, based on venue-defined thresholds
- comfort_flag: categorical estimate such as calm, moderate, loud, shaped for comfort rather than engineering perfection
- bleed_risk: mapping from this zone to adjacent zones, low, medium, high, describing risk of unwanted transmission
- occupancy_context: empty, light, busy
- confidence: signal of how reliable the current estimates are
- last_update: timestamp
- provenance: which sources contributed, and whether they are authoritative or advisory
- evidence_pointer: optional reference to the measurement or log record used for audit and commissioning review

Occupancy levels have a significant impact on these numbers, and on the acoustic behavior of a venue. A loud zone while empty is usually an operational failure. This model is intentionally practical. It trades academic nuance for operational usefulness, because that is what venues need.

Update frequency

Acoustic state should update on two time scales, and the venue should be explicit about the difference.

1. Fast loop updates: seconds-level updates used for immediate comfort decisions, such as whether a zone should be considered loud right now.
2. Slow loop updates: minutes-level aggregates used for policy tuning, trend analysis, and understanding persistent bleed issues.

A venue that only measures sound in slow averages will miss the spikes that create fatigue. A venue that only measures spikes without aggregation will drown operators in noise. The correct posture is a layered update loop: fast enough for moment-to-moment comfort, and slow enough for pattern learning.

How acoustic state is used in decisions

Acoustic state should influence decisions at both micro and macro scale, always within explicit bounds.

Micro scale decisions
- If comfort_flag is loud, the system can offer Personal Channel options more prominently: captions, personal audio, or a quiet narration mix.
- If intelligibility_flag is poor, the venue can shift interpretive delivery toward text or personal audio rather than forcing comprehension through a noisy room.
- If occupancy_context is empty, the system can reduce program audio as a hygiene policy, preventing bleed into neighboring spaces.

Macro scale decisions
- Routing suggestions can treat loud zones as higher "cost" for visitors who have opted into calm or quiet modes.
- Load balancing can avoid sending additional crowds into zones that are already acoustically saturated, even if physical capacity exists.
- Operating mode changes, normal to busy, can tighten acoustic constraints so clarity is prioritized during crowd surges.

This is the practical meaning of acoustic layer as state. It is that the venue treats sound as part of operational truth, not that the venue becomes obsessed with decibels

Comfort mode triggers, explicitly consent-driven

Comfort adaptations must not depend on covert inference. They must be driven by explicit, dignified choices. A practical implementation uses three trigger types:

1. Visitor-selected comfort modes.
   Visitors opt in to calm or quiet modes through the Personal Channel, or a simple venue UI. This is the primary trigger for comfort-aware routing and delivery.

2. Contextual prompts, not automatic flips.
   If a zone becomes loud and the visitor has not selected a comfort mode, the system may offer a prompt such as "Want a quieter narration mix?" The visitor accepts or declines. The system does not silently decide on their behalf.

3. Venue policy triggers for empty zones.
   When zones are empty, the venue can reduce audio output by policy to prevent bleed. This is operational hygiene, not personalization.

These triggers keep the system helpful without becoming intrusive.

Operator controls for zone volume policies

Operators need control because acoustics is not static. Programming changes, events happen, and guest demographics shift. A serious venue, therefore, provides controls that are simple, bounded, and auditable.

At a minimum, operators must be able to:

- Set target program level ranges per zone and per operating mode.
- Set bleed-sensitive adjacency pairs where stricter limits apply.
- Set quiet-hours policies for zones near sensitive exhibits or learning spaces.
- Enable or disable Lifecycle Automation Layer audio reduction per zone.
- Override automation temporarily for special events, with a clear expiry.
- View current acoustic state and recent changes in a clear, zone-based view.

These controls are how the acoustic layer becomes operable.

Digital audio as a system layer

Digital audio is no longer a peripheral. In modern venues it is a software-defined plane, and it should be treated as a first-class system layer. The practical shift is that audio is now networked, multi-channel, and policy-shaped. A single node can render many zones, support multiple languages and mixes, and do so with consistent timing.

Interoperable standards, such as AES67, make this practical across mixed-vendor systems.

Spatial audio belongs in the same category when it is governed. It can reduce fatigue, improve intelligibility, and support guidance cues. Soundscapes also belong here. Variation can keep environments from becoming repetitive, but only when constrained so the venue remains coherent and calm.

Special events add another requirement: controlled ingest. Venues may need to incorporate time-bounded external audio sources for a specific celebration, or branded moment. In a reference-grade system, these sources are treated as inputs, not dependencies. They are cached, substituted under degraded connectivity, and the venue retains a safe fallback posture.

The design rule is the same as everywhere else in this book. If you can change audio dynamically, you must be able to prove it is correct. Verification and drift detection are part of the acoustic layer, not optional add-ons.

Lifecycle Automation Layer and sound bleed, expressed as policy

The Lifecycle Automation Layer should include an explicit acoustic policy, not only a power policy. A clean policy looks like this:

- If a zone is unoccupied longer than a defined threshold, reduce program audio to a defined bleed-safe level.
- If adjacent zones are occupied, prioritize bleed reduction to protect intelligibility elsewhere.
- Restore program audio before visitors can perceive a transition when occupancy returns.
- Verify restoration as part of the same loop that verifies display restoration.

This turns sound bleed control into consistent venue behavior rather than a manual habit.

Acoustic acceptance tests to require

Use acceptance tests that prove intelligibility and stability, not just SPL.

- Intelligibility test method declared and repeatable (owner-approved).
- Targets per zone, and a documented procedure for re-verification after changes.
- Fault handling: loss of a channel or device does not create unsafe peaks or silent failures.
- Maintenance plan: filters, fans, and temperature drift are monitored and alarmed.

Acceptance tests for intelligibility and bleed

A reference-grade venue needs acceptance tests that can be executed and repeated. Acoustics should be commissioned like any other system layer. At minimum, a venue should pass these tests:

Intelligibility tests

- In each zone, speech and narrative content are intelligible at expected crowd conditions, not only in an empty room.
- Quiet-mix and Personal-Channel modes remain synchronized and intelligible under typical interference and device variation.
- In busy operating mode, intelligibility does not collapse below the venue's defined threshold.

Bleed tests

- Adjacent zones do not significantly interfere with one another under defined program levels.
- When a zone becomes empty, Lifecycle Automation Layer audio reduction measurably reduces bleed into adjacent occupied zones.
- When a zone becomes occupied again, restoration occurs before the visitor can perceive a missing, or broken, audio layer.

Governance tests

- Comfort-aware routing and delivery are triggered only by explicit visitor choice, or by venue policy for empty zones, never by covert inference, and never when acoustic-state confidence is below the venue's declared threshold.
- Operator overrides take effect immediately, and are logged, with clear expiry.

If a venue can pass these tests, the acoustic layer is not merely "considered." It is implemented.

Acoustic state is part of venue reality

When you start treating audio as a system layer, you also start treating sound as state. A space has an acoustic state the way it has an occupancy state. It can be calm or loud. It can be intelligible or muddy. It can be empty or crowded. It can be in a condition where a certain kind of narrative works, or a condition where that narrative will be lost.

At micro scale, acoustic state can influence what you deliver. If a room is already loud, the system might default to captions, or Personal Channel delivery. If a visitor indicates they are easily overstimulated, the system can offer a quieter mix option without asking the entire room to change.

At macro scale, acoustic state can influence flow. If a gallery is already operating at high sensory load, a venue-scale system can direct visitors who need calm toward quieter routes or quieter exhibits, without segregating them, and without calling attention to the fact that accommodation is happening.

This is not about to turn acoustics into a medical model. It is about to treat comfort as part of hospitality.

A practical acoustic posture for modern venues

If you want a single operational stance that fits museums, themed entertainment, cruise ships, and retail, it is this:

1. Design the room mix for coherence. One story per zone. Intelligibility over loudness.
2. Offer personal mixes as opt-in inclusion: language, narration-only, and comfort modes.
3. Treat sound bleed as a system-level constraint. It is part of experience quality.
4. Use occupancy-aware behaviors to keep the venue calm. The Lifecycle Automation Layer is experience design, too.
5. Verify what you manage. If you can change audio dynamically, you must be able to confirm it is correct.
6. Measure drift, not only failure. Trend signals belong to preventive maintenance, because reliability is part of experience quality.

The power of the acoustic layer is that it touches everything without demanding attention. Get it right, and visitors will not compliment your audio. They will simply stay longer, understand more, feel less tired, and leave with a sense that the place was designed for humans rather than for equipment.

In the next chapter, we move from sound to the physical substrate that makes adaptive systems operable under real constraints: networks, timing, and resilience. The more capable a venue becomes, the less it can afford fragile dependence on perfect conditions.

9

NETWORKING, TIMING, AND RESILIENCE

Minimum network separation

At minimum, separate three concerns so one cannot collapse the others:

- Guest network (best effort).
- Operations network (monitoring, logging, updates, administrative control).
- Show and safety network (time-critical control and verification).

Hard safety is not a network feature

Life-safety systems and emergency stop chains are not "on the network." They must remain functional if all compute, all software, and all networks fail.[xvi]

The venue intelligence stack may observe these safety states and react by suppressing automation, changing messaging, and removing routes, but it must never be required for:

- an emergency stop to work,
- the stopped state to remain latched until manual reset, or
- the building's life-safety behavior to execute.

Design intent: software can assist, but it cannot sit in the causal chain of stopping, alarming, or evacuating.

Document the boundaries and test failure modes.

Implementation pointers: see Appendix D.

Governance depends on network boundaries

In a governed venue, segmentation is not only a security measure. It is a governance measure.

If a system can change what people see, hear, or where they go, then the pathways that carry control, logging, and verification must remain reliable and protected from guest traffic and from best-effort pathways.

Treat the guest network as untrusted by default. Treat the operations network as the evidence path (monitoring, logs, update control, and administrative authority). Treat the show and safety network as the deterministic path (time-critical cues, safety-adjacent controls, and outcome verification signals).

This boundary is also what keeps adaptive components, including conversational and agent-like interfaces, from becoming accidental control planes.

> *A venue is not a website.*

That sounds obvious, until you watch how many modern experiences plans quietly assume connectivity will behave like the internet at home: always on, always fast, and always forgiving. In a venue, none of that is guaranteed. Networks are shared, radio conditions are messy, devices roam, guests bring their own interference, and the building itself can be hostile to signals in ways nobody notices until opening week.

The consequence is simple. When the network falters, the experience falters. If you want hyper-personalization to feel calm and inevitable rather than fragile and performative, networking, timing, and resilience cannot be treated as infrastructure footnotes. They are part of the narrative contract you make with visitors. You are promising that the place will respond coherently, and coherence requires that the system can move information through space and time reliably, even when the venue is stressed.

This chapter is about that discipline. It treats networking, timing, and resilience as experience principles with implementable patterns that operators, owners, and cities can demand. It is not an IT lecture.

The myth of perfect connectivity

A venue network is not one network. It is several. There is the operational network, which must remain stable and predictable. There is the show control network, which must behave deterministically enough that the room feels synchronized. There is the guest network, which is noisy by nature and should never be allowed to dictate the behavior of the core system.

When teams "simplify" by collapsing everything into one or two flat networks, they are not simplifying. They are deferring complexity until it becomes failure. A well-designed venue assumes connectivity will be imperfect, then builds so imperfection degrades gracefully.

That posture matters even more in personalization systems because personalization increases the number of moving parts: more nodes, more sensors, more surfaces, more Personal Channels, and more state. The answer is not to fear that complexity. The answer is to contain it with an architecture that respects priorities and keeps the critical path protected.

Latency is experience quality

Visitors do not perceive latency as latency. They perceive it as hesitation. A button press that lags becomes doubt. A media cue that fires late becomes awkwardness. A synchronized moment that slips out of alignment becomes a loss of magic. The system can be technically correct and emotionally wrong.

That is why it helps to think in experience budgets rather than raw bandwidth. Immediate interaction, such as a button, a gesture, or a kiosk selection, must feel instant. Background adaptation, such as updating recommendations or preparing a surface for the next group, can tolerate more delay as long as it does not intrude.

The discipline is to decide which behaviors must be low-latency, then ensure they do not depend on fragile paths. This is one reason provisioning sufficient compute matters beyond performance. When you have sufficient compute at the edge, the system can make decisions locally without waiting for a round trip to a distant service. The experience stays responsive even when the network is imperfect.

Timing is choreography

If the network moves information, timing makes moments land together. Venues are full of moments that must be synchronized to feel intentional: a line of displays that change together, a sound cue that aligns with a reveal, a lighting transition that supports a narrative beat, and a Personal Channel that remains synchronized to a shared room show.

The visitor experiences this as choreography. Either the place is together, or it is not.

Two principles keep choreography stable. First, do not assume that "close enough" is always close enough. Small drift becomes visible quickly when outputs share line-of-sight, or when adjacent zones bleed into one another. Second, treat time as a shared resource that must be managed rather than assumed. If systems cannot agree on "now," they cannot reliably agree on "together."

That is why serious venues treat synchronization as a base design requirement, not an integration afterthought.

Local-first is a reliability stance, and a privacy stance

There is a fashionable tendency to treat cloud services as the default. In venues, default cloud is often a category error. This does not mean cloud is forbidden. It means cloud belongs in the right place.

If an experience requires low-latency interaction, immediate responsiveness, or synchronized cues, the decision loop should run locally. If the experience requires continuity across multiple exhibits, it can still be local, because continuity does not inherently require remote services. If the experience requires venue-scale operational truth, it should remain on-premise, because the venue must remain operable even when the outside world is not.

Cloud is appropriate for tasks that are not in the critical path of the moment: aggregated reporting, content distribution workflows, update pipelines, and long-horizon analytics that can be buffered and reconciled. Local-first also supports privacy-forward design. When decisions can be executed locally, the system can deliver meaningful personalization while minimizing what must leave the venue. The more capable the venue becomes, the more valuable this posture becomes.

Graceful degradation is hospitality

Resilience is not only redundancy. It is the ability to fail without humiliating the visitor. A resilient venue system has fallback behaviors that preserve dignity. If a Personal Channel cannot connect, the core experience still works. If a network link drops, local nodes continue operating in a stable mode. If a service degrades, the system reduces capability but does not collapse into confusion. If a sensor becomes unreliable, the system stops trusting it and continues with safer assumptions.

Visitors forgive limitations. They do not forgive chaos.

Security is part of experience quality

Security in venues fails in two opposite ways. One failure is neglect: everything is connected, everything is reachable, and nobody wants to talk about it until a problem arrives. The other failure is overreaction: fear-driven lockdown that makes systems difficult to operate, update, or support.

A practical posture sits between those extremes. Segmentation, access control, and clear boundaries between guest traffic and operational control are designed in from the beginning. Remote access, when used, is deliberate, auditable, and scoped.

Hostile input is a design assumption

Any venue system that accepts natural-language input, public device input, QR-driven deep links, or other open-ended user-controlled inputs must assume hostile or malformed inputs will occur.

The practical posture is separation of concerns: conversational and advisory outputs do not become executable actions directly. Actions remain constrained by explicit permissions, bounded scopes, and auditability, and they execute only through the controlled planes defined above.

This is not paranoia. It is what keeps a helpful system from becoming a fragile system.

Security is part of resilience because security incidents create the same outcome as technical failures: loss of trust and loss of coherence.

Observability replaces hope

As systems become more distributed, the venue's relationship with truth becomes more important. It is not enough to have logs. Operators need a clear view of the state that matters, and the ability to answer practical questions quickly. They need to know whether the system is actually doing what they think it is doing. They need to locate the failure quickly when something is wrong. They need to see drift toward failure before it becomes a visitor-facing breakdown.

This is why verification and drift detection appear repeatedly throughout the book. In venue systems, hope is not an operating strategy.

Evidence, not screenshots

Logging is necessary, but raw logs are not automatically evidence. Evidence is reviewable, bounded, and reconstructable.

For venue-scale systems, the minimum bar is that an operator can answer, with artifacts: what the system believed, what constraints were checked, what was attempted, what was verified, and what was done when verification failed.

This is why later appendices define a logging minimum, an evidence pack manifest, and a monitoring plan. Those are not paperwork. They are how a venue remains defensible after upgrades, vendor changes, and incidents.

The chapter's practical commitments

If you take only a few commitments from this chapter, take these. They are the difference between systems that feel alive and systems that feel brittle.

1. Separate networks by purpose: guest, operations, and show control.
2. Put low-latency decision loops close to the experience, and provision the right compute there.
3. Treat time as a shared resource, and design synchronization explicitly.
4. Assume imperfection, and design graceful degradation so the experience simplifies rather than collapses.
5. Make the system observable, verify outcomes, and detect drift early.
6. Use cloud where it belongs, off the critical path, and never as the heartbeat of the venue.

In the next chapter, we move from network posture to the control posture that turns moments into reliable performance: show control, orchestration, and power as experience quality.

10

SHOW CONTROL, ORCHESTRATION, AND POWER AS EXPERIENCE QUALITY

A venue makes promises it rarely states out loud. It promises that a button will do what it says it will do, that a story will begin when a visitor expects it to begin, and that it will not be interrupted by the building itself. It promises that a reveal will land, that a transition will feel intentional, and that a space will remain coherent even when it is busy, noisy, or imperfect.

For much of the industry's history, show control meant something precise: deterministic cues that fire in a reliable sequence. A door opens. A projector starts. A light fades. A sound hits. When the system is well built, the room behaves like a composed piece of music. It is one of the great achievements of physical experience design, and it remains essential.

The change, and the reason this chapter exists, is that modern venues now ask the environment to do more than play a sequence. They ask the environment to adapt. They ask it to present a different language without visual compromise, deliver a quieter mix to one visitor without changing the room for everyone, and alter signage, pace, and guidance because a gallery is congested, a show is about to start, a route is closed, or a school group arrived early. They ask it to behave less like a fixed performance, and more like a living place.

That is not a betrayal of show control. It is the natural next demand placed on systems that have become capable. It does require a distinction that many teams do not make early enough: the difference between show control and orchestration.

Where governance sits

In governed venues, orchestration is not the final authority. A governance gate sits between orchestration and execution.

The orchestration layer proposes actions based on state and intent. The governance layer evaluates those proposals against explicit rules: safety, accessibility, privacy, consent, and operational constraints. The result is one of three outcomes: approve, modify, or reject.

This is how a venue stays adaptive without becoming arbitrary.

Show control is the score, orchestration is the conductor

Show control is best understood as the score. It is the designed sequence of states and cues that makes an experience land. It is the part of the system where you want determinism, repeatability, and crisp timing. It is where you decide that a light fade is twelve seconds, not ten, because twelve feels right. It is where you decide that an audio cue must precede a visual by a fraction, because the brain perceives it as more natural. It is where you encode intentionality.

Orchestration is best understood as the conductor. It does not replace the score. It decides which score to play, under what conditions, and how to keep the environment coherent when the situation changes. Orchestration is what allows a venue to remain hospitable under variability: crowd patterns shift, language mixes change, accessibility needs differ, and physical systems occasionally misbehave.

In Stage 0, the score can be enough. In Stage 1, you begin to need light orchestration, because multiple exhibits must remain consistent with a visitor's preferences. In Stage 2, orchestration becomes central, because the venue is coordinating behavior across zones, and those behaviors have externalities. This is why the book separates these concepts. It is operational, not theoretical.

If you confuse show control with orchestration, you will either build a brittle system that cannot adapt, or a chaotic system that adapts without preserving intent. Both failures look the same to the visitor. The place stops feeling designed.

Deterministic where it must be, adaptive where it can be

A modern venue needs two planes of behavior, and it needs them to coexist without conflict.

One plane is deterministic. This is where show control lives. It includes cues that must fire on time, sequences that must remain consistent, safety behaviors that must be predictable, and synchronization that must remain stable.

The other plane is adaptive. This is where orchestration lives. It includes selecting language modes, choosing narrative depth, offering Personal Channel options, adjusting guidance based on congestion, scheduling content based on time of day, programming, or audience mix, and managing energy and equipment life based on occupancy.

The art is deciding what belongs in which plane. If you push too much into the deterministic plane, the environment cannot respond to reality. If you push too much into the adaptive plane, the environment feels inconsistent, and sometimes arbitrary. Visitors will forgive a fixed show that is excellent. They will not forgive a system that behaves as if it is guessing.

The stance is simple: be deterministic where you must be, be adaptive where you can be, and design so the adaptive plane cannot corrupt the score.

A rule that prevents fragile autonomy

Treat adaptive outputs as proposals, not commands.

This is especially important when any part of the system uses generative techniques. Confident text is not proof. A proposal must be translated into bounded, permitted actions, checked against constraints, and logged with decision evidence.

If a proposal cannot be verified or cannot be executed within policy, the correct behavior is refusal or escalation, not improvisation.

That stance keeps the room composed.

Power as experience quality

In the AV world, power is often treated as a backstage concern, something you handle and then forget. That is a mistake in modern venues, because power behavior changes experience quality in direct and visible ways. A projector left running unnecessarily ages faster. A display left bright in an empty gallery contributes to a sense of waste. When audio is paired with visuals, it also contributes to sensory clutter, and to bleed into adjacent spaces. A space that remains loud when empty is not merely inefficient. It is unfriendly to adjacent spaces, because it turns the building into a chorus of competing narratives.

When a gallery is empty, there is no reason for it to broadcast at full volume. Lowering volume reduces sound bleed and makes neighboring experiences more intelligible. Turning off certain devices extends their life and reduces heat. Doing these things automatically, and restoring them seamlessly before any visitor notices, is a form of hospitality. It makes the venue calmer, and it makes operations less fragile.

This kind of behavior belongs in orchestration because it is driven by state: occupancy, time, and anticipated flow. But the moment you allow power to be dynamic, you acquire a new responsibility. You must verify.

The difference between
"we told it to turn on" and "it is actually on"

One of the quiet truths of complex venues is that commands are not outcomes.

Confabulation is why this matters

A system can also fail in a quieter way: it can produce a coherent, confident account of what happened that is not true.

In AI safety and governance language, this is confabulation: confidently stated output that is wrong or ungrounded. In venue operations, the analogue is "phantom success," where a system reports success because it issued a command, not because it verified an outcome.

The remedy is the same in both cases: require evidence. Verify the physical result. If verification is missing or inconclusive, the system must surface uncertainty, fall back to a safe mode, or escalate to an operator.

A system can send a command to a device, and the device can fail to comply for mundane reasons. A cable is loose. A setting is wrong. A temperature threshold is reached. A fault state is latched. A human intervention happened, and nobody logged it. The venue does not care why. The visitor only sees the result: a dead screen, a black wall, or a space that feels broken.

This is why verification belongs in the same conceptual breath as orchestration. If you have an occupancy-aware mode that lowers volume, powers down projectors, or changes display states, you also need a mechanism to confirm that restoration has actually occurred when the zone returns to use. The venue should not rely on hope. It should rely on observation.

That observation can be automated. A camera can confirm that screens are showing an image, and that the image is moving when it should be moving. A system can confirm that audio is present where it should be present.

A space can be treated as something you verify, not merely something you command. This is maintenance of truth in the physical world.

Later chapters treat this as a full operational loop with preventive maintenance, trend analytics, and automated checks. The point here is the architectural posture: once the environment is dynamic, verification becomes part of the experience, because experience quality depends on it.

Orchestration needs observability

As venues become more capable, failure becomes less dramatic and more subtle. A projector does not only fail. It drifts. A fan does not only stop. It speeds up over time. A device does not only go dark. It becomes slightly dimmer, slightly warmer, slightly more erratic, and then one day it stops cooperating.

The most operationally mature venues do not wait for the dramatic moment. They look for trends. They treat rising temperature, increasing fan speed, and other slow indicators as signals that maintenance is needed. They clean filters before failure. They schedule service before the public sees the fault. They keep the show by preventing the emergency.

This kind of observability is not a luxury. It is what makes complex systems sustainable, because complexity without visibility becomes fatigue, and fatigue becomes neglect.

Measure trust failures, not only outages

Treat confabulation incidents the way you treat drift and fault trends: as measurable operational signals.

Track when the system produced an ungrounded claim, when it cited missing evidence, when it had to abstain, and when it required human escalation. Trend these signals across updates. A system that becomes "more fluent" but less grounded has degraded, even if it is still available.

This is one reason commissioning must include evidence requirements and post-update regression checks, not only functional "it runs" tests.

A boundary that keeps teams aligned

Projects drift when different disciplines believe they own the system. The exhibit team thinks in stories. The AV team thinks in signal chains. The IT team thinks in networks. The operator thinks in staffing and uptime. The visitor experiences all of them as one.

Show control and orchestration provide shared language that keeps those disciplines aligned. Show control preserves intent. Orchestration preserves coherence under reality. Power management preserves sustainability and calm. Verification preserves trust. Observability preserves longevity. When those are treated as separate concerns, venues become brittle. When they are treated as one designed system, venues become resilient.

Commitments for teams building modern venues

If you want a practical list to carry into design meetings, carry this:

1. Define the deterministic plane explicitly and protect it.
2. Define the adaptive plane explicitly and constrain it.
3. Make occupancy-aware behavior part of experience design, not an afterthought.
4. Treat sound bleed as a system-level constraint and manage it operationally.
5. Verify outcomes, not just commands.
6. Design for drift detection, not only failure detection.
7. Ensure operators can override, intervene, and understand what happened - and show clearly when a system is in override mode.

These commitments are how modern venues keep their promises. In the next chapter, we move from orchestration as concept to identity, privacy, and continuity, because as soon as a venue begins to adapt across multiple exhibits, it must decide what it will remember, what it will forget, and how it will remain trustworthy while doing so.

Part III

MICRO HYPER-PERSONALIZATION ACROSS MULTIPLE EXHIBITS

Micro hyper-personalization is where most venues should start: real continuity across multiple exhibits, with benefits that visitors can feel immediately.

This part focuses on the personal channel and the mechanics of doing "small-scale" personalization correctly, meaning accessible, multilingual, private by design, and stable under real-world conditions.

Chapter 11 frames the personal channel as the accessibility and language backbone.

Chapter 12 makes the case for offline-first behavior, so the experience degrades gracefully instead of collapsing.

Chapter 13 tackles synchronization across audio, captions, sign languages, and Braille so accessibility is not second-rate.

Chapter 14 addresses multilingual delivery at scale while resisting bias and uneven quality.

Chapter 15 argues for visitor models that help, models that improve guidance, comfort, and outcomes, rather than surveillance theater.

Chapter 16 covers virtual docents and embodied guides as interfaces, not mascots.

Chapter 17 turns the pieces into a deployable pattern across multiple exhibits.

If you are trying to ship something in a realistic timeline, this part is the highest return. It is also the fastest way to earn trust, because visitors notice when a venue remembers what matters, without feeling watched.

11

THE PERSONAL CHANNEL FOR ACCESSIBILITY AND LANGUAGE

The Personal Channel began with a practical discomfort that too many venues learned to tolerate. Hearing support was often treated as a special case: a service desk request, a hardware loan, or a compliance checkbox. When it existed, it was frequently awkward and conspicuous. The equipment did not fit, the equipment was not charged, the staff did not understand it, and the guests who most needed clarity often received friction.

Personal Channel minimum viable spec

- Language selection and captions available on demand, with clear fallback.
- Synchronization commitments defined per environment class.
- Offline-first behavior: core content still plays when connectivity degrades.
- Privacy-forward session continuity: do not create permanent identity by default.
- Confabulation controls: do not present ungrounded output as fact; require approved sources when claims matter; refuse to invent when grounding is insufficient.

At the same time, the outside world moved in the opposite direction. The phone became the default interface for personal settings, personal media, and personal control, and hearing technology began to follow the same trajectory. When the U.S. Food and Drug Administration's over-the-counter (OTC) hearing aid category went into effect on October 17, 2022[xvii], it signaled more than a regulatory update. It marked a shift toward a consumer-led, device-centric model shaped by personal devices, apps, and everyday usability.

The Personal Channel is what happens when you take that shift seriously inside a venue. If a visitor already carries a capable personal device, and the venue can provide the right content synchronized in time, then that device becomes an accessibility channel the venue can deploy at scale. Not because it is fashionable, but because it is already in the visitor's pocket, already configured for their life, and increasingly integrated with the assistive technologies they depend on.

Confabulation is one of the quiet ways accessibility fails.

If a system fills gaps with plausible language, the visitor is forced to do extra work: to detect errors, to resolve contradictions, and to decide what to trust. In accessibility delivery, that is not only frustrating. It can be unsafe.

The Personal Channel therefore carries a higher standard than a generic assistant. When claims matter, the channel must be able to operate in explicit delivery modes, require approved sources for factual claims, and refuse to invent when grounding is insufficient. The correct fallback is a stricter mode, a safe default, or human escalation, not improvisation.

This is not a niche need. The National Institute on Deafness and Other Communication Disorders reports that approximately 15% of American adults ages 18 and over report some trouble hearing[xviii]. That statistic captures only one dimension of the need. Language diversity, cognitive comfort, and sensory variability widen the circle. In practice, almost every venue serves people who participate more fully when the experience can be delivered through a Personal Channel in the right mode, at the right intensity, and in the right language.

A hearing story becomes an inclusion story

It is tempting to treat accessibility as a checklist. It is more accurate to treat it as a system, because timing, modes, and dignity have to work together in the real room.

The Personal Channel started with a hearing question: how do you synchronize a personal audio stream to the main show so a visitor using hearing aids, earbuds, or other personal devices receives the same narrative at the same moment, without delay, without drift, and without needing to sit in a special location?

Solving that problem changes what becomes practical. Synchronized audio makes synchronized captions feasible. Captions are not only for deaf visitors. They also help second-language visitors, noisy spaces, families who do not want a loud room, and anyone who benefits from redundancy between eyes and ears.

Adaptive Delivery Across Modalities

Personalization that stops at the screen is incomplete.

A governed system must treat visual, audio, and non-visual delivery as equal, first-class outputs of the same personalization pipeline. Accessibility, language, and dignity depend far more on modality than on content volume, and modality must be adaptable at runtime.

Visual delivery as a spatial system

In a compute-based architecture, visual delivery is no longer bound to a single audience per surface. When visitor position and orientation can be sensed, the system can deliver different visual layers to different regions of the same physical space.

This allows patterns that are difficult or impossible in static systems:

- Different languages rendered simultaneously to different zones in front of a display.
- Visual emphasis adjusted by proximity, dwell, or interaction history.
- Contextual overlays introduced or removed without interrupting the core presentation.

Language selection does not need to be a menu. It can be spatial behavior.

The system can also make that behavior legible. Projection mapping, lighting cues, surface markers, or subtle graphical indicators can show which areas are currently serving which language or mode. New visitors can orient themselves physically rather than cognitively.

Audio delivery as a controllable field

Audio must be treated as adaptive infrastructure, not as a single shared channel.

When compute nodes output serious digital audio directly, it becomes possible to:

- Deliver multiple languages concurrently through spatially controlled audio zones.
- Steer intelligible audio to defined areas without flooding adjacent zones.
- Adjust balance, loudness, and intelligibility dynamically based on crowd conditions.

This is not about novelty. It is about avoiding contention. A multilingual venue that forces all visitors to share one audio channel is choosing conflict by design.

Non-visual navigation and communication as a core channel

Non-visual delivery is not an accommodation. It is a primary channel.

Guidance, wayfinding, descriptive narration, and accessibility communications must be:

- Language-aware,
- Verbosity-controlled,
- Pace-controlled,
- Context-aware,
- Governed by the same policy constraints as visual media.

A visitor navigating non-visually should not be served a degraded experience. They should be served a different expression of the same canon.

This requires that non-visual systems participate fully in orchestration and recognition state. If visual delivery can adapt but non-visual delivery cannot, the system fails its own inclusivity claims.

Emerging display capabilities and position-aware delivery

As display technologies evolve, including displays that allow multiple viewers to perceive different images simultaneously, the need for deterministic, position-aware rendering becomes more pronounced.

These technologies are not speculative. They are appearing incrementally. The limiting factor is not display hardware. It is whether the system can ingest position and context signals, reason about them, and render the correct layers under governance.

This again points back to compute at the node and orchestration above it. Without those layers, adaptive modalities remain isolated experiments.

Adaptive delivery across modalities is not about adding channels. It is about ensuring that every channel participates in the same governed system.

Once text becomes a synchronized mode, sign-language video becomes feasible as well. Sign languages are diverse, regional, and culturally specific, which is precisely why fixed in-room solutions struggle. A Personal Channel can present the appropriate sign-language video for the visitor who requests it, without forcing the entire room into a one-size-fits-all compromise.

The same posture extends to Braille outputs. The important point for venue design is not any single device model. It is that personal devices increasingly act as bridges to assistive outputs, and a Personal Channel makes that bridge usable in a venue context where fixed exhibit hardware rarely can.

Language dignity without visual clutter

The Personal Channel is also a clean answer to a multilingual reality that fixed signage can only partially address. In a theatre or show environment, the room often needs one coherent language mix for a given cycle, because clarity, energy, and shared experience matter. But the room is never perfectly uniform, and printed panels cannot keep up with the language mix that actually walks through the doors.

A Personal Channel solves this without turning the room into a compromise. The venue presents a coherent main narrative, while a visitor receives their preferred language through their own earbuds, hearing aids, or device speaker, synchronized to the shared show. The room remains visually calm, and the visitor remains included without forcing the venue to redesign the space every time demographics shift.

This is also where comfort becomes a form of access. Not everyone is excluded by missing ramps. Many people are excluded by sensory overload. A layered effects soundtrack can be thrilling for one person and exhausting for another, and that difference often determines whether someone stays, learns, and returns. An opt-in quiet mix, narration-forward, reduced effects, and lower intensity can expand participation without flattening the core show for everyone.

Neurodiversity and cognitive comfort

Neurodiversity is part of accessibility in public space. Many visitors participate more fully when an experience offers predictable pacing, reduced sensory intensity, and clear agency over how content is delivered. This is not a special-case feature. It is a normal mode set delivered through the Personal Channel.

Practical requirements that make this real:

- Comfort modes are explicit and opt-in: calm or quiet modes are chosen by the visitor (or guardian), not inferred covertly.
- Predictable interaction pacing: avoid surprise timers, sudden mode shifts, and "gotcha" interactions.
- Pause, resume, and exit without penalty: allow a visitor to pause an interaction, step away, and continue, or skip, without losing the thread of the visit.
- Reduced sensory intensity options: narration-forward mixes, reduced effects, reduced motion, and reduced flashing for the same core story.
- Consistency across stops: if a visitor selects a comfort mode once, it should persist across the micro cluster, and not reset at every station.

Age and interest dignity

Language is only one axis. Age band and interest lens matter just as much. A five-year-old and a seven-year-old do not share vocabulary, and an adult is not served by forced simplifications.

A Personal Channel lets a venue offer explicit choices for language, age band, and interests, then honor those choices consistently across stops without adding visual clutter to the room. Interests can be expressed as a small set of venue-appropriate lenses, such as art, history, and science, or within science: physics, mechanical systems, and biology. Age can be handled as a coarse band selected by the visitor, or a guardian, treated as a preference rather than identity, and allowed to expire at session end.

A privacy-forward system by construction

The Personal Channel has another advantage that matters more as venues become more sophisticated. It is naturally privacy-forward. When a visitor receives language, captions, sign-language video, or a quiet mix through their own device, the venue does not need to collect a dossier. Many benefits can be delivered with anonymous session behavior and short-lived preferences, designed so personalization is offered as a service rather than extracted as data.

This posture is ethical, and practical. It reduces liability, improves trust, and increases adoption. Visitors use systems they trust. Later chapters address identity approaches that support higher-trust interactions without oversharing, including models where the venue receives a minimal proof rather than a full document. For the Personal Channel, the principle is simpler: default to helpfulness without intrusion.

Why compute still matters here

A Personal Channel is only as good as its timing, and timing is only as good as the system that holds it. If room audio and personal audio drift, the brain notices. If captions lag, the visitor loses trust. If sign-language video falls behind the moment it describes, the experience becomes work.

This is why the Personal Channel sits naturally on a compute-first substrate. Compute is not only about speed. It is about control.

When the critical loop runs locally, the venue can hold a stable timing contract, enforce delivery modes, and degrade safely under stress. When the loop depends on remote round trips, the Personal Channel becomes fragile, and fragile accessibility turns help into work at exactly the moment the visitor most needs the system to be calm and reliable. Synchronization, switching, and multimodal delivery are timing commitments. The system must provision the right compute where it is needed so the Personal Channel remains stable under real-world conditions: crowds, interference, and the minor imperfections that every public venue contains. The deeper point is strategic. When timing is reliable, inclusion stops being a special-case promise and becomes a baseline expectation.

What the Personal Channel teaches the rest of the book

This chapter is about more than a platform. It is about a design stance. You can build experiences that assume a single average visitor and retrofit exceptions, or you can build experiences that assume human diversity from the outset and provide modes that make that diversity participate.

The Personal Channel makes the second approach achievable at scale. It aligns with the reality of modern devices, it aligns with privacy-forward operation, and it aligns with the practical needs of operators who cannot manage a warehouse of specialized loan equipment.

Most importantly, it aligns with the public purpose and the moral center of public venues.

A museum, a theme park, a cruise ship, a retail environment, and a city all exist, in part, to serve the public as it actually is: multilingual, varied in ability, varied in comfort, varied in pace, and worthy of dignity without negotiation.

The Personal Channel is one expression of that commitment.

The next chapter

In the next chapter, we take the Personal Channel one step closer to the floor. Inclusive delivery has to survive the real building: crowded days, messy radio environments, and moments when connectivity degrades and staff cannot afford to troubleshoot in public.

That is where Offline-First Personalization comes in. Offline-first means language, depth, and comfort modes remain immediate and coherent because the critical loop runs locally, close to the visitor, and close to the exhibit. It is the bridge between micro hyper-personalization and the macro WorldModel Venue, and it is one of the most practical ways to keep privacy-forward design enforceable in the field.

12

OFFLINE-FIRST PERSONALIZATION

A venue that depends on perfect connectivity is not modern. It is optimistic.

Public spaces are messy. Radio conditions change by the hour. Devices roam. Crowds interfere. Back-of-house networks are never as tidy as diagrams suggest. Even when everything is engineered well, there will be moments when the system must keep working without asking the outside world for permission.

Offline-first personalization exists for that reality. It is a posture for personalized delivery in which the intelligence required for a coherent visitor experience lives close to the visitor, close to the exhibit, and close to the venue's operational truth, so the visitor's day is not held hostage by network conditions.

This chapter is about that posture, and why it is the missing bridge between micro hyper-personalization and macro WorldModel Venues. It is also about why offline-first is an upgrade rather than a downgrade. It improves privacy, improves resilience, and makes personalization feel immediate rather than delayed.

The problem offline-first personalization solves is not "AI." It is latency and trust. Most people frame AI in venues as a model choice. That is the wrong starting point. The first question is operational.

When a visitor asks for something, can the venue respond immediately, consistently, and safely?

If the answer depends on a remote round trip, you have built a system that will fail at the worst moment: when the venue is busiest, when connectivity is noisiest, and when staff do not have time to babysit the system. Offline-first is therefore not a technical preference. It is the posture that makes personalization feel like hospitality rather than a demo.

Offline-first is also a governance posture.

When decisions are made locally, the venue can apply constraints before outputs are delivered, and it can avoid the most common operational failure of generative systems: confabulation (hallucination) under uncertainty. If the system cannot reach a remote service, it must not "fill in the blanks" with confident guesses. It must fall back to stricter modes, safe defaults, or human escalation.

In this book, Offline-First Personalization is also the form of Virtual Docent Layer designed to operate in that posture: local enough to be fast, local enough to be reliable, and local enough to be privacy-forward.

OnBoard as a pattern, not a device

Do not over-literalize the word OnBoard. OnBoard is a pattern: the core functions that must be reliable are executed locally within the venue's trusted perimeter, and often on the visitor's own device when that is appropriate.

This includes mode enforcement.

If the venue declares that a surface is curated-only, or Body-of-Knowledge only, that promise must still hold during degraded connectivity. Offline-first does not mean "do anything you can." It means "do what is permitted, safely, and consistently, even when the outside world is unavailable."

In practice, OnBoard can mean a local exhibit node that delivers personalized content without external dependency. It can mean a venue-local service that continues under degraded internet and responds with low latency. It can mean a visitor-device mode where preferences and comfort settings are held locally. It can also mean a synchronization system that keeps state coherent between local nodes and the venue's central operational truth.

OnBoard is often paired with simple physical entry points. A QR code or NFC tag at each stop can boot the Personal Channel and bind the session to the correct content and timing domain, which is why OnBoard supports retrofit with minimal physical change.

The crucial point is architectural. The critical path does not require the cloud. Cloud remains useful for distribution and reporting. It is not allowed to be the heartbeat.

It is also not allowed to be the arbiter of safety-adjacent guidance, accessibility instructions, or other consequential visitor-facing claims.

Why this posture is privacy-forward by construction

Privacy becomes difficult when the venue must send everything away. Privacy becomes manageable when the venue can compute locally.

Offline-first personalization supports privacy by default because preferences can be held as session state rather than as identity, personalization can be executed without exporting sensitive signals, and the venue can deliver value without collecting a dossier to "make the model work."

This is one of the main reasons OnBoard belongs in this book as a first-class idea. It makes the privacy covenant enforceable in the physical world. The venue can say, honestly, that it does not need to ship a visitor's day to a remote service to be helpful.

That changes trust. Trust changes adoption. Adoption changes outcomes.

The three OnBoard functions that matter most

1. Local delivery of language, depth, and modes

A visitor chooses language. The system must honor it without delay.

A visitor chooses an age band. The system must honor vocabulary, pacing, and explanation style without asking again.

A visitor chooses an interest lens. The system must honor it consistently, so a science-forward visitor is not forced through an art-history framing, and vice versa.

A visitor chooses Streaker, Stroller, or Student depth mode. The system must honor it across stops.

A visitor chooses captions, personal audio, sign-language video, Braille output, reduced light or screen brightness level, or a quieter mix. The system must honor it without requiring staff intervention.

These are not optional enhancements. They are the baseline of inclusive personalization. If they become unreliable, the venue becomes frustrating in exactly the way visitors now refuse to tolerate.

OnBoard means these choices are served locally, at exhibit nodes, and through the Personal Channel, so they remain stable even when the world outside the building is not.

2. Coherence across multiple exhibits

Stage 1 micro hyper-personalization depends on continuity. Continuity does not require identity. It requires a stable, privacy-forward way to carry declared preferences across exhibits during a visit.

Offline-first personalization supports this by treating continuity as a local session contract: language stays consistent, depth mode stays consistent, comfort modes stay consistent, and interest vectors shape what is delivered next.

This is where the museum stops feeling like islands and begins to feel like a system. It is also where OnBoard becomes a bridge to the WorldModel™ framework. The same continuity that helps one visitor can become aggregated state that helps the venue operate, without turning continuity into surveillance.

3. Safe fallback when everything degrades

OnBoard enables that by keeping a local minimum viable experience available.

If the Personal Channel cannot connect, the exhibit still functions.

If a cloud service is unavailable, the exhibit still delivers language and depth.

If a venue-wide orchestration component is degraded, local nodes still run safely and coherently within their permitted envelope.

This is what makes the venue resilient without becoming simplistic. Visitors forgive limitations. They do not forgive chaos.

OnBoard and Personal Channel are the natural pairing

The Personal Channel delivers synchronized audio and other modalities to the visitor's device. Offline-first personalization is the local intelligence posture that makes delivery reliable and immediate. Together, they solve a hard problem elegantly: the shared room remains coherent, the visitor receives individualized modes privately, and the system remains functional under imperfect connectivity.

This pairing also supports the book's insistence on inclusion beyond minimum compliance. Comfort modes, quiet mixes, long-tail language delivery, and accessible content layers can be delivered as normal options rather than special cases. When the system is local-first, it is easier to keep those options consistent.

OnBoard and Embodied Docent Interfaces

Many venues want embodied interfaces: a character, a specialist, a greeter, a curator-like presence. Embodied Docent Interfaces are one delivery form of Virtual Docent Layer, and OnBoard is what makes that form stable.

An avatar that is slow, inconsistent, or dependent on remote services becomes a novelty that irritates staff and undermines trust. An avatar that is local, responsive, and constrained by curated knowledge behaves like a real guide.

The value of an avatar is not that it looks human. The value is that it behaves as a consistent, inclusive interface to the venue's curated knowledge, in the visitor's language, at the visitor's preferred depth, under the venue's governance constraints. OnBoard makes that behavior reliable because the critical loop stays local.

Implementation posture, publish-safe and concrete

A reference book has a practical obligation: it must help teams make real decisions, not merely agree with principles. At the same time, it has to remain publish-safe and avoid wandering into patent-sensitive mechanics. The goal here is to strike that balance. The OnBoard implementation posture is described at the level of operational truth: what must run locally, what can be managed remotely, what must be measurable, and what constraints keep the system reliable, inclusive, and governable.

Content packaging and versioning

OnBoard systems require content to be packaged in a way that can run locally: language packs, depth layers (headline, context, deep), modality assets (captions, sign-language video, quiet mixes), and curated knowledge snippets used by Virtual Docent Layer dialogue.

Publishing must be versioned, staged, and capable of rollback. If the venue cannot roll back an update quickly, it will become afraid to update, and the system will stagnate.

Runtime modes and policy control

Offline-first personalization does not require one fixed delivery style. In the implementation examples discussed in this book, Virtual Docent Layer OnBoard can deliver fully curated fixed layers, Body of Knowledge layers, or a hybrid under owner-controlled policy.

A practical posture is to treat delivery mode as policy state, selectable by exhibit, station, or topic group. The venue can define which layers are fixed versus Body of Knowledge driven and can schedule those choices by time window or operating mode. This enables disciplined hybridity: a headline layer that remains fully curated, a context layer that is constrained, and a deep layer that is either curated, exploratory within boundaries, or a mix.

One operational benefit is repeat visitation. When deeper layers can vary within curated boundaries, a returning visitor can receive a fresh path through the same subject without the venue rewriting the exhibit, and without the system becoming unbounded.

Synchronization with venue operational truth

Offline-first acceptance criteria

Offline-first is not a slogan. It is a commissioning requirement.

- Define what still works with WAN loss and prove it during commissioning.
- Define what degrades gracefully, and how the visitor is informed.
- Define recovery behavior: state reconciliation without corrupting logs or consent state.

OnBoard does not mean isolated. Local nodes must align with venue truth: what exhibits are open, what routes are available, what mode the venue is in (normal, busy, incident), and what constraints are currently active.

The pattern is that local nodes consume a minimal subset of WorldModel state needed for their behavior and publish local events back to the truth bus. They are participants in shared truth, not separate worlds.

Local execution boundaries

OnBoard is responsible for immediate delivery. It is not responsible for unbounded decisions. Macro decisions remain governed at venue scale. OnBoard nodes execute within the envelope of permitted actions and modes. This separation prevents a fleet of independent agents with inconsistent behavior.

The bridge to phone-agent negotiation

Chapter 28 describes the phone as an agent. Offline-first personalization is a present-day steppingstone to that world.

Today, OnBoard emphasizes local execution for resilience and privacy. Tomorrow, the visitor's phone agent will negotiate language and comfort preferences, eligibility proofs as minimal yes or no disclosures, and what the venue is permitted to remember. OnBoard aligns with that future because it already treats the phone and local compute as trusted, privacy-forward layers. It makes the venue comfortable with the idea that the visitor's device is not an enemy, but a boundary. A venue that learns to operate OnBoard systems now will be ready for negotiated experiences later without needing to change its ethics or its architecture.

Acceptance tests that make OnBoard real

If OnBoard is not tested, it becomes marketing language. The acceptance tests are straightforward.

Responsiveness

- Language selection changes output immediately at the exhibit.
- Depth mode changes output immediately without restarting flows.
- Comfort mode changes output immediately without breaking synchronization.

Offline resilience

- With external internet unavailable, language, depth, and modality delivery still works.
- With partial venue network degradation, local exhibit behavior remains coherent and safe.

Continuity

- Preferences persist across the micro cluster without requiring identity.
- Preferences can be reset easily and visibly, with no hidden retention.

Governance alignment

- In incident mode, OnBoard respects tightened constraints immediately.
- OnBoard never executes macro actions outside permitted envelopes.

Publishing discipline

- Content updates can be staged and rolled back.
- Post-update verification confirms synchronization and modality delivery remain correct.

A venue that can pass these tests has an OnBoard implementation, not a concept.

The chapter's conclusion

Offline-First Personalization is the resilience and privacy bridge that makes personalization feel like hospitality rather than a network dependency. It keeps the critical path local: language, depth, comfort, and coherence across multiple exhibits. It pairs naturally with the Personal Channel. It supports embodied interfaces as stable, inclusive guides. It integrates with WorldModel truth rather than competing with it, and it prepares the venue for a future of phone agents and negotiated experiences.

Most importantly, it makes a promise the public will increasingly demand: this place can serve me well without asking me to surrender myself.

In the next chapter, we move from posture to precision. Offline-first personalization keeps delivery local, resilient, and privacy-forward, but inclusive delivery is only real when it is synchronized. A Personal Channel that is even slightly late turns help into work. Captions that lag break trust. Sign-language video that drifts out of alignment destroys comprehension. Braille output that arrives after the moment arrives is not access.

Chapter 13 is about synchronized audio, captions, sign languages, and Braille, and about the engineering discipline required to deliver them as reliable modes, not as special cases.

13

Synchronized Audio, Captions, Sign Languages, and Braille

Accessibility failures in venues are often the result of architecture rather than intent. A system is designed to deliver one experience through one primary channel, at one pace, and inclusion is added later as an attachment. A headset cart appears at the service desk. Captions arrive late and never quite match the show. A bilingual label gets added and the room becomes visually crowded. Sign language interpretation is offered for special events only, which is better than nothing, but still signals that full participation is occasional.

The Personal Channel changes the pattern because it changes where adaptation lives. The room can remain coherent while individuals receive the mode they need privately. This protects the shared experience from permanent compromises and expands who can participate fully.

That outcome depends on synchronization. Visitors care about synchronization in one practical way: the story lands at the right moment, in the right language, with the chosen modality, without additional work. This chapter defines the timing discipline that makes the Personal Channel dignified, reliable, and scalable.

The timing problem everyone feels

When audio is delayed by a second, the brain notices. When captions lag behind speech, the visitor stops trusting them. When sign language video drifts out of alignment with the moment it describes, the visitor is forced to work twice, first to interpret the language, then to reconstruct context. When Braille arrives late, reading becomes guessing.

Synchronization is therefore not a technical flourish. It is a minimum requirement for dignity.

Dignity also depends on correctness.

A perfectly synchronized message that is wrong is still a failure. This chapter focuses on timing discipline, but the delivery modes defined earlier still apply: when factual claims matter, the Personal Channel must remain inside approved sources, and it must refuse to invent when grounding is insufficient. Synchronization prevents drift in time. Governance prevents drift in truth.

In venues, synchronization is also harder than it looks. Public space is messy. Networks drift, devices roam, crowds interfere, and wireless conditions vary by the hour. Even in controlled environments, clocks slide unless time is treated as a managed layer. This is why the preceding chapter on networking and timing sits directly underneath this one. If the system cannot hold a stable sense of "now," it cannot deliver stable inclusion.

When you design the Personal Channel, you are designing a timing contract. You are promising that the visitor who opts into a mode will not be penalized for asking.

Part of that contract is graceful degradation. If timing cannot be held, the system must fall back deliberately, announce the limitation in a respectful way, and avoid turning accessibility into troubleshooting.

Hearing makes the requirement visible

A Personal Channel often begins with hearing support because hearing makes timing failure obvious. Venue shows depend on cadence: pauses, crescendos, reveals, and humor that arrives on a beat. If a visitor receives the show through hearing aids or earbuds and it is not synchronized to the room, the moment becomes uncomfortable. People withdraw, stop engaging, and stop trusting the system. Some remove the device, and the accessibility layer has not only failed, it has exhausted the person it was meant to serve.

The wider world has been moving toward personal hearing technology that is configured through phones and apps. The U.S. Food and Drug Administration's over-the-counter hearing aid category, effective October 17, 2022, is one marker of that direction[xix]. The practical implication for venues is behavioral rather than regulatory. Visitors increasingly expect hearing support to be personal, device-centric, and under their control.

Captions as a universal usability layer

In practice, they are a usability layer that helps many people in many conditions. They support deaf and hard-of-hearing visitors. They also support second-language visitors, visitors in noisy spaces, families who prefer quieter rooms, and visitors who benefit from redundancy between eyes and ears. Captions reduce anxiety as well. When a visitor can read and hear, they stop fearing they missed the key sentence.

The design choice that matters is delivery. Captions forced into the shared room as a permanent overlay can degrade the experience for everyone by competing for attention. They also push teams toward compromises that reduce readability, such as cramped typography, shortened sentences, or reduced language richness.

The Personal Channel supports a cleaner posture. Captions become a mode a visitor chooses. The room stays visually calm. The visitor who wants captions receives them at full clarity. The visitor who does not want them does not have to ignore them.

Sign languages are plural

Many people still assume sign language is universal or that it is signed English. Neither is true. Sign languages are natural languages with their own structure, and there are hundreds in use worldwide. The United Nations notes that, collectively, Deaf communities use more than 300 different sign languages.[xx]

That plurality matters for venues because it exposes the limitation of fixed solutions. A "sign language option" is not one thing. It is a family of options shaped by region, community, and culture. The Personal Channel makes that diversity manageable because it removes pressure to pick one sign language for the room. A visitor can choose the sign language that fits them, and the system can deliver a synchronized sign-language video feed aligned to the show.

The requirement remains timing integrity. For sign language, alignment is not a cosmetic detail. It is the story.

Braille already lives in the personal-device ecosystem

Braille is sometimes assumed to be purely physical, separate from personal devices. In practice, mainstream platforms already support refreshable Braille displays as external devices.[xxi] Apple documents using Voice Over with a Bluetooth refreshable Braille display on iPad, allowing the user to navigate and read what is on screen[xxii]. Android documentation likewise describes connecting to a Braille display through Talk Back Braille display settings[xxiii].

This matters for venues because it changes what is architecturally plausible. A venue does not need to invent Braille hardware. It needs to ensure its content can be rendered through the Personal Channel in a form that a visitor's own Braille device can consume, if they choose.

The Personal Channel becomes a bridge into the accessibility ecosystem visitors already use, rather than a parallel universe of specialized loan equipment.

The room language and the personal language

Multilingual delivery is often treated as a translation problem. In shared shows it is primarily an orchestration problem. A theatre cannot play five languages simultaneously without becoming noise, so it chooses a coherent room mix, often based on schedule or audience composition.

That choice is reasonable, and incomplete. There will always be visitors who prefer a different language. The Personal Channel solves this with dignity. The room remains coherent, while the visitor receives their preferred language through earbuds, hearing aids, or a device speaker, synchronized to the same show timing.

The same posture supports comfort. A visitor who is easily overstimulated should not be forced out by intensity. A narration-forward, quieter mix delivered through the Personal Channel can preserve the story without the full sensory load. Inclusion becomes practical when agency is preserved.

Why compute still matters

Multi-modal delivery is a real-time workload. If you are delivering synchronized audio in multiple languages, captions in multiple languages, sign-language video options, and Braille-friendly text streams, you are running a timing pipeline under field conditions. That pipeline needs sufficient compute at the right nodes, and it needs architecture that keeps time stable under crowd load and imperfect connectivity.

This is why the compute-first posture matters beyond performance. It makes inclusion operable. When a node can be upgraded by provisioning more compute, a venue can expand capability over time. More languages, richer caption options, additional sign-language options, and improved comfort mixes become feasible upgrades rather than rebuild triggers. A venue that cannot expand the Personal Channel without disruption eventually stops expanding it, and inclusion stagnates.

The promise of synchronized modes

The Personal Channel does not replace good spatial design, and it does not excuse poor acoustics. It exists to keep the room coherent while making the experience available to more people, in modes they control.

It is also one of the most future-proof layers a venue can build because it aligns with the direction of everyday technology: phones as personal interfaces, phones as accessibility hubs, and increasingly phones as trust devices capable of minimal disclosure.

In the next chapter we turn to multilingual strategy and language dignity at scale, because once language and modality become runtime choices, venues stop building bilingual compromises and start building environments that can reflect the real demographics of a city, a tourist destination, or a global audience.

14

MULTILINGUAL AT SCALE, WITHOUT BIAS

If you want to understand the future of venues, begin with an ordinary sign. The bilingual panel still appears in countless galleries, lobbies, and visitor centers. English is large and visually integrated. Spanish is smaller, squeezed beneath it. Sometimes a third language arrives as a compromise in a corner.

The panel is meant to signal welcome. The underlying structure often signals hierarchy.

The problem is architecture rather than intention. Fixed multilingual signage turns language into a physical scarcity problem. A wall has finite space, so languages compete for prominence. The result is predictable: one language feels primary, and the others feel like accommodations.

In many places, the "two-language solution" also fails as a description of reality. The public a venue serves is not bilingual. It is multilingual, and often dynamic.

Confabulation has a specific failure mode in multilingual systems: fluent wrongness.

It shows up as a translation that reads cleanly but subtly changes meaning, mistranslates technical terms, misstates proper nouns, or introduces details that were never present in the approved sources.

The consequence is trust. If the same exhibit says materially different things in different languages, the venue looks careless, and the visitor is given a second-class version of the story. Multilingual at scale is therefore a truth-consistency problem, as well as a translation problem.

Los Angeles is a useful example because the City has formalized language access as a tiered operating responsibility.[xxiv] The Citywide Language Access Plan establishes "threshold languages" organized into Tier 1, Tier 2, Tier 3, and Emergency Response Languages, with different service obligations across tiers. In Appendix A of the City's Language Access Annual Report, Tier 1 includes Spanish, Korean, Armenian, Chinese, Filipino/Tagalog, and Farsi (Persian), and the City describes Tier 1 responsibilities that include translation of vital documents and readiness to provide interpretation upon request. Tier 2, which should be supported at minimum by publicly available systems, adds additional languages.

Tourist destinations are even more demanding. The language mix is diverse and changes with seasons, school calendars, conventions, festivals, and global travel conditions. What you need on a Tuesday is not what you need on a peak weekend.

Multilingual support is no longer a design garnish. It is an operating condition of public space. Like most operating conditions, it becomes manageable when you stop treating it as decoration and treat it as a system.

Language Coverage Is a Strategy, Not a List

Multilingual support is often discussed as a checklist item. In practice, it is a strategic design decision that shapes who a venue actually serves.

Public spaces are inhabited by populations that speak far more languages than are typically offered in labels, tours, or media. Historically, this gap was unavoidable. Fixed-media localization is expensive, slow to update, and operationally fragile. Every additional language multiplied production cost and long-term maintenance burden.

A governed, compute-based delivery pipeline changes this equation.

The language gap as a design signal

The relevant question is not "how many languages do we offer."

The relevant question is "how many languages are spoken by the people who arrive."

That gap is the measure of exclusion.

Modern systems make it possible to address that gap deliberately rather than defensively. This does not mean every language must be supported equally or immediately. It means the system must be capable of supporting them without architectural penalty.

Tiered language coverage

Language coverage should be designed in tiers, aligned to audience, mission, and operational reality.

- Tier 1: Always-on languages.

- Tier 2: Event-enabled languages.
  Additional languages activated by schedule, season, school programming, special events, or known audience composition.

- Tier 3: Long-tail languages.
  Languages supported through controlled workflows and appropriate channels, including BYOD, text-first delivery, and other modalities where practical.

Tier 3 is where a venue demonstrates whether it is serious about serving its real audience.

Indigenous and non-mainstream languages

Long-tail support must explicitly include languages that fall outside typical "top language" lists.

In the United States, this includes Native American languages and other community languages. These are living languages tied to living cultures. Supporting them requires care, respect, and governance. Naming conventions, terminology choices, and update workflows must be deliberate. Where appropriate, consultation is part of correctness, not ceremony.

The system architecture must be capable of this support even when the operational decision is to phase it.

Scaling without asset explosion

The purpose of tiered language design is not to promise infinite translation. It is to avoid architectural dead ends.

A venue that can support fifty languages can often support more. A venue that can support one hundred can often support one hundred fifty. The limiting factors become governance, workflows, and priorities, not the media pipeline itself.

This book treats multilingual capability as a normal requirement of modern venues. The decision is not whether to support language diversity. The decision is how to do it responsibly.

Language is experience delivery

Teams often approach multilingual support as if the core question is linguistic: how do we translate the words. Translation matters, but the harder problem is delivery. The venue has to deliver language without distorting readability, pacing, tone, or dignity.

When you print language into a panel, you lock typography, hierarchy, line length, and reading order into a physical artifact. You make updates expensive, which means your language layer becomes stale. Staleness is a quiet form of disrespect.

A multilingual system treats language as runtime. Words are rendered rather than printed into a layout. Audio is selected rather than embedded as a single track. Captions are a mode a visitor can choose, rather than a permanent overlay that competes with the exhibit.

Once language becomes runtime, the correct planning question changes. The venue stops asking, "Which two languages do we print?" It starts asking, "How do we deliver language as a dignified mode, under real conditions, at real scale?"

Dignity is a design requirement

Bias enters multilingual design in quiet ways. It enters through font size, placement, and which language gets full narrative depth versus abbreviated summaries. It also enters through workflow decisions, such as which languages are considered "worth review," and which are treated as perpetual exceptions.

A venue does not need to solve politics to solve dignity. Dignity is achieved when language is delivered with parity in three dimensions:

1. Presence: the language exists as a first-class option, not a hidden sub-menu.
2. Quality: the language carries the same intent and clarity, not a rushed approximation.
3. Access: the language is available through at least one low-friction path, without special pleading.

Dignity also improves when the venue stops forcing language into physical scarcity. A wall has finite space. A software-defined surface, treated as a canvas, turns scarcity into configuration.

In operational terms, language dignity means two things.

First, every supported language gets the same quality bar: readability, pacing, and completeness, not "best effort."

Second, meaning must remain consistent across languages. If a claim matters in one language, it must mean the same thing in the others, or the system must fall back to a curated, known-good phrasing rather than improvising.

Synchronization acceptance criteria

If you claim synchronized delivery, specify how you will measure it.

- Measurement method declared (for example, flash-and-blip), and repeatable.
- Thresholds declared per zone and per environment class.
- A method for re-verification after updates or device replacement.

Three delivery planes: room, surface, and Personal Channel

A fourth AR plane is discussed later in this book, but the core strategy is already complete with three.

The room plane

The room plane is the shared experience, the show and soundscape that plays into the space as a whole. Shared space requires shared choices. A theatre cannot play five languages aloud at once. A gallery soundscape cannot become a cacophony of competing narrations.

The room plane therefore often carries a dominant language selection, which can change by schedule, time of day, or program. This becomes exclusionary only when the venue offers no dignified path for other languages.

The surface plane

The surface plane is the built environment's visible communication layer: walls, displays, projections, and canvases. This is where programmable surfaces change the practical possibilities. A surface can render language based on context, adapt typography for readability, and switch languages cleanly for a new group without turning the space into a permanent compromise.

The surface plane is also where operational messaging lives. Closures, reroutes, safety notices, and schedule changes are not special cases. They are venue reality. A programmable surface is how you keep operational truth aligned with what the visitor sees.

The Personal Channel plane

The Personal Channel plane is the visitor's device and personal audio. It is where language and accessibility become individually controllable without disturbing the shared experience. This is the most powerful plane for long-tail language support because it is not constrained by the room's need for coherence or the wall's need for restraint.

A visitor who prefers Mandarin can receive Mandarin through earbuds while the room plays Spanish or English for the majority. A visitor who needs captions can receive them privately. A visitor who wants sign-language video can receive it without requiring the room to become a compromise for everyone.

The Personal Channel also aligns naturally with privacy-forward design. Many language benefits do not require identity. They require preference selection, which can be anonymous and short-lived.

A serious multilingual strategy uses all three planes deliberately, rather than forcing everything into printed panels or forcing everything into an app.

A language strategy that scales

Most venues fail multilingual support because they treat it as a list. A sustainable approach treats it as a plan with tiers, workflows, and accountability.

1. Define language tiers by the venue's reality

Most venues need at least three tiers:

- Core languages: languages that appear constantly, in meaningful numbers.
- High-frequency languages: languages common enough to justify full support in vital content and wayfinding.
- Long-tail languages: languages that appear unpredictably but deserve dignified support at least through Personal Channel narration and essential wayfinding.

2. Separate vital content from deep content

Not all content carries the same risk. A venue should classify content into at least two categories:

- Vital content: safety, wayfinding, critical rules, accessibility instructions, and essential context.
- Deep content: interpretive depth, optional narratives, and extended learning.

Vital content must be accurate, fast to update, and consistently available. Deep content can expand over time and vary within curated bounds. Programmable surfaces are especially valuable for vital content because they can update quickly as conditions change. The Personal Channel is especially valuable for deep content because it can scale breadth without cluttering public space.

3. Build a translation and review workflow that respects time

Translation is not only words. It is voice, tone, reading level, and cultural fit. A label translated without regard for readability becomes a barrier even when it is technically correct. A workflow that scales typically includes translation and localization, review by a competent human reviewer who understands the audience, versioning so changes are trackable, rollback so mistakes do not become permanent, and scheduled refresh because language evolves and venues must keep pace.

4. Treat multilingual support as a living capability

A venue that treats language as a one-time deliverable is designed to be behind its own audience. A venue that treats language as a living capability can evolve with the city around it and with visitor patterns that change every season.

This is where provisioning sufficient compute becomes a practical lever. When nodes can be upgraded by provisioning more compute, you can expand language modes, caption options, and delivery sophistication without reconstructing the venue. Evolution becomes routine rather than traumatic.

The boundary: multilingual without creepiness

Multilingual support often tempts recognition, because recognition looks like the fastest way to infer language. Inference is not always the right answer. A privacy-forward venue prefers explicit choice when choice is easy. A visitor selecting language is simple, dignified, and auditable, and it avoids the failure mode where systems guess wrong in a way that feels intrusive.

When recognition is used, the posture remains consistent: default to anonymous behavior, minimize data, keep retention short, and treat trust as a requirement, not a marketing line.

Language dignity is achieved most reliably when the visitor feels in control.

A short standard for language dignity

If you want a practical standard you can apply to any project plan, use these questions:

1. Can a visitor receive the experience in their language without being treated like an exception?
2. Is language delivered as runtime, rather than hard-baked into design compromises?
3. Can the venue add or update languages without a fabrication cycle?
4. Does the plan support long-tail languages through a Personal Channel without forcing the room to become noise?
5. Are vital messages covered with a robust workflow and a fast update path?
6. Is the approach privacy-forward, with explicit choice as the default?

If you can answer yes, the venue is building a multilingual capability rather than producing translated artifacts.

Acceptance tests to require for multilingual at scale

1. Equality of presence: language selection is available without friction, and no language is visually or structurally treated as a secondary option.
2. Meaning consistency: a defined set of "must-match" statements produces consistent meaning across all supported languages, including exhibit titles, names, dates, safety instructions, accessibility instructions, prices, hours, and eligibility rules.
3. Proper nouns and controlled terms: names, locations, and exhibit-specific terminology remain stable across languages and do not drift into approximations.
4. Safe fallback: when a translation is missing or ambiguous, the system does not invent. It falls back to a curated, approved message, or offers a clear alternate mode.

In the next chapter we move from language to the visitor model itself, because multilingual delivery is only one dimension of personalization. The deeper question is how a venue learns what a visitor wants, what a visitor can tolerate, and how a visitor prefers to move through a day, while remaining privacy-forward and inclusive by design.

15

VISITOR MODELS
THAT ACTUALLY HELP

The easiest way to lose a visitor is to pretend they all arrived for the same reason.

They did not.

Some people came to be moved. Some came to be impressed. Some came because a child needed something to do for two hours. Some came because a school assignment required it. Some came because the venue is part of a day that includes lunch, a meeting, a drive, a flight, a wedding, or a hotel check-in. Some came because they live nearby and have been three times already. Some came because they have not been able to enjoy places like this for years, and they are quietly hoping this one will be different.

A venue that designs for a single "average visitor" is not being inclusive. It is designing for a fiction.

This chapter argues for a different posture: visitor models that are useful, non-invasive, and kind. The point is to help the system behave like a good host, offering the right depth, in the right mode, at the right moment, without demanding a dossier in exchange.

A visitor model must also be epistemically disciplined.

Confabulation is not only a content risk. It is a modeling risk. Systems that fill gaps with plausible output can quietly invent visitor attributes, needs, or intent, and then act as if those guesses are facts.

The rule for this book is strict: a visitor model may use declared preferences and observed session signals, but it must not invent personal attributes. When evidence is missing, the correct behavior is to ask, offer choices, or fall back to safe defaults, not to guess.

Conversational language is part of the visitor model.

When a venue uses conversational agents or avatars, language is not only a delivery attribute. It is part of the interaction contract. A visitor who addresses a system in a given language should be answered in that language. Where a visitor has explicitly opted into a membership or profile, the system may default to the visitor's preferred language without requiring repeated selection.

This behavior does not require identity. It requires context. Language can be inferred from the interaction itself, selected explicitly, or associated with an opted-in profile. The important point is that conversational systems must respond in the language the visitor is using, not merely present translated media alongside a fixed interaction layer.

Treating conversational language as first-class avoids a common failure mode, where exhibits are multilingual but dialogue is not. In a governed system, conversation, narration, and guidance all participate in the same language and modality rules.

Conversation follows the same governance posture as everything else.

When an agent speaks, it can confabulate. That is why conversational systems must remain inside declared delivery modes. Where factual claims matter, conversation stays within approved sources. Where visitor attributes matter, the system does not infer and then pretend it knows. It asks, it offers choices, or it uses session-scoped signals with clear uncertainty.

You do not need identity to model intent

There is a persistent misunderstanding in experience technology that leads to poor design and worse trust. It is the belief that personalization requires knowing who someone is.

Most of the time, what the venue needs is intent, not identity.

Intent is lightweight. It is how much depth someone wants, how they want it delivered, and how much friction they will tolerate before they disengage. Intent can be declared anonymously in seconds. It can be adjusted as the day unfolds. It can be held in a short-lived session context. If someone wants continuity across visits, it can also be held on their personal device.

Identity is heavier. It creates liability. It requires retention rules, access control, and trust that is difficult to rebuild once broken. Identity should be invited only when the benefit is clear, and when the visitor chooses it.

The posture used throughout this book remains consistent: available without identity and invite opt-in only when the visitor receives obvious value. Visitor modeling begins there.

A practical evidence rule keeps visitor models both useful and defensible.

- Tier 1: Declared preferences the visitor chose. Language, depth mode, interests, modality, and comfort choices. This tier is the most reliable and the least controversial.
- Tier 2: Observed session signals. Interaction choices, dwell, and what the visitor explicitly engages with during this visit. This tier should be short-lived and should not be treated as identity.
- Tier 3: Inferred attributes. Use sparingly, label as uncertain, and do not let inference override Tier 1 preferences or accessibility and safety constraints. When in doubt, ask or fall back to a conservative default.

Streakers, Strollers, Students: a model grounded in reading behavior

A visitor model only helps if it matches what people actually do. The most reliable behavioral signal is reading behavior, because reading behavior reveals intent under time. These modes show up across museums, themed environments, retail, corporate centers, and city-scale installations.

Streakers

Streakers consume the venue the way some people consume news. They read headlines. They want the main idea, the core point, the essential story, and then they move on. This is often a rational choice under constraints: limited time, a crowded day, a group to keep up with, or a preference for breadth over depth.

Design implication: Streakers need meaning fast without guilt. They benefit from clear one-minute stories, strong visual anchors, and optionality that never punishes skipping.

Strollers

Strollers read headlines, then they read a little more. They pause long enough to take in bylines, captions, and context. They want orientation before they commit to depth. They are the audience that benefits most from layered design, because they will step down into deeper content when the invitation is clear and the friction is low.

Design implication: Strollers need a clean path from headline to context. They respond well to "expand" moments, short explanations that lead naturally to deeper layers without forcing commitment.

Students

Students read everything they can lay their hands on. Sometimes they are literally students. Often, they are simply in a learning posture. They want the full narrative, the technical detail, the footnotes, and the adjacent sources. Summaries are not enough. They want the material.

Design implication: Students need deep layers that are real, not performative. They need extended narratives, optional technical depth, and alternate modalities that let them stay longer without fatigue. They also need the venue to respect their pace by not forcing motion.

A critical note: these are modes, not identities

These archetypes are not permanent labels. A person can be a Streaker in one gallery, a Stroller in the next, and a Student for the one topic that truly matters to them. The value of the model is that it gives visitors permission to choose depth, and it gives the venue a way to respond without guessing.

Do not backfill these modes from demographics or recognition.

If the system wants a depth mode, it should ask politely, or offer a one-tap choice. Guessing is not personalization. It is misclassification.

The three-layer content structure that makes the model real

If you want Streakers, Strollers, and Students to be more than a clever idea, each exhibit needs a disciplined content structure. The structure does not need to be complicated. It does need to be consistent.

Every exhibit should have three layers:

1. Headline layer: the one-minute story, the core meaning, the reason this matters.
2. Context layer: byline and first-paragraph depth, enough to orient and connect.
3. Deep layer: full narrative, technical detail, adjacent stories, and sources.

This structure also reduces confabulation.

If the deep layer includes sources and adjacent material, the system can stay grounded when it expands detail. If a venue allows bounded generation anywhere in this stack, it belongs in the deep layer only, and only within approved sources and constraints. The headline and context layers should remain stable enough that they can be trusted under load.

This structure is the backbone of micro hyper-personalization. It lets a venue serve different depths without building separate exhibits for different people. It also scales cleanly into macro behavior later, because it gives the venue a disciplined way to modulate load, pacing, and comfort without treating any visitor as a problem.

Interests as vectors

Depth is one dimension. Interests are another. If you ask someone, "What are your interests?" you will often get a vague answer because the question is too broad. The practical approach is to ask it in a way that fits the venue.

A museum can offer options such as art, history, science, people, and place. An automotive exhibit can offer design, aerodynamics, engines, mechanics, racing, sustainability, and culture. A retail environment can offer speed, discovery, health, premium, family, and value. A themed attraction can offer story, thrills, behind-the-scenes, characters, spectacle, and comfort.

These are not psychological traits. They are navigation tools. They allow the venue to pull the right narrative thread for the person who asks for it.

Interests should be chosen, not guessed, because guessed interests are a common confabulation pathway that feels like the venue is projecting onto the visitor.

The rule is restraint. Keep interest choices few, plain, and meaningful. Visitors should be able to choose without thinking hard. The system should do the heavy lifting.

Modality and comfort

Some visitors want audio. Some want text. Some want captions. Some want sign language. Some want a quieter mix. Some want reduced sensory load. Some want simple language. Some want technical language. Some want to hear it in the room, and some want it privately through earbuds.

When the venue offers modality as a normal mode, people stop struggling in silence. This is also where the acoustic layer becomes operational. A visitor who is easily overstimulated does not need to be managed. They need an option that lowers intensity without lowering dignity. A narration-forward personal mix is often enough to turn exclusion into participation.

Inclusion at this level is not a special feature. It is a normal set of choices.

How visitor models help operations without turning guests into inventory

Visitor models are sometimes misunderstood as marketing tools. In this architecture, they are operational tools in the best sense. They help a venue run smoothly while serving people better.

On a crowded day, you cannot create more square footage. What you can do is reduce decision friction and distribute people gently. Streakers benefit from clear, low-friction highlight paths. Strollers benefit from well-marked expansions that let them go deeper without getting lost. Students benefit from deep zones and structured resources that reward the choice to linger.

In Stage 2 venues, these modes can also inform routing suggestions and load balancing under governance constraints, because the venue can offer paths that fit intent while preserving fairness and safety. In Stage 1 deployments, the same model reduces choice overload and reduces the feeling of being lost. The visitor model becomes a way to be kind at scale.

A privacy-forward posture, always

Visitor models fail when they become excuses for collection. A visitor can choose depth, modality, and interests anonymously. The venue can hold that as a short-lived session context. If the visitor wants continuity across exhibits, it can be maintained without real-world identity. If the visitor wants continuity across visits, it can be offered as opt-in, with clear benefits and clear boundaries.

The model is about service, not surveillance.

A clean default is session scope.

Keep inferred signals short-lived. Keep retention minimal. Let the visitor opt into continuity when they want it, and let them reset easily. If the system cannot explain why it retains something, it should not retain it.

A template you can use immediately

If you are planning a micro deployment across multiple exhibits, you can start with a visitor model that fits on one screen and works without an account.

1. Depth mode: Streaker, Stroller, Student.
2. Interests: pick 3 to 6 that fit your venue, such as History, Engineering, Art, Science, and Mechanics.
3. Modality: room audio, personal audio, captions, sign language, quiet mix.

That is enough to transform the experience, because it tells the system how to be helpful.

Acceptance tests to require for visitor models

1. Works without identity. A visitor can select depth, interests, language, and modality without creating an account, and the experience remains complete.
2. No invented personal attributes. When evidence is missing, the system asks, offers choices, or falls back. It does not guess and present the guess as fact.
3. Preferences behave like a contract. Declared preferences persist across the micro cluster, are honored consistently, and can be reset easily and visibly.
4. Session signals are bounded. Observed session signals are short-lived, are not silently converted into persistent profiles, and do not override accessibility and safety constraints.
5. Confabulation controls are demonstrable. If the system cannot ground a claim or a recommendation, it abstains, falls back to a safer mode, or escalates, rather than improvising.

In the next chapter, we move from visitor models to multi-exhibit coherence and the hyper-personalized Virtual Docent, because once you know how to serve different depths and different interests, the next question is how to do it consistently across an entire venue day while staying privacy-forward and operationally sane.

16

VIRTUAL DOCENTS AND EMBODIED GUIDES

The oldest and most effective personalization technology in any museum is a good human being.

A docent who can read the room, adjust the pace, sense confusion before it hardens into disengagement, and choose the right story for the right person at the right moment has always been the gold standard. The docent does not deliver information. The docent delivers belonging. They make the visitor feel that the museum is not a warehouse of objects, but a place that can meet them.

For a long time, we treated that kind of personalization as inherently scarce. It depended on staffing, schedules, training, and the simple fact that one human can only be in one place at a time.

Virtual Docent Layer begins from a different premise: that we can make the docent's best qualities more available, without sacrificing the virtues that make docents valuable in the first place. The aim is not to replace the human guide. The aim is to extend the museum's ability to be helpful, coherent, multilingual, and inclusive, across multiple exhibits, for more visitors, more of the time.

This chapter is about Virtual Docent Layer as a micro-scale hyper-personalization layer across multiple exhibits. It is not a chapter about "AI as novelty." It is a chapter about how to build a system that behaves like a good guide, and how to do it in a way that respects privacy, respects curatorial intent, and remains operable in the real world.

Confabulation is the primary reason a Virtual Docent Layer must be governed.

A conversational system can produce fluent, plausible statements that are not grounded in the venue's approved sources, or in the venue's operational truth. In museums and cultural institutions, that failure mode is especially corrosive because it erodes the institution's authority quietly, one confident error at a time.

The posture used throughout this book applies here without exception: treat conversational output as a proposal. Delivery becomes a decision only after constraints are applied, approved sources are respected where factual claims matter, and safe fallback behavior exists when grounding is insufficient.

The shift from labels to dialogue

A fixed label is a monologue. It makes one bet about what a visitor needs.

Dialogue is different. Dialogue begins with a question, sometimes spoken, sometimes implicit. A visitor's curiosity is rarely shaped like the label that was written months ago. Curiosity is shaped by who the visitor is today, who they are with, how much time they have, what they already know, what they do not know, and what they are ready to care about.

Virtual Docent Layer treats the visitor's questions, and the visitor's chosen modes, as part of the experience. It can answer the person who wants the headline. It can reward the person who wants depth. It can adapt to the visitor who wants the story framed through design, or through history, or through engineering, depending on the subject matter of the exhibit and the interests the visitor chooses.

This is the point where the Streaker, the Stroller, and the Student become practical as a reading mode that the visitor can inhabit without apology.

1. A Streaker wants the headline, the essential meaning, and then they move. Virtual Docent Layer should be able to deliver that in a clean minute, with no scolding and no friction.
2. A Stroller wants the headline and then a little more, enough to orient and connect. Virtual Docent Layer should be able to provide that second layer gracefully, like a docent who knows when to stop talking.
3. A Student wants everything. The full story, the context, the technical details, the adjacent links. Virtual Docent Layer should not pretend to deliver depth if it cannot. It should deliver depth that is real, curated, and consistent with the institution's voice.

That last word matters. Voice.

A venue, a company, or a park has an identity. It has a tone, a seriousness, a warmth, a restraint. When a conversational layer is added, it must not dilute that identity. A Virtual Docent Layer must sound like it belongs.

Curatorial intent as the center of gravity

The quickest way to create chaos in a venue is to let a system become its own author.

Confabulation is one form of becoming an author by accident.

It does not arrive as vandalism. It arrives as plausible improvisation, and it is often delivered with confidence. That is why the conversational layer must remain anchored in curated delivery, or in a Body of Knowledge that the institution owns, governs, and updates deliberately.

If the institution cannot define what the Virtual Docent Layer is allowed to know, what it is allowed to say, and what it must refuse to invent, then the layer will eventually drift away from the venue's intent, even when everyone involved has good intentions.

Museums, cultural attractions, themed environments, and brand centers are not built on improvisation. They are built on intent. They exist because someone decided that certain stories matter, that certain objects deserve context, and that certain narratives should be told with care.

The Virtual Docent Layer must therefore be anchored in curated knowledge or be told to use curated text. It can do both, and it can do either based on a schedule.

Curated delivery, Body of Knowledge delivery, and hybrid depth

A conversational layer becomes useful only when the venue can choose how much variability it will allow. In the implementation examples referenced in this book, Virtual Docent Layer can operate in curated fixed mode, Body of Knowledge mode, or a governed hybrid, and the venue can select the posture by schedule.

There are three legitimate delivery modes, and venues should treat them as policy rather than as an accident of tooling.

These modes are also the venue's confabulation containment system.

Curated mode minimizes confabulation by design, because the system is selecting from institution-approved text.

Body of Knowledge mode makes confabulation containable, because the system remains inside approved sources and can be required to point back to those sources when claims matter.

Hybrid mode is how most professional deployments stay both helpful and disciplined. The system can be generous at deeper layers while keeping headline and core framing stable, accurate, and institutionally consistent.

If a venue cannot state which mode is active, by zone, by topic, or by schedule, then the venue cannot reliably commission the system, and cannot defend it after launch.

Some venues also permit bounded generation as a governed extension, but many deployments treat it as optional and keep conversational delivery within curated, Body-of-Knowledge, and hybrid modes.

1. **Curated fixed delivery** is the most controlled. The venue writes and approves the content, and the system delivers it consistently, like a docent following an approved script.

2. **Body of Knowledge delivery** draws from the venue's curated knowledge base to generate responses upon visitor registration, or in the moment. Where a claim matters, the system must be able to point to approved source material, and where grounding is insufficient, it must refuse to invent and fall back to a stricter mode or a human handoff. This is valuable when visitors ask unpredictable questions, and when the venue wants return visits to feel fresh. It must remain constrained by the same curatorial boundaries, tone rules, and safety policies as any other interpretive layer.

3. **Hybrid delivery** is where many professional deployments will land. The most reliable pattern is to mix modes by depth layer, because depth layers have different risk and different purpose:
   - Headline layer, fixed:
     the one-minute meaning is consistent, accurate, and approved.
   - Context layer, fixed or variable:
     the byline and first-paragraph layer can be either a curated expansion or a constrained synthesis from the Body of Knowledge.
   - Deep layer, fixed, Body of Knowledge, or both:
     deep content can be delivered as curated essays, or as guided exploration from the Body of Knowledge, depending on how much variability the institution wants to allow.

This structure keeps institutional voice stable at the top of the experience while still allowing the venue to reward curiosity at deeper layers. It also makes controlled novelty possible. In Body of Knowledge mode, the information a visitor receives can vary across visits because the questions asked and the paths chosen differ, while the approved sources remain bounded. That variation, when governed, is a practical reason return visits are more likely.

In practice, this means the venue defines what Virtual Docent Layer is allowed to know, how it is allowed to say it, what topics require careful framing, and what the visitor should be guided toward when the conversation risks drifting into noise. You can think of this as a Body of Knowledge, in the curatorial sense. It is the living library that Virtual Docent Layer draws from, shaped by the institution's scholarship and values.

This is also how the system stays patent-safe, trustable, and consistent.

A reference book like this cannot, and should not, publish implementation shortcuts. What it can do is clarify the architectural principle: the conversational layer must be governed by the venue's intent, and that intent must be explicit.

Virtual Docent Layer is not valuable because it can speak. It is valuable because it can speak within the museum's boundaries and still feel natural.

Multi-exhibit coherence, the moment it becomes real

A single conversational kiosk can be charming. It can also be a novelty.

Virtual Docent Layer becomes genuinely transformative when it is coherent across multiple exhibits.

This is Stage 1 hyper-personalization. The visitor moves through a sequence, and the system remembers what the visitor asked for, in a privacy-forward way. It remembers language preference, chosen interests, preferred depth, and chosen modalities. It does not need a full identity to do this. It needs continuity of intent.

Acceptance tests to require for multi-exhibit coherence

1. Preference persistence: language, chosen depth mode, chosen modality, and chosen interests persist across the micro cluster without requiring identity.
2. No re-asking fatigue: the visitor is not forced to repeat the same settings at every station unless they choose to change them.
3. Consistent voice: the headline layer remains stable and institutionally consistent across stops, even when deeper layers vary.
4. Confabulation containment: when the system cannot ground an answer in approved sources, it abstains or falls back, rather than producing a fluent guess.
5. Safe handoff: when a question exceeds policy scope, the system can hand off to staff in a way that is opt-in, and does not create a visitor dossier.

The effect is subtle and powerful.

A visitor who chose "design" in an automotive gallery should not be forced to reassert that preference at every station. A visitor who asked for the headline should not be dragged into deep exposition in the next room. A visitor who enabled captions should not have to hunt for the setting again. A visitor who chose a quiet narration mix should not be punished by a default that spikes sensory load every time they move.

Coherence is what makes a venue feel intelligent. It is what turns a set of installations into a system.

This is also where the phone, the Personal Channel, and Programmable Canvases begin to collaborate rather than compete. The visitor can engage in dialogue where it makes sense, see rendered language on surfaces where it makes sense, and receive a personalized modality on their own device where it makes sense. The environment stays beautiful. The visitor stays in control.

Multilingual dialogue without bilingual clutter

Multilingual strategies often fail at the point of conversation.

Translating labels is one thing. Delivering dialogue in a way that feels dignified is another. Dialogue also has a higher integrity requirement than static panels.

A visitor who asks a question in one language should receive the same meaning, not a different story. Proper nouns, dates, and core interpretive claims must remain stable across languages. When the system cannot confidently deliver a correct translation within approved sources, it must fall back to curated phrasing or a known-good alternate, not invent a fluent approximation.

This is one of the practical ways confabulation shows up in multilingual systems: not as nonsense, but as subtle meaning drift. If a venue tries to force multilingual conversation into static signage, it becomes visually noisy and still incomplete.

Virtual Docent Layer supports a better posture, but only if language is treated as a runtime choice, not a printed hierarchy.

A shared space may run a coherent room language for a given show cycle. That can be the right choice for clarity and shared experience. At the same time, a visitor should still be able to ask questions and receive answers in their preferred language through a Personal Channel, without making them feel like an exception.

The goal is a venue that feels welcoming to the long tail of languages without turning every wall into a compromise.

Inclusion that feels normal

Virtual Docent Layer should not only be multilingual. It should be inclusive in the broader sense that this book has pressed from the beginning.

A visitor who prefers simple language should not be treated as a problem. A visitor who wants technical language should not be flattened. A visitor who is easily overstimulated should be able to choose a quieter mode and keep it. A visitor who uses captions, sign language, or Braille-connected access should experience those modalities as normal options, not as exceptions that require staff intervention.

This is one reason the system approach matters. Virtual Docent Layer is not the only layer. It is one layer among several that together make inclusion practical: the acoustic layer, the Personal Channel, programmable surfaces, and the operational discipline that keeps everything reliable.

In a well-designed venue, inclusion is not a retrofit. It is a posture.

Returning visitors, the gentlest personalization

The most persuasive argument for hyper-personalization in cultural venues is not novelty. It is return.

People return to places that feel alive, that reward attention, and that do not treat every visit as the same script. Virtual Docent Layer can support this in a modest, privacy-forward way.

The system does not need to know a visitor's full identity to avoid repeating itself within a visit. It can vary explanations across exhibits, offer new angles when the visitor chooses, and guide toward adjacent topics based on declared interests.

If a visitor chooses opt-in continuity across visits, the experience can become richer, but the boundary remains: opt-in only, minimal data, clear benefit, and clear control. A docent remembers because they care. A system should remember only because the visitor asked it to.

Embodied Docent Interfaces:
Embodied interfaces without the gimmick

An avatar is not a feature. It is an interface choice.

When an avatar works, it lowers the barrier to asking the first question. It invites engagement. It reduces intimidation in spaces where visitors worry about sounding uninformed. It can also humanize guidance in large environments where a purely utilitarian kiosk feels cold.

When an avatar fails, it becomes a novelty that people try once, then avoid. The most damaging failure mode is not visual. It is behavioral.

An avatar that confabulates will eventually say something that sounds authoritative and is wrong. In a cultural institution, that is not a minor defect. It breaks trust in the institution's voice.

So embodied guides must be held to the same standard as the rest of this book: constrained by role, constrained by mode, anchored in approved sources, and required to refuse to invent when grounding is insufficient.

The failure mode is usually not graphics. It is behavior: inconsistency, latency, unbounded improvisation, or a tone that does not match the venue.

Embodied Docent Interfaces are one delivery form of Virtual Docent Layer, and they should be treated as such. They are not allowed to become their own authors. They must remain governed, curated, multilingual, and inclusive.

A practical way to think about avatars is as role-based specialists and generalists:

- a curator-like specialist for a specific gallery or topic
- a generalist greeter at entry who can orient, reduce anxiety, and suggest first steps
- a historical character interface, when appropriate, that remains constrained by scholarly boundaries
- a cruise ship director-style interface that can answer practical questions and hand off to staff when needed
- a corporate center guide that adapts to stakeholder roles and time budgets without becoming salesy

An avatar can also function as a front-of-house amplifier by supporting staff, without trying to replace them. When a visitor opts in, an avatar can notify staff that someone has arrived and would like help, or that a group is present in a zone and needs assistance. This must remain opt-in and privacy-forward, because the goal is service, not surveillance.

Avatars must also inherit the full inclusion posture:

- captions and readable text as normal
- sign language availability as a mode
- language as runtime, not baked into a single voice
- comfort modes, including quieter delivery and reduced intensity when requested

If the avatar cannot respect these requirements, it should not exist. A venue does not need an embodied interface everywhere. It needs embodied interfaces only where they increase approachability without increasing risk.

The governing rule is simple:

An avatar is permitted only when it is a faithful, constrained expression of the venue's intent, and when it improves hospitality more than it increases complexity.

Finally, avatars inherit every inclusion requirement: captions, readable text, multilingual delivery as runtime, sign-language availability where appropriate, and comfort modes, including quieter delivery and reduced intensity when requested. If these modes are not supported, the avatar is a gimmick, not an interface.

Avatars can also support staff without replacing them, but only by consent. If a visitor opts in, the avatar can offer a handoff, such as notifying staff that assistance was requested, or that a group has arrived and would like help. This must remain service-first and privacy-forward, and it must never become a hidden tracking system.

Hybrid delivery applies here, too. Many venues will choose fixed headline content for an avatar, while allowing deeper exploration through a Body of Knowledge under curated constraints. Others will run fully fixed delivery during certain hours and allow more exploratory depth during special programs. That choice should be schedule-controlled, auditable, and reversible.

Role defines what the avatar is for: a greeter, a specialist for a gallery, a character with scholarly constraints, a director-style helper, or a corporate guide that adapts to stakeholder roles. Boundaries define what it is not allowed to do: it does not invent facts, it does not improvise outside approved scope, and it does not use covert profiling to be persuasive.

A personal vignette makes the purpose of embodied guides obvious. I once walked through the Smithsonian's National Air and Space Museum beside Dick Gordon, the Apollo 12 command module pilot. The value was not that he recited labels. The value was that he connected artifacts to lived experience and context no wall text can hold.

That is the promise of avatars when they are treated as interfaces to curated knowledge. A visitor should be able to choose an astronaut, a test pilot, or an aircraft designer as their guide, then receive a narrative at the right depth, in the right language, and tuned by age and interest, without turning the museum into a scripted ride.

An embodied guide should behave as an interface to the venue's intent, not as an author. In the implementation examples referenced in this book, this is delivered through Virtual Docent Layer Avatars as a constrained interface layer, not as an unbounded author. The safest way to think about it is role plus boundaries.

What avatars are not allowed to do

- override safety, show integrity, or operator holds.
- require identity for basic help, language, or accessibility.
- infer sensitive attributes or create covert profiles.
- act as an unbounded author or improvise facts beyond curated sources and approved Body of Knowledge scope.
- present ungrounded output as fact, including invented names, dates, citations, policies, or instructions.

What avatars are allowed to:

- log operational evidence, such as mode changes and verification results, without creating a visitor dossier.
- hand off to staff when a question exceeds policy scope, when safety requires it, or when the visitor requests human help.
- provide captions, Personal Channel audio, quiet mixes, and other accessibility modes as opt-in choices.
- guide visitors through space using governed route truth, including accessibility routes, without inventing paths.
- deliver a fixed headline layer that is institutionally approved, then expand into deeper layers only within approved scope.
- offer a visitor explicit choices for language, age, and interest, then honor those choices consistently across stops.

Avatars are permitted only when they behave as constrained interfaces to venue intent and curated knowledge. In practical terms, that means they are allowed to do specific hospitality tasks, and they are not allowed to improvise outside policy.

The operator's view: calm beats clever

A system that delights visitors and exhausts operators is not a success. It is a liability that will be throttled until it becomes irrelevant.

Virtual Docent Layer must therefore be operable by design. That means:

- the venue can update and expand the Body of Knowledge without drama
- tone, verbosity, reading level, and sensitive-topic boundaries are policy, not emergency fixes
- the system can be monitored for health and integrity
- confabulation incidents are observable in aggregate, including when the system had to abstain, fall back, or hand off, so the venue can improve the Body of Knowledge without turning logs into dossiers
- the venue can see question themes in aggregate to improve interpretation, without turning analytics into a dossier factory
- and the system degrades gracefully under imperfect conditions rather than failing theatrically

Visitor questions are not noise. They are a map of curiosity and friction. A venue that learns from them becomes better over time because it listens with restraint.

Key takeaways

Virtual Docent Layer is valuable only when it is a constrained interface to curatorial intent, not an unbounded author.

Confabulation is a predictable failure mode in conversational systems. The remedy is mode discipline, approved-source boundaries, safe refusal when grounding is insufficient, and an operator-visible evidence posture.

The Virtual Docent Layer is not an ornament. It is a way to make a venue feel more human by making it more responsive, more multilingual, and more respectful of how different people learn.

It makes micro hyper-personalization coherent across multiple exhibits. It extends the docent's best qualities into more moments of the day. It does so with privacy as a constraint, not as a promise, and with operability as a requirement, not as an afterthought.

Embodied Docent Interfaces can make the venue more approachable in the right contexts while remaining governed, inclusive, and calm.

In the next chapter, we move from experience behavior to deployment discipline: how you roll micro hyper-personalization out across multiple exhibits in a way that is staged, maintainable, and reliable, so the system stays useful long after the novelty has worn off.

17

DEPLOYING MICRO HYPER-PERSONALIZATION ACROSS MULTIPLE EXHIBITS

A single exhibit can be excellent and still feel lonely. It can be beautifully designed, impeccably built, and technically flawless, and yet fail to change how a visitor experiences the venue as a whole. The visitor has a good moment, then walks away, and the building returns to its old habits. The day becomes a sequence of unrelated islands.

Micro hyper-personalization begins when those islands start speaking to one another through continuity. The visitor chooses a language once, chooses a depth once, chooses a mode once, and then the venue behaves as if it remembers. It does not force repetition. It does not demand identity. It does not treat inclusion as a special request. It stays coherent as the visitor moves.

This chapter explains how to deploy that coherence in a way that is operable, privacy-forward, and expandable. It is written for the reality of most projects. You do not start with a thousand endpoints and a perfect budget. You start with three to five exhibits, you prove value, and then you grow.

One operational discipline belongs in this chapter explicitly: confabulation control.

If any part of the micro deployment uses generated or synthesized language, then "it sounds good" is not a pass condition. Confabulation is a predictable failure mode, and it gets worse under load, missing context, and degraded connectivity. That is why deployment discipline must include mode discipline, approved-source boundaries where claims matter, and safe refusal when grounding is insufficient.

Treat this as commissioning posture, not as ideology. If the system cannot demonstrate that it refuses to invent when it is unsure, it is not ready to scale.

Begin with the smallest deployment that supports continuity

The temptation in modern systems is to begin with ambition. Ambition is not the enemy. Indistinct ambition is.

The right starting point is a small cluster of exhibits that forms a natural narrative loop: a gallery wing, a sequence of interactive stations, a single themed area, a brand center zone, or a retail experience that includes entry, discovery, and checkout.

You want a loop because a loop makes continuity visible. If the visitor sees the same preference honored at the second stop, the value becomes obvious.

If they only see it at the fifteenth stop, they may never notice it, and you will have built the right capability in the wrong order.

The first micro deployment should therefore be small, and complete. It should demonstrate language carried across multiple exhibits, depth carried across multiple exhibits (Streaker, Stroller, Student), at least one inclusive modality that is available by default as a mode (captions, personal audio, or a quiet mix), and a coherent tone so the venue feels like one place rather than a set of widgets.

That is enough to change a day.

Treat preferences as a contract

Micro hyper-personalization does not require identity. It requires agreement. A visitor does not need to tell you their name to tell you what they want. They can choose language, or a ranked list of languages, depth mode (Streaker, Stroller, Student), interests expressed as a small set of venue-appropriate choices, and comfort modes such as captions, personal audio, sign-language video, or quieter narration.

Treat those choices as a contract and the deployment stays clean. The visitor understands what they are getting. The venue understands what it must honor. The relationship feels fair.

Treat the same choices as a profile you are collecting and the system will drift. There is a second drift that is quieter and more corrosive: invented preferences.

A system that fills gaps with plausible guesses can begin to behave as if it knows what a visitor wants when it does not. That is confabulation expressed as personalization. The visitor experiences it as projection, not hospitality.

The fix is simple and scalable: preferences are chosen, not inferred. When evidence is missing, the system asks politely, offers a one-tap choice, or falls back to safe defaults. It does not guess and then behave as if the guess is a fact.

The deployment grows in scope, retention grows by default, inference creeps in, and trust erodes. A contract posture keeps the system humane, and it keeps the model scalable because a small, stable set of choices is easier to carry across exhibits and across time.

Build the experience in layers, and make each layer complete

Micro hyper-personalization succeeds when each exhibit has depth in layers rather than length in paragraphs. The Streaker, Stroller, Student model only works when content is written to support it. Each exhibit needs a headline layer that can be delivered cleanly in about a minute, a context layer that connects and orients, and a deep layer that rewards sustained attention without pretending.

If you skip the deep layer, Students will feel dismissed. If you skip the headline layer, Streakers will feel punished. If the context layer is thin, Strollers will not bridge into depth, and the venue will lose the broad middle of its audience.

A layered structure also reduces confabulation because it gives the system stable headline and context layers, and constrains any expansion into deep layers where approved sources and evidence rules can be enforced. When content is modular, you can revise one layer without rewriting the entire exhibit.

Decide what belongs in the room, on the surface, and in the Personal Channel

A micro deployment becomes elegant when it respects the three delivery planes.

The room plane is shared and should remain coherent. The surface plane is architectural and should be readable without turning every wall into a compromise. The Personal Channel plane is individual, and it is where long-tail languages, private accessibility modes, and comfort choices can exist without cluttering the environment.

Do not force everything into one plane. A gallery with a strong visual rhythm can keep the room calm and use the Personal Channel for optional depth, alternate languages, captions, and quiet mixes. A space that needs clear wayfinding can use surfaces as primary truth and keep personal devices as optional reinforcement. An exhibit that requires strong shared audio can keep the room mix coherent and let the Personal Channel provide language alternatives and accessibility modes.

This is how you avoid the bilingual wall problem without turning the venue into a phone-dependent experience. Design for phone-era expectations without requiring a phone.

Provision compute as a capacity decision

Micro deployments fail when teams treat hardware as permanent. It is not. The correct posture is to treat each exhibit node as a capacity envelope. The practical question is not which brand you bought. The practical question is how much compute you need in that node to deliver the experience reliably.

When compute is treated as capacity, replacement becomes straightforward. You swap a node, restore configuration, and move on. Upgrades become incremental. If a cluster needs more capability, you provision more compute at those nodes and leave the rest of the system untouched. Expansion becomes repeatable. Adding exhibits becomes repeating a pattern rather than inventing a new stack.

This is the foundation that allows micro deployments to expand toward macro behavior later. You do not need to predict every future function. You need a substrate that can absorb future functions without reconstruction.

Make operability part of the initial design

A micro deployment is not a prototype. It is the beginning of a venue system. Operations must be designed in from the outset, even at small scale. The operator needs clarity about what is supposed to happen, visibility into what is actually happening, and a way to recover when reality disagrees.

Visibility must include integrity, not only uptime.

Operators need to see when the system had to abstain, fall back to a stricter mode, or escalate because grounding was insufficient. Those events are not "failures." They are proof that the system is governed.

If the venue cannot observe confabulation containment in aggregate, it cannot improve the Body of Knowledge, and it cannot defend the system when something goes wrong.

Pilot plan template

A pilot is successful only if it produces operational truth. It is not easy to deploy a pilot, as the entire governance stack must be implemented for it to be relevant.

- Pick three exhibits, one simple, one medium, and one high-traffic.
- Define the stage target for each, the acceptance tests, and rollback plan.
- Instrument logs and drift signals before go-live, not after.
- Run a confabulation challenge set: questions that tempt the system to guess, and verify it abstains or falls back rather than inventing.
- Run for a full operating cycle, then decide what scales.

The simplest way to achieve that is to treat the deployment as a living product with a small set of recurring rituals: content versioning so you can change and roll back, health checks so failures are detected quickly, and a maintenance posture that prevents drift from becoming downtime.

Even a three-exhibit deployment benefits from this discipline because it teaches the organization how to operate the system before the system scales.

Guard privacy as if it were a design material

Most of the value in micro hyper-personalization can be delivered anonymously. That should be the default. Language choice, depth mode, and interest vectors do not require a name, a face, or a persistent identity. They require only session context and a way to carry that context across multiple exhibits.

If you need continuity across a single visit, use short-lived context. If you need continuity across multiple visits, invite opt-in, and make the value explicit. Never make identity the price of basic participation.

This is strategic. In public spaces, trust is the limiting factor. A privacy-forward system can be expanded. A system that feels hungry will be resisted, constrained, and quietly abandoned. Micro deployments are where trust is won.

Scaling from three exhibits to twenty

Scaling a micro deployment is not a matter of adding endpoints. It is a matter of preserving the same promises as the surface area grows. If you want to scale cleanly, keep four elements stable.

1. Stable promise: the visitor's chosen language, depth mode, and selected modalities remain honored across the cluster.
2. Stable patterns: each exhibit implements the same layered content structure and the same preference contract.
3. Stable operations: content is versioned, health is observable, recovery is defined.
4. Stable expansion logic: new exhibits plug into the same capability map: language, depth, modality, and interest vectors.
5. Stable integrity: confabulation controls remain consistent across the cluster, including approved-source boundaries where claims matter, abstain behavior when grounding is insufficient, and operator-visible evidence of fallbacks and escalations.

When these are stable, scaling becomes almost boring. Boring is good in operations. The venue feels alive to the visitor and predictable to the operator.

Cross-vertical examples

A museum wing can begin with three stations: an orientation wall, a hero artifact, and a deep-dive interactive. The visitor chooses language and depth once, and the wing responds consistently. A Streaker gets a tight narrative arc. A Student gets a library.

An automotive brand center can offer interest vectors such as design, aerodynamics, engines, and mechanics. The visitor does not need an account to select a path. The system behaves as if it understands what they came for.

A retail or supermarket experience can use the same discipline because the need is the same: reduce friction, increase relevance, and respect privacy. The system can adapt language, guidance, and assistance without demanding identity, and invite opt-in only when there is a clear benefit.

The technology differs by vertical. The discipline does not.

The end-state of micro is not micro

Micro hyper-personalization is not a consolation prize. It is the foundation. It is where a venue learns to manage language as runtime, deliver inclusion through modes rather than retrofits, provision the right compute rather than build one-off stacks, and operate a living system.

Most importantly, it is where a venue learns how to stay coherent without becoming invasive. Once you can do that across twenty exhibits, the next question becomes inevitable. If the venue can honor what a visitor wants across exhibits, can it also help them move through the venue day, reducing friction, balancing load, and improving comfort without sacrificing privacy or fairness?

That is the macro question.

In the next part of the book, we begin answering it, beginning with privacy as the central design pressure, because the more capable the environment becomes, the more important it is to remain worthy of trust.

Part IV

ANALYTICS, PRIVACY,
AND
ANONYMOUS RECOGNITION

You cannot build modern personalization without measurement, and you cannot earn trust if measurement turns into extraction. This part separates venue intelligence from people tracking, and it makes privacy an explicit design goal, not a policy footnote.

Chapter 18 states the posture: privacy is the product, and the product is the relationship.

Chapter 19 explains anonymous recognition in plain language, including what it can do, what it must not do, and why the distinction matters operationally.

Chapter 20 draws a critical boundary: measuring a place is not the same as measuring people, and confusing the two creates both ethical and business risk.

Chapter 21 addresses retail and the phygital moment, where convenience pressure often tempts teams into shortcuts that are difficult to justify later.

Read this part before you choose vendors, before you sign off on sensing, and before you define KPIs. It gives you language for requirements like minimization, consent, separability, and auditability, which are the difference between a system that scales and a system that becomes a liability.

18

PRIVACY IS THE PRODUCT

Every generation of venue technology begins with an argument about possibility. What can we do now that we could not do before? How responsive can an environment become? How quickly can content adapt? How precisely can we serve different languages, needs, interests, and paces without turning the room into a compromise?

The harder argument should arrive at the same time. Should we?

It is tempting to treat privacy as a legal constraint that appears near the end of a project, usually as a meeting, a policy document, and a list of prohibitions. That approach is risky, and it misses the operational reality of public environments. Privacy is the condition that makes capability deployable.

A system can be brilliant and still fail if it is not trusted. Trust is not an abstract feeling. Trust shows up as behavior. Visitors opt in, or they refuse. Staff use the system, or they work around it. Operators defend it when something goes wrong, or they quietly limit it until it becomes irrelevant. Institutions allow it to scale, or they freeze it at pilot size. A venue is either able to say, credibly, that it built something for the public, or it is not.

This chapter makes one claim, then turns it into design discipline. Privacy is the product, in the sense that privacy determines whether everything else can live.

The difference between being helpful and being hungry

Modern systems can be extraordinarily helpful. They can deliver the right language at the right moment, offer the right depth without forcing visitors to hunt, and provide captions, sign-language video, personal audio, and quieter mixes without making anyone feel conspicuous. They can make a large environment feel navigable rather than overwhelming. They can reduce friction, confusion, and fatigue.

The same systems can also feel hungry. Hungry systems collect because they can. They infer because it is clever. They store because storage is cheap. They correlate because correlations are tempting. They expand scope because it is easier than drawing boundaries.

A hungry system may not intend harm. Harm still arrives through secondary effects: unnecessary exposure, unnecessary risk, and unnecessary mistrust.

The central design question is therefore practical. Can the venue remain helpful without becoming hungry?

Privacy-forward design means restraint. It is an explicit choice to draw boundaries early, then enforce them mechanically, not abstinence.

Restraint also prevents a common mistake in AI deployments: trying to cure confabulation by collecting more.

When a system produces wrong output, teams often reach for more data, more identity, and more correlation, hoping "context" will fix accuracy. In public venues that move is usually backwards. It increases privacy risk, expands attack surface, and creates stronger illusions of certainty without actually improving truth.

The correct remedy is governance: approved-source boundaries where claims matter, evidence requirements for consequential statements, safe refusal when grounding is insufficient, and verification against operational truth. Accuracy is a discipline, not a data grab.

Privacy as architecture

When privacy is treated as policy only, policy is violated by accident. A well-meaning team collects more than intended because it was easier. A vendor component logs more than expected because that was the default. A new feature requires a new data source, and retention quietly expands because nobody wants to break the feature. A small exception becomes permanent, and the venue gradually becomes a place that remembers too much.

The way out is architectural. Privacy must be built into the structure of the system, so the easy path is the safe path. The rules are straightforward, and they are worth stating plainly.

One additional rule makes the rest of the chapter easier to implement: evidence without dossiers.

A governed venue should be able to prove what it did and why, without retaining biographies of visitors.

Decision records, mode state, consent state, and outcome verification are compatible with privacy when they are bounded, minimized, access-controlled, and retention-limited by design. If an audit artifact cannot be justified as necessary for safety, accountability, or operations, it should not exist.

1. **Rule 1: Default anonymous**
   Most of what a venue needs to be helpful does not require identity. Language choice does not require identity. Depth choice does not require identity. Interest vectors do not require identity. Comfort modes do not require identity. Many forms of analytics do not require identity. Most operational decisions at both micro and macro scale can be made from state: occupancy, congestion, schedules, closures, acoustic state, and capacity.

   Default anonymous is not a moral posture. It is a design posture that keeps systems lightweight, trustworthy, and scalable.

2. **Rule 2: Invite opt-in only when the benefit is obvious**
   Identity is not forbidden. It is expensive. It creates responsibility. It creates risk. It creates a burden of explanation. If you ask for identity, you must justify it, protect it, and minimize it.[xxv]

   Identity belongs where the value is clear: continuity across visits, membership benefits, saved itineraries, accessibility persistence, entitlements, paid upgrades, or other conveniences the visitor wants. Opt-in should not be a gate for basic participation. A museum should not require identity to read a label. A theme park should not require identity to find a restroom. A public venue should not require identity to feel included.

3. **Rule 3: Minimize what you store, and shorten how long you keep it**
   In public environments, "we might need it later" is not a reason to keep data. It is a reason to design a system that does not depend on keeping it. This is also an epistemic rule.

   If a system cannot justify retaining a signal, it also cannot justify treating that signal as a stable truth. Short retention, limited scope of correlation, and explicit purpose prevent the system from treating weak inferences as facts.

That matters because confabulation is not only model output. It can also occur at the system level when correlations are treated as certainty and then acted upon as if they were proven. A privacy-forward venue collects only what it needs, for the moment it needs it, and then it lets it go.

When something must be retained, it is retained for a specific purpose, for a specific time, with a deletion posture that is part of the system's rhythm. Many venues quietly fail here. They build systems that can collect and never build the discipline to forget. Forgetting is what makes systems safe in public space.

End-user responsibility is enabled not by disclaimers, but by systems that make consent boundaries visible, retention limits explicit, and forgetting enforceable as part of the operational rhythm.

4. **Rule 4: Keep sensitive decision loops local when possible**
   The more execution you can keep inside the venue, the less you must expose outside it. Local-first is a resilience stance, and it is a privacy stance. When language, personalization, and many analytic signals can be processed locally, the venue can deliver value while minimizing what leaves the premises. It also avoids turning the visitor experience into a hostage of external dependencies.

   Local execution does not eliminate risk. It reduces surface area and makes governance more realistic because the venue controls the operating environment.

5. **Rule 5: Make the system intelligible to the public**
   A system that cannot be explained cannot be trusted. This does not mean publishing internal diagrams. It means the visitor can understand, in plain language, what the system is doing, what it is not doing, what choices the visitor has, and what happens when the visitor chooses not to participate.

   When privacy is clear, opt-in rises.
   When privacy is ambiguous, even a helpful system feels suspicious.

If the visitor does not opt in, the venue should still work beautifully.

Multi-Domain Correlation Under Explicit Consent

Advanced personalization frequently requires correlating signals across domains. These domains may include people, devices, vehicles, transactions, and environmental state.

Correlation is where systems become powerful. It is also where they become dangerous.

For that reason, correlation must be governed explicitly.

Correlation must also be treated as a source of false confidence.

Cross-domain correlation can produce patterns that look persuasive and still be wrong, incomplete, or unfair. A governed system therefore treats correlated signals as proposals, not as truths. Where consequences are real, correlation must be bounded by consent, must be auditable, and must not be used to invent attributes or intent that the visitor did not choose to disclose.

Default anonymity and consent boundaries

The default posture of the system is anonymous continuity. A venue may recognize that someone has been here before, or that a presence is persistent, without knowing who that person is.

Identity binding, when used, must be:

- Explicit,
- Revocable,
- Membership-based,
- Purpose-limited.

Consent is not a checkbox. It is an ongoing relationship with defined boundaries.

Vehicles and arrival context

Correlating a person with a vehicle can enable arrival detection, accessibility assistance, concierge workflows, and staff notification. These are legitimate use cases. They are also sensitive.

Such correlation must be constrained to contexts where:

- The visitor has explicitly opted in,
- The purpose is clear and communicated,
- The behavior remains auditable.

The system must be capable of the correlation without assuming it.

Transactions and contextual assistance

Transactional correlation is more sensitive still.

Correlating a person with a purchase must never occur by default. It must be tied to an explicit opt-in to a membership or service relationship. Where visitors have chosen such a relationship, transactional context can support practical assistance. A café can suggest recipes. A shop can offer guidance. A venue can personalize follow-up.

Without opt-in, transactional data remains unassociated.

This distinction matters because it preserves trust while enabling value.

Why this belongs in the architecture

These capabilities are named here not to encourage surveillance, but to define the boundaries of what a governed system can do.

Privacy-first does not mean capability-limited. It means consent-first, purpose-limited, and auditable. Systems that cannot express those boundaries architecturally cannot enforce them operationally.

Anonymous recognition

There is a phrase in this book that matters because it resolves a false binary: anonymous recognition.

People often assume a venue must choose between two extremes, recognition or privacy. That binary is false.

Anonymous recognition means the system can recognize a context, a returning session, or a repeated pattern without knowing the person's real-world identity. It is how a venue can support continuity and analytics without turning every visitor into a named record.

This concept must be handled carefully. Anonymous does not automatically mean harmless. Linkability can still exist. A token can still become a shadow identity if it is retained too long or correlated too broadly. That is why anonymous recognition must be constrained by strict minimization and short retention.

When done well, anonymous recognition becomes a practical bridge between hospitality and restraint. It allows the venue to do things visitors want without demanding identity.

- Keep a visitor's language preference consistent across multiple exhibits during a visit.
- Keep depth mode consistent during a visit.
- Avoid repeating the same explanation in the same day.
- Provide analytics better than raw headcounts without building dossiers.
- Support operational decisions based on patterns rather than individuals.

Later chapters describe anonymous recognition more precisely and show how to prevent it from drifting into identity by accident. The important point here is conceptual. A venue can be responsive without being invasive.

Privacy becomes more important as capability increases

At Stage 0, privacy is mostly about restraint. Do not over-collect. Do not force accounts. Do not require identity for language selection.

At Stage 1, privacy becomes continuity without creepiness. Preferences persist across multiple exhibits, but the venue does not become a collector. The system behaves like a good guide who remembers what you asked for and forgets what you did not ask it to remember.

At Stage 2, privacy becomes central because power increases. When a system can guide flow, influence queues, coordinate scheduling, and shape how people distribute through space, the temptation to use identity as a lever grows. Externalities appear, and public scrutiny increases.

The response is governance. A system that affects public behavior must be constrained by explicit rules, including rules about data, retention, and consent. A macro system that is not privacy-forward will not scale. If it scales, it will eventually be shut down politically, legally, operationally, or through quiet abandonment. That outcome is not a loss of technology. It is a loss of trust.

Trust is more fragile than code.

Privacy makes inclusion easier

Privacy-forward design often produces better inclusion. When you avoid identity by default, you avoid the friction that comes with it. Visitors who would never register still receive multilingual support. Visitors who would never share personal details still receive captions, sign-language video, and personal audio. Visitors who are cautious still receive comfort modes. A family can enjoy the venue without negotiating privacy policies at the entrance.

When identity is invited only for clear benefits, the visitor understands the exchange. Participation becomes a choice rather than resignation. This is what turns personalization from something done to people into something done for them.

In inclusive venues, participation should not require surrender.

The same posture improves integrity because visitors are not forced into systems that guess; instead, the venue offers explicit modes and choices, and confabulation is contained by design.

Human Authority Is Not an Exception

In a WorldModel system, human authority is not a fallback. It is a core input.

Automation exists to reduce cognitive load, not to displace judgment. Operators, hosts, and staff remain meaningfully accountable for safety, hospitality, and ethics, through explicit checkpoints, escalation paths, and auditable decision traces. The system's role is to surface better options, enforce constraints, and verify outcomes, not to overrule human intent.

This is also how confabulation is handled in practice: when uncertainty rises, the system escalates or abstains, and human judgment remains a primary input, not a failure mode.

Humans as state, not overrides

Traditional automation treats human intervention as an exception. WorldModel treats it as state.

Operator actions, holds, approvals, and refusals are authoritative signals. They are logged, versioned, and propagated through the same state loop as automated events.

This ensures that human decisions:

- Are visible to the system,
- Are respected by automation,
- Are not silently undone by later logic.

Escalation and refusal are valid outcomes

A governed system must allow actions to be refused.

If a proposed action conflicts with policy, accessibility requirements, safety constraints, or human judgment, refusal is a legitimate and expected outcome. The system must record why the refusal occurred and adjust future proposals accordingly.

Automation that cannot accept refusal will eventually be bypassed.

Why this distinction matters

Public venues are not factories. They are social spaces with moral, cultural, and emotional dimensions.[xxvi]

A system that assumes it knows better than staff will fail, even if it is technically correct. A system that collaborates with staff earns trust and longevity.

WorldModel systems are designed to assist humans, not replace them.

Operating commitments

Privacy, rather than a feeling, is a set of decisions repeated consistently. If you are building a system described in this book, commit to the following posture:

1. Design so the venue works beautifully without identity.
2. Invite opt-in only when the value is obvious, and never as a gate.
3. Treat anonymous recognition as a tool, constrained by short retention and clear purpose.
4. Minimize collection and maximize deletion. Forgetting is a feature.
5. Keep sensitive decision loops local, when possible, for resilience and privacy.
6. Treat confabulation containment as an operational metric: track abstain and fallback rates, escalation frequency, and post-update regressions, and review them as part of the same cadence used for drift and incidents.
7. Make the system explainable in plain language and keep choices visible.
8. Ensure every privacy posture has an operational counterpart: logs, audits, and override.

The next chapter takes the key concept introduced here and turns it into practical mechanics. If we want continuity and useful analytics, how do we recognize what matters without identifying people, and how do we keep that discipline from drifting over time?

That is the promise of anonymous recognition, and it is where privacy becomes a design material rather than a slogan.

19

ANONYMOUS RECOGNITION, EXPLAINED CLEARLY

Anonymous recognition is one of those phrases that sounds like marketing until you put it under a bright light.

People hear the word recognition and assume surveillance. They hear the word anonymous and assume hand-waving. The result is a familiar argument: either a venue recognizes people, or it respects privacy. Either it can be helpful, or it can be restrained.

That binary is false.

Anonymous recognition is a practical middle layer. It allows a system to recognize a context, a returning session, or a repeated pattern without learning, storing, or requiring a person's real-world identity. It is one of the tools that makes privacy-forward personalization possible at scale, not a loophole around privacy.

This chapter exists for one reason: to make the concept precise enough that designers, operators, and owners can use it responsibly.

We will be plain about what it is, what it is not, what it can enable, and what boundaries keep it from drifting into identity by accident.

One more boundary matters in practice: do not let the system invent identity stories.

Confabulation can appear here as "plausible explanations" about who a visitor is, why they did something, or what they want, derived from weak signals and then presented as if certain. Anonymous recognition must remain a state tool, not a narrative engine. When evidence is missing, the correct posture is uncertainty and restraint, not a fluent guess.

Recognition is not identification

> *Identification answers the question,*
> *"Who is this person."*

In many venue use cases, you do not need to know who someone is. You need to know whether the visitor's experience should remain coherent across multiple exhibits, whether a preference was already selected, whether a particular piece of content was already delivered, or whether a space is repeatedly confusing people in a way that should inform design.

Those needs are real. If you deny them, the venue becomes clumsy. Visitors are forced to repeat themselves. Operators lose the ability to improve. The day becomes less coherent than it could be.

Anonymous recognition aims to satisfy those needs while refusing the step into real-world identity.

It answers, "have we seen this pattern" without answering "who is this person."

What anonymous recognition means in a venue

In practice, anonymous recognition means three things at once.

- The system can maintain a continuity token that persists long enough to be useful.
- The token is not a person's name, and it is not directly linked to a person's real-world identity.
- The system is designed to minimize linkability, minimize retention, and constrain correlation, so the token does not quietly become a shadow identity.

This is where many systems fail. They begin anonymous, then drift.

They drift because retention becomes convenient. They drift because logs become sticky. They drift because analytics wants longer horizons. They drift because teams forget that "anonymous" is a posture that must be maintained, not a label you apply once.

The way to prevent drift is to make the constraints explicit and enforce them.

One constraint must be explicit: do not backfill identity-like attributes from anonymous continuity.

A system may hold declared preferences and short-lived session signals, but it must not infer and then treat inferred attributes as facts. Inference becomes shadow identity faster than most teams expect, especially when it is logged, correlated, and retained.

The three varieties of anonymous recognition

Anonymous recognition is not one technique. It is a family of postures. The safest way to discuss it is to separate it into three varieties, because each has different risk, different value, and different governance requirements.

1. **Session continuity**

   This is the simplest form. A visitor's choices, language, depth mode, comfort settings, and interest vectors are held for the duration of a visit session. When the session ends, the context ends. The visitor can leave without anything being retained

   This is the right default for most micro deployments across multiple exhibits. It supports coherence without implying memory across days

   Session continuity can be implemented through mechanisms the visitor already accepts, such as a session QR code, a local token on a personal device, or a short-lived anonymous identifier that is not tied to a real person. The important point is not the mechanism. The point is the retention boundary.

   The system forgets when the visit is over.

2. **Anonymous return recognition**

This is the form that becomes valuable for repeat visitors without requiring them to register.

A venue can sometimes recognize that a visitor is returning in a way that allows non-creepy improvements, such as:

- avoiding repeating a welcome tutorial,
- offering a different angle on a story within the same day,
- recognizing that someone already selected a language preference recently,
- or providing better analytics about repeat patterns at the level of aggregates.

Return recognition is still anonymous, but it creates longer linkability.

That linkability must be constrained. Retention should be short and purposeful. Correlation should be limited. The system should avoid accumulating a stable behavioral fingerprint that becomes identity-like.

In other words, if you want anonymous return recognition, you must also accept a stricter governance posture.

3. **Pattern recognition without person recognition**

This is the most underappreciated form, and often the most useful.

Instead of recognizing individuals at all, the system recognizes patterns in the environment:

- occupancy patterns,
- congestion patterns,
- repeated confusion at a particular decision point,
- repeated dwell spikes in a particular zone,
- repeated disengagement at a particular exhibit segment.

This form of recognition can be done without tracking individuals through time. It can be done through aggregated counts, anonymous flow measures, and short-lived observations that decay into statistics rather than records.

Pattern recognition of this kind is the basis of humane operations. It helps staff without turning guests into inventory. It is also often sufficient for macro decisions, because macro decisions are usually about state, not about identity.

A venue that invests in environmental pattern recognition can achieve many of the benefits people wrongly assume require identity.

The drift problem: when anonymous becomes identity by accident

It is worth naming the main failure mode clearly, because it is common and avoidable.

Anonymous recognition becomes identity by accident when four things happen:

1. The token becomes stable over long periods.
2. The token is correlated across many contexts.
3. The correlated record is retained.
4. The system begins to infer attributes, preferences, and behaviors over time.

This is also where confabulation enters, even without a chatbot.

Once inference is treated as certainty, the system can begin to act as if it "knows" a visitor. That is a governance failure. Anonymous recognition must treat inference as tentative, bounded, and disposable, or it will become identity-like in effect, even if no name is stored.

At that point, you have built a shadow identity, even if you never stored a name. This is why anonymity cannot be declared. It must be maintained.

Maintained means:

- short retention,
- limited scope of correlation,
- deletion as a first-class behavior,
- and auditability of how tokens are used.

If those boundaries sound strict, that is because public trust is strict. The venue is a public environment. People do not expect to be remembered without permission.

We do not use 'anonymous' as a loophole. In a WorldModel venue, anonymity is an engineered state. The system is designed to 'flush' session data by default, retaining only the aggregate metrics needed to improve the facility (flow, heat, wear) without retaining the biography of the visitor. If the system cannot demonstrate that it forgets, it cannot claim to be anonymous.

How anonymous recognition supports the visitor without demanding trust up front

Failure modes to prevent

Anonymous recognition fails when correlation creeps in through side channels.

- Retention creep: logs stored longer than declared limits.
- Correlation creep: identifiers that become stable in practice.
- Shadow logs: analytics that bypass the evidence and consent rules.
- Interpretation creep: systems that generate fluent explanations or summaries about individuals or intent from anonymous signals, and treat those outputs as truth.

Mitigation must be explicit, auditable, and tested.

When anonymous recognition is done well, it produces a particular kind of experience quality. It feels as if the venue is paying attention, but it does not feel as if the venue is watching.

That is the line.

A visitor can select a language once and have it persist through multiple exhibits. A visitor can select a depth mode once and not be punished for it later. A visitor can enable captions once and have them remain available. A visitor can choose a quieter mode and keep it consistent.

The visitor receives coherence. The venue receives operability. Neither requires identity.

This is the positive promise of anonymous recognition. It makes personalization practical without turning a public place into a data collection apparatus.

Where recognition belongs, and where it does not

A responsible system draws boundaries not only around what it collects, but around what it uses recognition for.

Anonymous recognition belongs in service of:

- experience continuity within a visit,
- accessibility and comfort persistence within a visit,
- aggregated analytics that improve operations,
- and limited, clearly bounded improvements that the visitor would obviously welcome.

Anonymous recognition does not belong in service of:

- covert targeting,
- undisclosed profiling,
- long-horizon behavioral records,
- or any use case where a reasonable visitor would say, "I did not agree to this."

This is the practical rule that keeps systems defensible, not a rhetorical stance.

How to explain it to the public, in one paragraph

A venue should be able to explain anonymous recognition in plain language, and the explanation should be short.

A visitor should not have to read a policy essay to understand the deal.

A good explanation sounds like this:

"We can remember your choices, like language and accessibility settings, during your visit, without knowing who you are. We do not need your name. We do not keep your information longer than necessary. You can enjoy the venue without opting in to anything. If you choose to opt-in for additional benefits, we will tell you exactly what that includes."

If a venue cannot say something like that confidently, it is not ready to use recognition.

The practical governance posture

Anonymous recognition is not the absence of governance. It is governance expressed through restraint.

A privacy-forward implementation posture can be reduced to a small set of commitments:

1. Default to session continuity and treat it as the baseline.
2. If you add anonymous return recognition, define purpose, retention, and scope, and enforce them technically.
3. Prefer environmental pattern recognition over person recognition when possible.
4. Minimize correlation and never correlate "just because you can."
5. Delete by default and prove that deletion occurs.
6. Make the system explainable and opt-out friendly.

These commitments are not bureaucracy. They are the reason a system can remain helpful without becoming hungry.

In the next chapter, we move from recognition to analytics, because recognition is not valuable on its own.

It is valuable only when it supports better decisions: better maintenance, better staffing, better content, better comfort, and better outcomes for the visitor.

The risk, as always, is that analytics becomes a justification for over-collection. The opportunity is that analytics becomes a way to improve the venue while remaining privacy-forward and humane.

20

ANALYTICS, AND THE DIFFERENCE BETWEEN MEASURING A PLACE AND MEASURING PEOPLE

A venue that does not measure is blind.

Early interactive installations often avoided analytics on principle. The work felt commercial, invasive, or off-mission. Operators still made decisions. They made them from anecdotes, staff impressions, sporadic headcounts, and the handful of incidents that happened to be memorable.

Modern analytics, done well, gives a venue a second sense. It reveals where visitors hesitate, where they disengage, where they cluster, where they return, and where the experience excludes quietly without intending to.

Analytics also creates a temptation. Measurement can slide into identification, retention, and correlation. That slide is the path from service to overreach, and it is why this book treats privacy as a first-class design requirement.

The purpose of this chapter is to define analytics as a first-class layer that improves operations, inclusion, and experience quality while remaining privacy-forward and defensible.

A second risk must be named: analytics can become a narrative engine.

Dashboards and reports tempt teams to summarize data into confident stories about visitors, intent, or causality. That is a form of confabulation at the analytics layer: fluent explanation that is not warranted by the measurements.

A defensible posture is to separate measurement from interpretation. Keep metrics traceable to the underlying signals, constrain narrative summaries to what the data supports, and do not allow prose to replace evidence in decisions that matter.

A practical distinction keeps the chapter grounded. A venue can measure a place without measuring people. A venue can measure people only under explicit consent, narrow purpose, and strict restraint. A mature venue knows where the boundary belongs.

What analytics is for

Before discussing sensors, models, or dashboards, it helps to define what analytics should deliver. The most useful venue analytics falls into five categories.

1. Flow and distribution
   Where people go, how long they take, where they queue, and where they cluster.
2. Engagement and dwell
   Where people linger, what they skip, what they return to, and what creates fatigue.
3. Clarity and friction

Metrics allowed without identity

Most operational metrics do not require identification.

- Flow and dwell by zone, hesitation points, queue times, and abandonment rates.
- Content engagement by segment without storing identity (session-scoped).
- Accessibility uptake metrics (captions used, language selected) without personal dossiers.
- Integrity metrics for governed systems, in aggregate: abstain and fallback rates, escalation frequency, and confabulation incident counts, without storing identity.

If a metric requires identity, treat it as governed and justify it.
Implementation pointers: see Appendix G.

Where people hesitate, turn around, ask staff, or abandon a path.

1. Operational health
   Whether systems are up, stable, synchronized, and delivering what they should.
   Operational health also includes integrity: whether generated guidance and summaries remained grounded, and how often the system had to abstain, fall back, or escalate.
2. Lifecycle and maintenance
   Whether equipment is drifting toward failure, and whether intervention can be scheduled before the public sees a fault.

Notice what this list does not start with. It does not begin with demographics, identities, or psychological claims. It begins with the mechanics of a place and the lived behavior of a crowd. Most venues can improve dramatically before identity is introduced at all.

The baseline most venues should start with

Anonymous place analytics is where responsible systems begin, and in many cases where they should remain. The most common anonymous measurements include counts, directional flow, dwell estimates, and occupancy by zone. These are the measurements that answer operational questions without turning visitors into records.

Even this baseline has engineering reality. People counting in the real world faces occlusion, lighting changes, camera placement constraints, and privacy constraints. The important point is accuracy that is sufficient for decisions and deployment that is explainable in plain language. When it is done well, it enables improvements visitors feel immediately without being asked to identify themselves.

Why device-based "unique visitors" is increasingly unreliable

For a period, venues estimated unique visitors by tracking devices. Wi-Fi and Bluetooth identifiers looked like a shortcut: devices broadcast identifiers, so the venue counts unique devices and infers repeat behavior.

Platform privacy changes have weakened this approach. Apple's "Private Wi-Fi Address" behavior uses a different Wi-Fi address per network and can rotate over time (Apple Support: "Use private Wi-Fi addresses on Apple devices").[xxvii] Android documents MAC randomization behavior for Wi-Fi as a privacy feature (Android Open Source Project: "MAC randomization behavior"). As a result, device identifiers are less stable by design[xxviii].

This shifts the practical conclusion. If uniqueness depends on device identifiers, the data will drift because the operating systems are actively reducing identifier stability. A privacy-forward venue should design around that trend rather than fighting it.

Face recognition and the stakes

There are contexts where a venue benefits from knowing whether it has seen the same person before without needing to know who they are. Unique visitors and repeat visitors can be operationally meaningful. Measuring whether an improvement increased return can be meaningful.

Face recognition can support stronger uniqueness than device-based methods because it matches faces rather than phones. It also raises stakes.

Accuracy varies by algorithm, image quality, and operational setting, and demographic differentials have been documented in NIST testing. NIST's Face Recognition Vendor Test program includes reporting on demographic effects [xxix](NISTIR 8280, 2019). A venue that uses face recognition for analytics must assume error modes, differentials, and sensitivity to image quality. Claims and policies have to reflect that reality.

Biometric processing also triggers legal and ethical obligations in many jurisdictions. Illinois' Biometric Information Privacy Act is a well-known example with strict notice and consent requirements[xxx], and it has continued to evolve through litigation and legislative changes. A venue that cannot operate with explicit consent, strict purpose, and strict retention should not use biometric uniqueness for analytics.

The operational takeaway is simple. Most venues do not need face recognition to improve service.

A privacy-forward analytics ladder

A useful way to structure decisions is an analytics ladder, similar to the Personalization Ladder. Each rung increases value and responsibility together.

At every rung, treat integrity as a measurable outcome: track confabulation containment and abstain behavior as operational signals, not as rare surprises.

Rung 1: Anonymous place analytics

Counts, flow, dwell, congestion, and hotspots, with no attempt to recognize individuals. This is where most venues should begin.

Rung 2: Anonymous recognition for continuity

Short-lived session context that preserves language, depth mode, and accessibility modes across multiple exhibits without identifying a person. This supports coherent experience delivery and improves session-level analytics without building long-horizon identity.

Rung 3: Consent-based uniqueness analytics

If a venue truly needs unique visitor accuracy beyond what anonymous methods provide, it can offer explicit opt-in as a benefit exchange, with clear explanation, narrow purpose, minimal retention, and a credible deletion posture. Biometric techniques belong only on this rung, and only with discipline.

At this rung, the system should be designed so the default experience remains fully usable without opting in, processing is local-first where possible, retention is minimal and purpose-bounded, and the venue can explain what it is doing in one short paragraph.

Analytics that improves inclusion

One of the most compelling uses of analytics is inclusion. Hesitation often indicates confusion. Confusion is not evenly distributed. It is shaped by language, cognitive load, sensory conditions, and unfamiliarity.

Place analytics can reveal:

- where multilingual messaging is insufficient,
- where sound bleed makes speech unintelligible,
- where signage density overwhelms, and
- where an experience is unintentionally optimized for the confident, the fluent, and the already initiated.

These insights allow a venue to justify investments in the acoustic layer, quiet modes, Personal Channel delivery, and better wayfinding without collecting personal identity. Used this way, analytics becomes a tool for dignity.

The other half of analytics: equipment health and preventive maintenance

Visitor analytics is only half the story. The other half is whether the venue can keep systems reliable over years.

Many assets drift before they fail. Projectors heat up. Filters clog. Fans work harder. Brightness shifts. Hardware errors become more frequent. If you wait for the public failure, you are operating reactively.

Preventive maintenance is the more humane model, and it is driven by trend analytics. Manufacturers provide maintenance guidance that ties overheating behavior to ventilation and filter conditions, including cleaning air intake vents and filters (for example, Epson support documentation). IBM describes condition monitoring as a predictive maintenance approach that uses real-time data to detect faults and anomalies before critical assets fail (IBM: "Condition monitoring").

In a venue context, a practical posture can be simple: trend rising temperature, trend rising fan speed, trend increasing fault warnings, correlate with occupancy and runtime, and alert when drift suggests a clogged filter or ventilation issue. The benefit is fewer broken moments.

Occupancy-aware automation
and why laser projectors changed the equation

Occupancy-aware operational automation became more practical as laser projection removed common warm-up and cool-down assumptions. Manufacturers describe laser projectors as turning on and off instantly with no warmup or cool-down period needed (for example, Epson's laser projector product pages). This supports strategies where unoccupied zones can be reduced and restored before visitors notice.

This capability becomes a foundation for Lifecycle Automation Layer and for verification layers such as Outcome Verification Layer. Analytics is what makes these systems behave intelligently rather than crudely. Occupancy signals and drift signals inform what to power, when to power it, and when maintenance is becoming urgent.

A rule that keeps analytics healthy

If you want one rule that protects both trust and utility, use this:

> *Measure what you need to improve the place,*
> *not what you can use to profile the public*

Most venues can achieve dramatic improvements with anonymous place analytics, short-lived continuity, and rigorous operations analytics. When identity is introduced, it should be an explicit opt-in benefit exchange and treated as a governed capability with strict purpose and strict retention.

In the next chapter, we turn to recognition capabilities that span beyond people counting and into cross-vertical patterns, including retail, vehicles, color, pattern, and point-of-sale correlation, using the same discipline: capability bounded by purpose, consent, and restraint.

21

RECOGNITION, RETAIL, AND THE PHYGITAL MOMENT

A visitor's day does not begin at the entrance. It begins in a car park, in a rideshare drop-off lane, at a bus stop, at a hotel lobby, or on a cruise ship deck where someone has decided that this hour will be spent somewhere real. By the time the guest reaches the first exhibit, the experience is already shaped by friction and ease: how long it took to arrive, whether they felt oriented, whether they could find a restroom without asking, whether the first sign spoke their language, and whether the space felt calm or chaotic.

Older venue systems treated that prelude as outside the story. Modern systems cannot. The prelude determines whether the visitor enters the day relaxed or already tired. This is where recognition, properly constrained, becomes useful.

Recognition is a way of translating the physical journey into state that a venue can respond to. The difficulty is that the word recognition is often heard as a synonym for surveillance, as if the only reason to recognize anything is to identify someone. That is the category error this book keeps trying to prevent. A venue does not need to identify people to be helpful. Most of the time, it needs to recognize context.

In this chapter, "recognition" is used in a state-first sense. Recognition is any mechanism that translates the physical world into bounded, time-limited state that the venue can respond to. That includes arrival context, queue and occupancy patterns, opt-in entitlements, and transaction state where the visitor explicitly requests service.

This matters because recognition is often conflated with identification. Identification is one possible use of recognition, but it is not a prerequisite for helpfulness, and it is not the default posture of a trustworthy public venue. The default posture is: recognize the place, recognize the moment, and recognize declared preferences, while resisting the slide into persistent identity, cross-domain profiling, or retention creep.

Academic note: confabulation is not limited to chat systems. In phygital environments, confabulation also appears as "confident service claims" that are not grounded in operational truth, for example, "this route is open," "this item is in stock," or "the wait is five minutes." In any service-first system, those claims must be treated as safety-adjacent outputs: grounded, verifiable, or refused.

Context is operational truth. A space is filling up. A route is congested. A group is arriving. A show is about to turn over. A queue is about to spike. A retail point is understaffed. A gallery is quiet and ready for discovery. The acoustic layer is already loud enough that offering a quieter mode would be kind.

The phygital future, the blending of physical and digital, becomes healthy when recognition is used in service of those outcomes and when the system refuses to become hungry. This chapter is about that refusal. It is about positive uses of recognition across verticals, including retail, supermarkets, automotive environments, and destination-scale venues, and about doing it with restraint, dignity, and clear boundaries.

The phygital moment is timing and relevance

Retail personalization is often framed as advertising. That framing is too small. In a physical environment, the most valuable personalization is usually timing and relevance. It is the right information at the right moment, delivered in a way that reduces friction rather than increasing pressure.

A visitor who cannot find the next gallery does not need a coupon. They need clarity. A family trying to keep a day coherent does not need a banner. They need a route that respects time and energy. A guest in a store often needs assistance: where something is, whether it is in stock, what it means, how to compare it, and how to move through checkout without a small stress becoming a large one.

A healthy phygital system treats advertising as a narrow subset of contextual assistance. Assistance can be delivered through surfaces, through the Personal Channel, and through staff tools, but only if the system recognizes enough context to choose well.

Treat operational outputs as bounded decision objects with evidence pointers.

In service-first phygital systems, the most dangerous failure mode is a fluent explanation that is not grounded. The correct posture is to treat service outputs as decision objects, not as free-form narrative. A decision object is a bounded response whose fields are tied to operational truth, for example:

- Recommended next action (bounded choice set).
- Confidence posture (high, medium, low).
- Evidence pointer (what state or source it relied on).
- Expiry or freshness window (how long this remains valid).
- Fallback behavior (what happens if the state becomes uncertain).

A system can still present this politely to humans, but internally, the output should remain structured, bounded, and traceable. This is how you prevent confabulation from becoming "confident help" that is wrong in the ways that matter.

Recognition beyond faces

When people hear recognition, they often think of faces. Faces are one modality, and they are one of the most sensitive modalities. This book treats face recognition as a high-stakes tool that should never be a default and never be a casual add-on.

A more useful taxonomy is recognition by modality and by consequence.

Low-consequence contextual recognition
(usually acceptable without identity):

- Zone occupancy and dwell patterns.
- Queue and congestion estimates.
- Environmental state: acoustic comfort, heat load, time-to-next show, and route availability.
- Device-local session continuity, when implemented as rotating, short-lived tokens.

Medium-consequence continuity recognition
(requires explicit boundaries):

- Returning-session continuity across multiple stops in the same visit.
- Group continuity (family or school group) where the group explicitly opts in to be treated as a unit.
- Staff role recognition for operator tools.

High-consequence identification signals
(require explicit opt-in and heightened governance):

- Face recognition.
- License plate recognition, persistent device identifiers, or any identifier that is stable across days.
- Any linkage that turns "help me now" into "track me later."

Governance rule: the more stable the identifier, and the more cross-domain the linkage, the more it must be treated as a governed capability with explicit purpose, short retention, and auditability. If the venue cannot state the purpose and retention in one sentence, it should not deploy the linkage.

The broader, and often more useful, vocabulary of recognition includes:

- Occupancy and flow recognition
- Group recognition, in the sense of "a cluster just entered," not "who they are"
- Object recognition, such as a stroller, a wheelchair, a school-group marker, or a tour-guide flag that changes pacing needs
- Vehicle recognition in a bounded sense, such as recognizing a car type, a color, a pattern, or a repeat arrival token without turning license plates into identity databases
- Color and pattern recognition for operational cues, brand activations, and experience continuity in contexts where it is relevant
- Point-of-sale recognition, meaning transaction state and entitlement, not extraction of full identity

This vocabulary matters because it is how venues become smarter without becoming invasive. It is also how you build systems that generalize across verticals. A museum that recognizes flow and dwell can reduce congestion and improve clarity. A brand center that recognizes vehicle context can adapt the narrative to what a visitor is actually looking at. A supermarket that recognizes queue conditions can change signage, open self-checkout, adjust staff prompts, and reduce stress. A cruise ship that recognizes deck congestion can redirect traffic, adjust show timing prompts, and keep the day calm.

None of those require knowing who a person is. They require recognizing what the environment is doing.

People and their cars:
a bounded bridge between arrival and experience

The relationship between people and their cars is a powerful cross-vertical signal, and it is also an area where systems can either be responsible or reckless. A car is not a person. A car can be correlated to a person, but it should not be treated as a shortcut to identity unless the visitor explicitly asks for that and the venue can operate under strict consent and minimal disclosure.

The positive use case is simpler. Vehicle context can help the arrival feel coherent.

A visitor arrives in an electric vehicle and is naturally curious about sustainability. The venue can offer a sustainability thread as an option, not as a push. A family parks in a far lot. The venue can surface shorter, calmer routes early, so the day begins well. A brand center recognizes a vehicle class entering a zone and switches a nearby canvas to a relevant story angle, about the object of attention rather than about the visitor.

In automotive culture, "what you arrived in" is often a legitimate part of why you came. People enjoy seeing their interests reflected back to them, and the vehicle can be a proxy for taste in a non-legal sense: design, engineering, performance, heritage. The boundary is important. Identity as taste is not identity as personal data. A system can support the first without collecting the second.

Recognition and point-of-sale as a service layer

It is where the guest reveals intent by committing to a choice. It is also where the venue can either create friction or remove it.

When recognition is tied to POS, the goal should be coherence and service rather than targeting:

- A guest purchases a ticket upgrade, and the system reflects entitlement immediately, without awkward handoffs.
- A family buys a timed entry, and guidance becomes specific and calm, helping them arrive on time without rushing.
- A guest purchases a product linked to an exhibit and the Personal Channel offers optional extended content that enriches the purchase without turning the store into a surveillance zone.
- A supermarket transaction triggers a privacy-forward receipt and guidance experience, helping the guest find the exit, pickup point, or the next service.

Commerce is part of the visitor's day. Commerce is also where privacy pressure is highest, because the temptation to build dossiers is strongest. A responsible venue links POS to experience state through minimal, purpose-bound tokens, and avoids creating a permanent record of "who bought what" unless the visitor opts into a benefit that requires it, such as loyalty continuity. Even then, the posture remains fixed: minimize, disclose, and retain only what is necessary.

Later chapters discuss proof models that allow entitlements without oversharing personal documents. The principle is the same here. A venue should receive proofs and states, not a person's entire identity.

Assistance that does not cheapen the room

Visitors can tolerate some commerce. They can also feel instantly when commerce is driving the room. When that happens, the place loses dignity, and the visitor loses trust.

The practical question is not whether a venue can use advertising. The question is what belongs in a place that claims to be cultural, experiential, or civic. The most effective commercial prompts often look like assistance because they are relevant, timely, and low-pressure.

Programmable canvases are not an excuse to plaster the venue with offers. They are an opportunity to make the environment more coherent and occasionally more commercially effective without degrading the space. Sponsor messages can be integrated with narrative moments rather than dropped as interruptions. Retail prompts can appear when they are helpful, such as "this item is available in the shop," without shouting it in every room. Guidance that reduces friction, such as "short wait in this gallery," "quiet route," or "next show in ten minutes," serves the visitor and the venue at the same time. Contextual education can link to a shop item in a way that feels like extension rather than extraction.

The design rule is simple. Canvases should default to experience clarity rather than marketing pressure. When a system must choose between operations and advertising, a healthy system chooses operations. This is commercially sensible over time. A venue that feels respectful generates return, and return is a reliable revenue channel.

Personalization and manipulation

Recognition can improve relevance. It can also be used to push. The difference is whether the visitor remains in control.

A helpful system uses recognition to reduce friction and offer options. A hungry system uses recognition to steer behavior without disclosure. A helpful system makes "no" easy. A hungry system hides controls and calls it "optimization."

A venue that wants to be trusted makes its posture obvious. You can participate without opting in. You can opt in for clear benefits. You can decline and the venue still works beautifully. The system collects minimally, retains briefly, and forgets on purpose. Guidance is transparent. Offers remain optional.

When these boundaries are held, recognition becomes a tool for hospitality rather than a tool for pressure.

Why compute matters here

Recognition is not only a sensor problem. It is a compute and architecture problem. If recognition decisions require long round trips, the experience becomes hesitant. If recognition depends on constant cloud connectivity, the venue becomes fragile and privacy becomes harder to defend.

If recognition is centralized without constraint, correlation becomes tempting, and correlation is where anonymous systems drift toward identity.

A compute-first posture is a privacy posture here. When recognition and classification can be done locally, the venue can limit what leaves the premises. It can also limit what is retained because local systems can decay session data quickly and convert what matters into aggregated statistics rather than long-horizon records.

Provisioning the right compute is therefore not only about performance. It is about where truth lives, where decisions happen, and how you prevent a helpful system from becoming hungry.

A cross-vertical pattern library

To keep this book useful across industries, it helps to name a few patterns that recur.

- Arrival handoff
  Recognize that a visitor has arrived, then reduce friction immediately: language selection, orientation, a calm first route, and a clear next step. This can be done anonymously and still feel personal.
- Context-aware canvas
  Use programmable surfaces to reflect the state of the place: guidance when needed, narrative when calm, sponsor messages only when appropriate, and language that respects the audience without hierarchy.
- Commerce state bridge
  Treat POS events as changes in state that unlock services, entitlements, and guidance without turning commerce into identity capture.
- Object-centric experience
  In automotive, retail, and brand environments, the object often carries meaning. Recognize the object's category, style, color, or pattern, and adapt the story around the object, not around the person.
- Comfort-first mode
  Use recognition of environmental conditions, such as noise and density, to offer quieter mixes, captions, personal language tracks, and calmer routes. Comfort is access.

These patterns remain useful because they are grounded in operational reality. They respond to the place, not to private biography.

The chapter's conclusion

Recognition governance checklist (practical, testable)

A recognition feature is acceptable only if you can answer "yes" to all of the following:

- Purpose is explicit, bounded, and written in plain language.
- The system works without identity by default, and identity is an opt-in benefit exchange.
- Retention is short, declared, and enforced by design (including deletion evidence).
- Linkability is bounded (no silent cross-visit tracking unless explicitly consented).
- Outputs that can influence behavior are grounded in operational truth or approved sources, and the system can refuse when grounding is insufficient.
- Operators can see the current posture (what is being recognized, for what purpose, and for how long), and can override or disable it, with audit logs.

If any item is "no," the feature is not ready for a public venue.

Recognition used badly destroys trust. Recognition used well creates coherence. The difference is restraint, architecture, and purpose.

When recognition is treated as a way to measure the place and serve the moment, systems can feel almost magically helpful without becoming invasive. Retail experiences feel like service rather than pressure. Museums can support multilingual dignity without clutter. Automotive experiences can feel relevant because they respect interests rather than harvesting identity. Destinations can become calmer because guidance reflects operational truth.

In the next chapter we move toward macro behavior, where recognition and analytics become inputs to venue-scale guidance, load balancing, and operational decisions. At that scale, governance becomes explicit because system choices begin to shape the day of everyone in the building. The goal remains the same: capability constrained by responsibility, expressed as hospitality.

PART V
MACRO
HYPER-PERSONALIZATION, WORLDMODEL VENUES, AND GOVERNANCE

Macro hyper-personalization is what happens when personalization stops being "at an exhibit" and becomes "how the venue behaves." At this scale, coordination is the hard problem: multiple systems, multiple stakeholders, and multiple simultaneous goals. This is where WorldModel™ Venues and governance are not optional.

Chapter 22 introduces macro-scale concierge interfaces, what it means to guide, translate, deconflict, and reassure at venue scale.

Chapter 23 defines the WorldModel framework as operational truth, a living state of the venue that reconciles sensors, intent, rules, and outcomes.

Chapter 24 moves from concept to implementation patterns.

Chapter 25 presents the four main tools, CGL™, EDE™, ICL™, and MAOL™ as the decision architecture, the functional separation that prevents one "smart" layer from breaking everything else.

Chapter 26 puts governance into practice, including boundaries, fallbacks, and what to verify.

If you are building a flagship venue, a campus, or any environment that must coordinate flows, accessibility, content, and safety, this part is the center of the book. Read it with operators in the room, because operability is the pass-fail test at macro scale.

22

VENUE CONCIERGE INTERFACES AT MACRO SCALE

Most venues are designed as if a visitor's day were a straight line: arrive, enter, absorb, exit. In reality, a visitor's day is a negotiation. It is negotiated with time, crowds, attention, fatigue, an unexpected closure, the child who suddenly needs a restroom, the show that starts earlier than expected, the exhibit that looks beautiful but feels too loud, and the queue that was supposed to be short but is not.

People do not come to a venue to solve logistics. They come to feel something, learn something, share something, and leave with a sense that the trip mattered. The macro turning point arrives when you accept a simple fact. In large venues, logistics becomes part of the experience because logistics decides whether people can reach what they want without being worn down along the way.

That is the moment where hyper-personalization stops being only a story selection problem and becomes an environment management problem. It is also the moment where a layer like Venue Concierge Interface becomes necessary, not as a gadget, but as the interface between human intent and venue reality.

Venue Concierge Interface is the point where the place begins to answer questions that are not about content alone: where do we go next, how do we avoid the crush, how do we keep the day calm, how do we keep access equitable, and how do we balance what visitors want with what the venue can responsibly support in the next thirty minutes.

A macro concierge has a higher burden than a conversational assistant, because its outputs are acted upon.

Confabulation must be treated as a primary macro risk. At macro scale, "sounding right" is not acceptable. If the concierge states that a route is open, that a show has capacity, that a queue is short, or that a quiet path exists, that statement must be grounded in operational truth and freshness, not inferred from a language model's fluency.

Practical rule: treat concierge output as a proposal. A proposal becomes a visitor-facing instruction only after it is reconciled against current state, checked against constraints, and assigned a confidence posture. When confidence is insufficient, the system must abstain, fall back to a safe default, or escalate to staff.

If the earlier parts of this book are about making individual moments more inclusive and more meaningful, this part is about making the whole place behave with the same hospitality.

The shift from exhibits to outcomes

It delivers language, depth, and modality with dignity. It prevents the fatigue of repetition and makes a sequence of exhibits feel coherent.

Macro hyper-personalization begins when the visitor's request is no longer "tell me the story," but "help me have the day." That shift changes the unit of design. An exhibit can be treated as a self-contained moment. A day cannot. A day is shaped by constraints that belong to the venue as a system: capacity, congestion, schedule, safety, staffing, accessibility routes, and the acoustic and sensory character of spaces that change as crowds move.

This is where the "small city" comparison becomes literal. A theme park, a large museum, a cruise ship, and a destination district all face the same class of problems: flow, routing, load balancing, scheduling, and public comfort. These are city problems, compressed into bounded environments.

The Venue Concierge Interface is the interface layer that makes those problems visible and negotiable. It is the surface through which a visitor, staff member, or system can express intent and receive an outcome without needing to understand the machinery underneath.

That sounds simple. It is not.

Evaluate concierge systems by the bounded decisions they produce, then render those decisions into language.

A macro concierge should not be evaluated primarily by how well it talks. It should be evaluated by how well it produces bounded decisions under governance. The underlying output should be a decision object with explicit fields (recommendation, constraints checked, evidence pointer, freshness, and confidence), and the spoken or displayed language should be a rendering of that object.

This is the cleanest way to increase reliability, reduce confabulation risk, and make audits possible without turning the venue into paperwork.

It is necessary.

What a concierge actually does

The word concierge is useful because it is human. A concierge does not merely provide information.

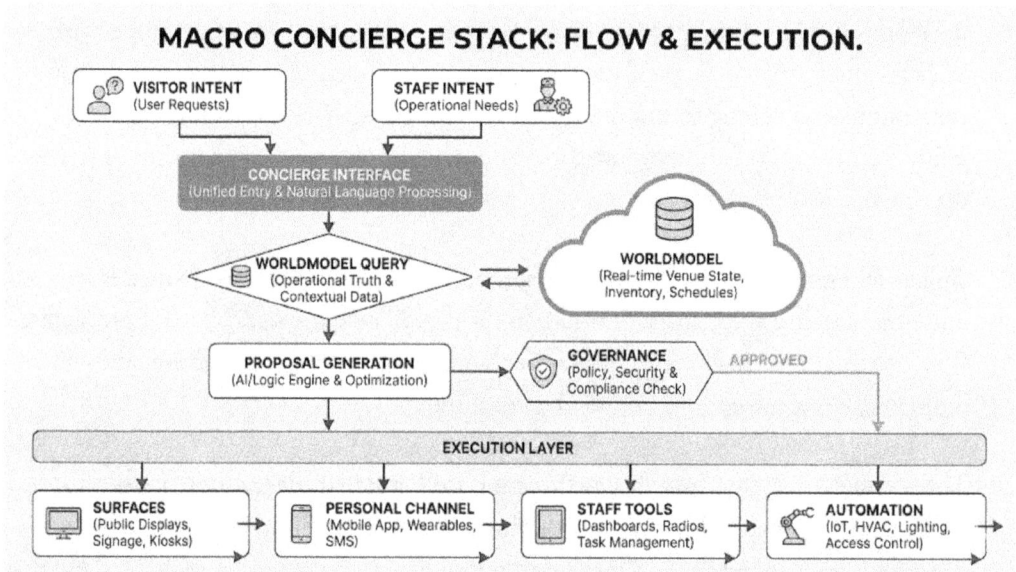

MACRO CONCIERGE STACK: FLOW & EXECUTION.

A visitor does not walk up and ask, "Optimize my path under occupancy and schedule constraints." They say, "We have an hour, we have kids, we want the highlights, and we do not want to stand in lines." A staff member does not say, "resolve competing objectives across zones." They say, "This gallery is getting congested, and we need to move people gently without making them feel managed."

A venue system does not say, "rebalance load." It says, "The next show is about to release a crowd into this corridor, one exhibit has gone offline, the restaurant is nearing capacity, and we need to keep the place calm."

The Venue Concierge Interface sits at that boundary. It takes human-shaped requests and system-shaped alerts and turns them into proposals that can be evaluated, constrained, and expressed through the appropriate channels: guidance on a phone, a change in signage on a programmable surface, a suggestion to staff, a schedule adjustment, a queue routing prompt, or sometimes a decision to do nothing. Restraint is often the best form of hospitality.

This is the turning point in the book's architecture. The unit of design moves from personalized media to an individual to governed outcomes for a place.

State: the difference between a map and reality

It is tempting to imagine wayfinding as a map problem. A map is static. A venue is a living system. Reality includes what the map does not hold well: closed doors, temporary reroutes, unpredictable queues, accessibility constraints, acoustic discomfort, staffing limits, and the difference between a path that is technically short and a path that feels exhausting because it is noisy, crowded, or confusing.

Venue Concierge Interface does not create truth. It queries it. It depends on a representation of operational reality that can answer questions such as: which routes are open right now, which zones are congested right now, which exhibits are operating normally right now, which showtimes are imminent and what that implies for flow, where the venue is loud and where it is calm, where accessibility routes are available and what their constraints are, and what the venue can support in this moment without creating new problems elsewhere.

This is where WorldModel™ architecture enters naturally, by necessity. A venue-scale concierge needs venue-scale operational truth. The visitor experiences it as guidance. The operator experiences it as calm.

Why governance is unavoidable

At exhibit scale, personalization is mostly private. One person receives a different narrative, and nobody else is harmed. At venue scale, decisions have externalities. A routing suggestion changes who goes where. A load-balancing action changes distribution.

A queue intervention changes who waits and who does not.

A schedule adjustment changes the day for many.

This is why governance becomes mandatory at this stage. The system is consequential.

A governed concierge is the difference between

- a place that offers helpful suggestions while remaining fair, safe, and inclusive, and
- a place that chases a single metric such as throughput and quietly degrades everything that makes the venue worth visiting.

Governance is the enabling structure that lets a concierge do more without becoming a liability. It constrains proposals under explicit objectives and hard constraints. It makes decisions auditable. It preserves operator override. It makes privacy and inclusion enforceable behaviors, not aspirations.

Example: informing rather than steering.

Consider an airport concourse where an escalator is inoperative and elevator capacity is limited. A metric-driven system might quietly reroute flows to protect throughput, shaping outcomes without making the tradeoff visible.

A governed concierge behaves differently. It discloses the constraint in plain language, offers options, and lets people choose: "Escalator unavailable. Stairs only on this route. Elevator service is limited and delays may occur. Alternative path available." This preserves autonomy and transparency. It also supports inclusion by making accessibility constraints explicit rather than forcing a hidden optimization.

The point is not that the system never recommends. The point is that governance determines when the system should inform, when it may recommend, and when it must take directive action for safety or compliance, with the rationale visible and auditable.

In a macro venue, governance does not suppress creativity. It protects creativity from operational collapse.

Privacy at macro scale: calm without creepiness

As environments grow more capable, the temptation to default to identity increases. It can feel as if identity would make everything easier: more precise personalization, more accurate uniqueness, cleaner correlation. It also creates the fastest path to mistrust.

A macro concierge should be privacy-forward by default for the same reason it should be governed by default. It operates in public space.

Most macro outcomes do not require identity. They require state. A venue can balance flow, manage queues, adjust signage, reduce sound bleed, and coordinate scheduling using anonymous inputs and aggregated measures. It can offer guidance to a visitor without knowing who they are. It can improve clarity without building dossiers.

Identity becomes appropriate only when a visitor explicitly asks for benefits that require continuity across time, such as membership privileges, accessibility persistence across visits, or entitlements. Even then, the posture remains minimal and consent-bound. The healthiest macro systems behave like good cities: helpful infrastructure without intrusive demands.

The acoustic layer as a macro constraint

Large venues often treat acoustics as local, something handled by room design and then forgotten. Macro operation makes that impossible. Sound bleed is an audio issue and a behavioral issue. A loud corridor pushes people away. A chaotic soundscape increases fatigue. A quiet gallery becomes a refuge and then, unmanaged, becomes congested because people flee to calm.

A macro concierge must consider acoustic state alongside occupancy. It needs to know where spaces are calm, where they are loud, and how that interacts with the visitor's comfort choices. It should avoid routing overstimulated visitors into the loudest paths, rather than by labeling them, by offering comfort-aware routes as a normal option.

The same logic extends to power and operations. When a zone is empty, lowering volume and reducing sensory output reduces bleed and preserves calm elsewhere. When a zone is about to be occupied, restoring it before visitor's notice is part of making the venue feel alive rather than mechanical. At macro scale, these become behaviors of the place.

A concierge is only as good as its restraint

The most common fear about venue-scale guidance is that it will feel like management. Nobody wants to feel herded. A good concierge respects agency. It suggests rather than commands. It explains when explanation matters. It provides choice rather than coercion. It allows a visitor to ignore guidance without punishment.

Tone matters. The system should sound like a host rather than an optimizer. It should speak in plain language. It should be concise by default and expandable by choice. It should be polite.

Restraint also has a technical meaning: do not invent operational facts.

A concierge must never claim certainty about operational reality when the system cannot prove it. If state is stale, contradictory, or missing, the correct response is to surface uncertainty, switch to conservative guidance, or route the request to staff. Visitors forgive "I cannot confirm that right now." They do not forgive being confidently misled.

When a concierge behaves this way, visitors begin to trust it. Trust increases usage. Usage makes the venue easier to operate. The day becomes better for everyone, including visitors who rarely engaged with the concierge at all. This is the paradox of well-designed macro systems. They improve the public experience most for people who rarely think about the technology.

Key takeaways

The Venue Concierge Interface is the macro turning point because it changes the unit of design. We stop designing only moments and begin designing outcomes. We stop treating guidance as signage and treat it as a negotiation between intent and reality. We stop treating a venue as a sequence of exhibits and treat it as a living environment with state, constraints, and responsibilities.

This is not primarily an AI story. It is an architecture story and a hospitality story. The intelligence is not the point. The point is that the place behaves coherently, inclusively, and calmly under real-world conditions.

In the next chapter, we formalize what that coherence requires: a representation of operational truth that allows the venue to reason, act, and evolve without becoming brittle. That is the role of WorldModel at venue scale, and it is where the architecture becomes explicit.

23

WorldModel as Operational Truth

If the Venue Concierge Interface is the turning point, the WorldModel™ framework is the ground it stands on.

A concierge can speak fluently and still be wrong in the ways that matter. It can send a visitor down a corridor that is closed, toward a gallery that is full, into a queue that is about to spike, past an area whose soundscape has become overwhelming, and into a moment that was meant to be beautiful but is currently broken. In a real venue, confidence is not enough. The system has to be correct about operational reality.

This is where confabulation becomes operational rather than rhetorical.

Confabulation is best understood here as any confident claim that is not grounded in verified sources or current state. In macro environments, confabulation is not limited to interpretive narratives. It includes operational assertions: open or closed, full or available, safe or unsafe, quiet or overwhelming, five minutes or forty minutes.

The venue-scale rule is simple: if a claim can change what people do, it must be grounded, verifiable, and time-bounded. Otherwise, the system must refuse, degrade to a conservative default, or escalate to human authority.

This is the central lesson of macro environments. When a place begins to guide, balance, schedule, and intervene, it must do so from a shared understanding of reality that is consistent, current, and defensible. Without that, you do not have a smart venue. You have a venue that improvises.

It is the idea that a complex environment can maintain a living representation of itself, expressed as operational truth: what is true now, what is constrained now, what is possible now, and what is likely to happen if the venue acts.

This chapter explains why operational truth is the core capability at venue scale. It also explains why a map is insufficient, why raw analytics is insufficient, and why a digital twin alone is insufficient. The most important capability is not speech or prediction. It is disciplined truth.

Named principle (citeable): Operational truth is bounded, versioned, and provenance-bearing.

"Truth" at venue scale is not a story. It is a bounded state object with:

- Versioning (so the system can reason about freshness),
- Provenance (so the system can justify why it believes a field),
- Uncertainty representation (so the system can abstain when it should),
- An evidence hierarchy (so conflicts resolve predictably, not rhetorically).

This is the difference between an environment that can speak and an environment that can be trusted.

The difference between a map and a model

A map describes where things are. A WorldModel describes how a place behaves.

A map may show a corridor between two galleries. A WorldModel framework represents whether the corridor is open, whether it is congested, whether it is accessible for a declared mobility need, whether it is loud enough to be uncomfortable, whether it is safe under current operating conditions, and whether routing additional people through it will create a downstream bottleneck.

A map is largely static. A WorldModel architecture is a living negotiation between structure and state. This is why macro systems cannot treat guidance as a geometry problem. Geometry is the skeleton. State is the body.

Why a digital twin is not enough

The phrase digital twin has become popular because it sounds like control. In practice, many digital twins are excellent at representing assets and geometry and less useful at representing lived experience. They often focus on what exists, not on what is happening, what is constrained, and what is safe to do next. Source: EY, "How digital twin technology and AI can reimagine theme park operations".[xxxi]

The WorldModel™ architecture is not a rejection of the twin. It includes the twin's strengths and extends them into decision integrity. It is the venue's operational truth system, rather than only a representation of the venue.

Operational truth means:

- state is current enough to drive decisions,
- state is consistent enough that subsystems agree,
- state is bounded enough to be governed, and
- state is verified enough that the venue can trust it.

A venue does not need a perfect simulation to be helpful. It needs a faithful, auditable present.[xxxii]

The WorldModel framework in plain language

WorldModel is the venue's shared state of reality, expressed in a form that systems and humans can use to make decisions coherently.

It is where the venue agrees on answers to questions such as: what is open, what is closed, what is full, what is quiet, what is loud, what is working, what is drifting toward failure, what is scheduled next, what is safe, what is accessible, and what is permitted under policy.

It also contains the constraints that keep the venue trustworthy: privacy posture, safety boundaries, accessibility requirements, fairness constraints, operational limits, and the discipline of not doing something merely because it is technically possible.

WorldModel is the venue's shared reality layer. It references content without being the content, supports recommendations without being the recommender, and can be informed by AI without becoming an AI personality. Its job is to prevent the venue from behaving like a set of disconnected apps.

How the World Model becomes WorldModel™

A governed architecture that transforms raw environmental intelligence into safe, value-aligned, human-centric operation.

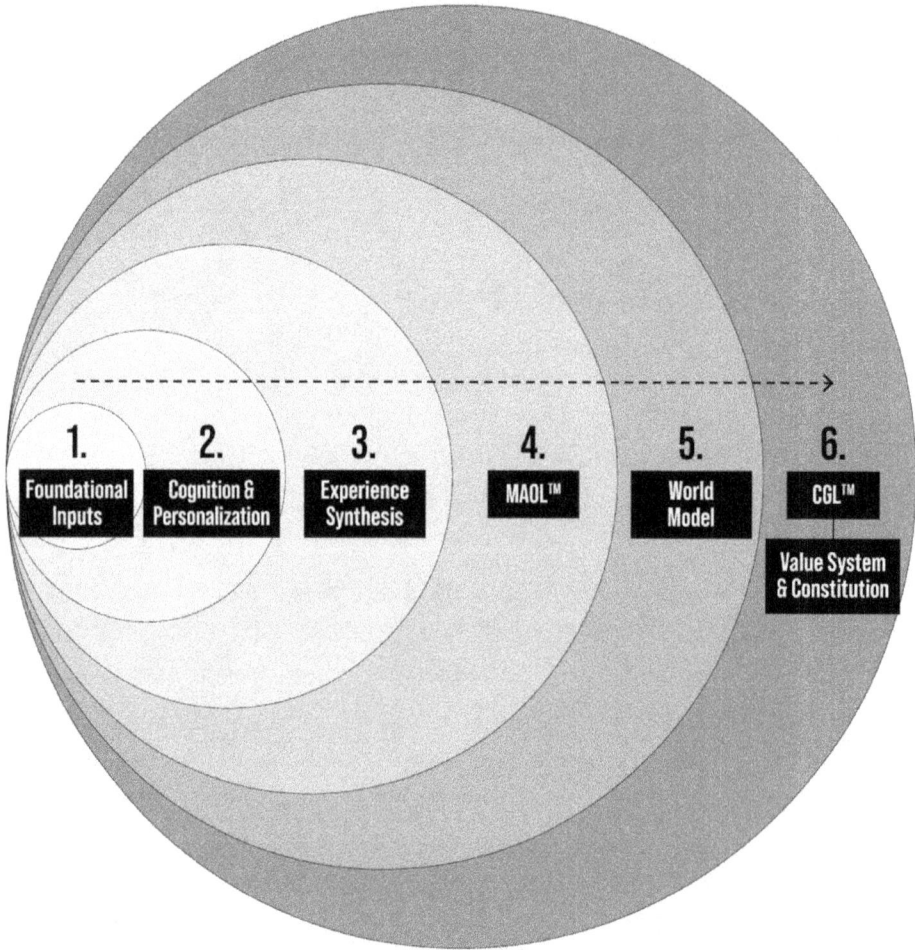

1.
Foundational Inputs

2.
Cognition & Personalization

3.
Experience Synthesis

4.
MAOL™

5.
World Model

6.
CGL™

Value System & Constitution

WorldModel™ is the complete, governed architecture that emerges when the World Model operates inside the Cognitive Governance Layer™

For a more detailed image of the WorldModel™, scan this QR code

1. Turn raw data into meaningful context

- Edge compute nodes
- Venue sensor
  (light, sound, occupancy, flow)
- Recognition layer
- Object recognition
- System/Show control
- Content nodes & media assets
- Visitor inputs
  (opt-in identity, preferences, history)
- Personal devices (OnBoard layer, Personal Layer, Concierge layer)

2. Transform signals into adaptive responses

- Personalization layer
- Contextual intent inference
- Accessibility adaptations
- Learning loops
- Eligibility & access logic
- Avatar interactions
- Object and visitor recognition tie to Recognition layer output form foundational input

3. Generate dynamic, human centric actions

- Real-time content assembly
- Dynamic routing & Wayfinding
  (non-visual navigation layer)
- AR layer
- Adaptive storytelling
- Preference modeling
- Pattern recognition
- Cross-device session continuity

4. Coordinate all venue agents into a unified intelligence

- Multi-agent collaboration
- Cross-venue continuity
- Zone-to-zone synchronization
- Task allocation
- Conflict resolution
- Role switching
  (guide, concierge, expert)
- Agent states and behavior patterns

5. Structured representation of environment, guests, and state

- Environmental Dynamics Engine
  (EDE™)
- Environmental dynamics
- Prediction surfaces
- State histories
- Temporal Governance Framework
  (TGF™)
- Identity continuity (logical model)
- Venue map & topology
- Identity Continuity Layer (ICL™)
- Spatial & semantic relationships

6. Enforces values, safety, constraints, and accountability

- BehaviorGuardrails
- Safety Interlocks
- Role Governance
- Delegation Rules
- Constraint-Driven Decision Making
- Oversight & Explainability
- Audit Trails & Compliance
- Domain rules & constraints

What a WorldModel framework must contain to be useful

In a venue-scale environment, state has layers. Some are obvious. Some are neglected until failure makes them unavoidable. A WorldModel framework good enough to operate a real place must, at minimum, represent these categories of truth.

Spatial and structural truth

Zones, rooms, corridors, entries, exits, vertical circulation, and the relationships between them. Structural truth also includes operational boundaries: where crowds can form, where staff can intervene, and where an accessibility route is materially different from a standard route.

Experience truth

What is happening experientially, right now:

- which exhibit is active,
- which show is running,
- what the next transition is,
- which narrative mode is selected for a given cycle,
- which language is the room language in a shared show, and
- which surfaces are in guidance mode versus story mode.

This is where micro and macro meet. The system can guide a day only if it knows which moments are actually available.

Operational truth

What the venue is doing, not what it intends:

- staffing posture,
- current incidents,
- temporary closures,
- queue management posture,
- scheduled events, and
- the realities that make a place feel smooth or chaotic.

Operational truth is also where policy constraints become enforceable rules, because macro decisions must be accountable.

How this differs from a digital twin

A digital twin can model a venue. The WorldModel framework is the operational truth the venue runs on.

- A twin can be descriptive. WorldModel must be decision-capable and governed.
- A twin can be updated later. WorldModel is updated continuously, in the loop.
- A twin can be optional. WorldModel becomes mandatory when automation has externalities.

Minimum viable WorldModel architecture implementation

If you cannot ship these, you do not have a WorldModel system in the sense used in this book.

- Shared state representation that multiple systems can read and update.
- Constraint evaluation and policy gating before action.
- Operator authority and override paths that are logged.
- Outcome verification, drift detection, and verified restoration.
- Versioning, rollback, and evidence pack delivery.

Acoustic truth

The acoustic layer is a constraint. The WorldModel framework needs acoustic state because acoustic comfort shapes behavior. Loud spaces repel, quiet spaces attract, and sound bleed destroys clarity. A macro system that routes people without acoustic awareness will push sensitive visitors into stressful paths and will overload calm spaces because they become refuges.

Acoustic truth includes ambient noise conditions, program levels by zone, bleed risk between adjacent zones, and comfort flags that inform guidance and mode selection.

Power and equipment truth

Power is cost and experience. It is also lifespan. WorldModel solutions must represent whether devices are running, warming, cooling, dimmed, muted, in lower-power states, or in error. It must represent thermal limits and maintenance drift because those constraints determine what is safe to run and what will fail if pushed.

This is where occupancy-aware behaviors become meaningful at the architectural level. The venue knows which zones are unoccupied, what that implies for sound bleed and equipment life, and what must be restored before a visitor can notice.

Without power truth, automation becomes guesswork. With it, automation becomes hospitality.

Recognition and analytics truth

Recognition is not one thing. It can mean anonymous session continuity, anonymous pattern recognition, or opt-in identity continuity. WorldModel must represent which posture is currently in effect, because privacy is part of operational truth.

Analytics truth includes aggregates that support decisions: occupancy counts, congestion patterns, dwell distribution, recurring confusion points, and trend indicators for equipment drift. It also includes what the system does not do. A privacy-forward venue must be able to state and enforce boundaries: no dossiers by default, retention bounded, recognition used for service rather than extraction.

Commerce and entitlement truth

Commerce is part of the day, and it must be coherent. WorldModel should represent entitlements and transactional state in the minimal form necessary to deliver service: tickets, timed entry, upgrades, reservations, and access rights.

This is not a call to merge the venue into a marketing database. It is the opposite. It is how you avoid oversharing by using minimal, purpose-bound tokens and yes-or-no outcomes where appropriate. The venue does not need a person's biography to know whether they can enter a timed experience. It needs entitlement state.

Safety truth

Safety is the ultimate constraint layer. At macro scale, safety includes crowd conditions, restricted zones, emergency procedures, and the policies that govern how the environment behaves under stress. A WorldModel Venue needs an explicit notion of operating mode, including normal mode and emergency modes, because the system must not behave the same way under radically different conditions.

Safety overrides are inviolable

Two safety boundaries must be stated plainly:

1. The life-safety boundary (building safety)
   Life-safety systems (fire alarm and signaling, emergency communications, egress controls, smoke control, and related functions as defined by applicable codes and the Authority Having Jurisdiction) have unconditional priority[xxxiii],[xxxiv] WorldModel consumes life-safety state as authoritative input. It does not arbitrate it.
2. The emergency stop boundary (hard stop)
   Emergency stop chains (E-stop, show-stop, ride-stop, or equivalent) are hard safety functions[xxxv],[xxxvi] They override all experience behaviors, all orchestration, and all agent outputs. The system may observe and log the stop state, but it must not be able to defeat, delay, or auto-reset it.

When either boundary is active, the environment must shift into a constrained posture:

- routing and recommendations remove affected zones immediately,
- any surface that normally promotes experiences may switch into a clarity-first mode,
- conversational interfaces suppress speculative answers, and
- the system escalates to staff rather than improvising.

The purpose of safety truth is predictability.

A place that changes behavior coherently under stress feels safe. A place that behaves unpredictably under stress feels dangerous, even when nothing is technically wrong.

Why truth must be shared

Large venues fail when different systems hold different truths. The signage says one thing. The app says another. Staff say a third. The queue behaves like a fourth. Visitors experience this as incompetence, even when each subsystem is individually competent. The failure is coordination.

The WorldModel framework is the shared substrate that prevents truth from fragmenting. It is the layer through which the venue maintains one operational reality, then expresses it through different channels: programmable surfaces, personal devices, staff tools, and orchestration systems.

This is also why the book insists on compute and local-first design. Shared truth cannot disappear because an external dependency is slow. External services can be used for non-critical tasks, but the core loop of operational truth must remain available when the building is full, the radio spectrum is messy, and the day is under pressure.

A smart venue that fails when stressed is theatrical, not smart.

The update loop: how truth stays alive

Truth is not a database. It is a loop. A venue-scale system must continuously:

1. observe the environment,
2. update state,
3. propose actions,
4. evaluate actions against constraints,
5. execute actions, and
6. verify outcomes.

This loop is the heart of a WorldModel Venue. It turns sensors into service and service into reliability. Verification is the piece many systems omit, and it is the reason they drift into unreliability. Commands are not outcomes. The system must confirm that what it thinks happened and detect drift early, not only when failure becomes public.

When you build the loop, the venue becomes calm under change. When you omit it, the venue becomes fragile under change.

The WorldModel framework as a promise about change

A museum does not rebuild itself every time its audience shifts. A theme park does not redesign its streets every time weather changes. A cruise ship does not rewire itself every time one venue becomes crowded. A destination district does not pour new concrete every time a corridor becomes popular.

Yet all these environments are being asked to behave as if they can respond faster than construction. WorldModel is how that promise becomes real. It is how a place can change its behavior without changing its bones.

It allows the venue to switch languages without reprinting, shift guidance without rehanging signs, reduce sound bleed without redesigning walls, extend equipment life without manual rounds, balance load without shouting at guests, and maintain inclusion without making it feel like exception handling. This is the flexibility that makes a physical place feel alive, aligned with reality rather than frozen in its build year.

What WorldModel solution is, and what it is not

The WorldModel™ architecture is not merely a database, a digital twin, or a concierge UI. In this book, it means one shared, continuously updated operational truth for a venue, used to guide action responsibly.

Comparison grid (conceptual, non-normative):

Concept	Typically good at	What WorldModel adds for venues
Digital twin	Assets, geometry, planning, simulation	Misses: the live operating day, and accountable decisions. Adds: shared operational state, constraints, and verification.
Event-sourced state store	Traceable state changes over time	Misses: venue-specific policy boundaries, and cross-modal integrity. Adds: governed decisions, operator authority, and outcome evidence.
Policy engine or rules engine	Consistent decisions under rules	Misses: a faithful state model, drift signals, and restoration under stress. Adds: state that reflects reality, drift detection, and verified restoration.
BMS, SCADA, and classical show control	Deterministic cues, timing, device control	Misses: personal channels, privacy posture, and negotiated intent. Adds: personal-channel integration, privacy-forward continuity, and governed autonomy.
Safety case and assurance	Proof a system is safe and fit for purpose	Misses: day-to-day coupling to experience quality, content operations, and drift. Adds: operational evidence packs, commissioning proof, and lifecycle monitoring.

The WorldModel architecture is:

- Shared operational truth, so different systems do not act on different versions of reality.
- A closed loop: observe, update state, propose actions, evaluate constraints, execute, and verify outcomes.
- Governed decision-making whenever choices create externalities for safety, accessibility, privacy, fairness, or integrity.
- A local-first reliability posture: the core loop remains available under load and under degraded connectivity.
- The WorldModel framework is not:
- A replacement for building design, staffing, or operations. It is a coordination layer that makes those investments behave coherently.
- A mandate for identification. Continuity can be delivered without identity, provided consent is explicit and retention is bounded.
- A promise that "AI can and will run the venue." Operators retain authority, and systems must degrade gracefully.

Minimum evidence, once governance applies

Minimum logging and traceability (summary):

- Decision events, identity or continuity state, and consent state.
- Data access events: what categories were accessed, and why.
- Safety events, refusals, guardrail triggers, and escalations.
- Human oversight events: approvals, overrides, and rationale.
- Incident events: suspected harm, complaint, breach, or anomalous behavior.
- Update events: configuration, model, and content updates, including rollback events.

Evidence pack contents (summary):

- AI System Inventory, Risk and Impact Assessment, and post-deployment Monitoring Plan.
- Model and configuration version history, incident register, and most recent assurance notes.

WorldModel Failure Modes

Understanding how a system fails is as important as understanding how it works. The following failure modes are not hypothetical. They are predictable outcomes when parts of the architecture are weakened or omitted.

- **Failure mode: Partial truth**
  Symptom:
  The system makes locally sensible decisions that create global problems.
  Cause:
  State is incomplete, stale, or unevenly trusted. Some subsystems update faster than others.
  Prevention:
  Explicit state provenance, confidence levels, and verification before consequential actions.
- **Failure mode: Orchestration without governance**
  Symptom:
  The system reacts quickly but violates policy, accessibility, or safety norms.
  Cause:
  Actions bypass constraint evaluation in the interest of speed or convenience.
  Prevention:
  Governance gating as a mandatory step, not a configuration option.
- **Failure mode: Silent degradation**
  Symptom:
  The system appears to function but gradually becomes less helpful or trustworthy.
  Cause:
  Lack of outcome verification and drift detection.
  Prevention:
  Outcome Verification Layer with tolerance thresholds and recovery behavior.
- **Failure mode: Over-eager personalization**
  Symptom:
  Visitors feel watched, guessed-at, or uncomfortable.
  Cause:
  Inference used where explicit choice would suffice.
  Prevention:
  Privacy-first defaults, explicit opt-in, and clear purpose limitation.

- **Failure mode: Automation optimism**
  Symptom:
  The system assumes actions succeeded and builds on incorrect assumptions.
  Cause:
  No verification loop.
  Prevention:
  Treat every action as unverified until proven otherwise.

These failure modes are not arguments against automation. They are arguments for architecture.

Procurement minimums that prevent fragile deployments

Owner procurement minimums (short list):

- Occupancy-aware power management, so equipment longevity does not require manual discipline.
- Verified restoration: the system proves a zone is exhibit-ready before visitors enter the experience area.
- Automated drift detection with preventive maintenance signals and escalation triggers.
- Frame-accurate synchronization across audio, video, captions, and interpretation where required.
- Offline-first capability for the Personal Channel when WAN or internet service is degraded.
- Privacy-forward session management: continuity without default identification.

For full language and acceptance criteria examples, see Appendix J. For logging and evidence requirements, see Appendix G.

The chapter's conclusion

The WorldModel framework is the operational truth layer that macro systems require. It is the difference between a venue that speaks and a venue that knows. Between a venue that reacts and a venue that behaves coherently. Between a venue that can guide and a venue that merely suggests.

It is also where the architecture becomes most disciplined because truth is where responsibility lives. If you represent state, you must represent it honestly. If you propose actions, you must constrain them. If you execute actions, you must verify them. If you collect data, you must minimize and forget.

A WorldModel Venue is not impressive because it is complex. It is impressive because it is restrained, coherent, and trustworthy under pressure.

In the next chapter we move from truth to decision, because once a venue has operational truth, it must still decide what to do with it. That is where governance becomes explicit, and where values stop being language and become enforceable constraints on behavior.

24

IMPLEMENTING WORLDMODEL VENUES

The State Loop, the Truth Bus, and the Governance Gate

A WorldModel Venue is not built by adding intelligence. It is built by formalizing truth. That claim can sound abstract until you watch venues fail at scale. They rarely fail because they lack sensors or compute. They fail because they allow different subsystems to hold different realities. The signage says one thing, the app says another, the staff says a third, and the building behaves like a fourth. Visitors experience this as chaos even when each subsystem is individually competent.

WorldModel architecture implementation prevents truth from fragmenting. It does so by maintaining a single operational state that is updated continuously, and by ensuring that every consequential action is derived from that state, constrained by policy, executed locally where needed, and verified against the physical world.

This chapter makes the implementation concrete without publishing proprietary recipes. It is written so a designer can specify it, an operator can demand it, and an integrator can build it, while staying safely above the line where a reference book becomes a blueprint.

The implementation in one sentence

> The WorldModel framework is a state store derived from an event stream, hardened by verification, and used by an orchestrator whose actions are permitted only through a governance gate.

The eight components every implementation needs

Different venues will choose different products. Some will centralize more. Some will distribute more. The implementation still reduces to the same eight components.

1. Sensors and authoritative sources
2. Ingress and normalization
3. The truth bus, an event stream
4. The WorldModel state store
5. The orchestration plane
6. The governance gate
7. The action plane
8. Outcome Verification Layer and drift detection

This is the reference architecture. It is what makes the venue coherent.

Let's break those down:

(1) Sensors and authoritative sources

> Sensors are not truth. They are evidence.

> This sentence is also a confabulation control rule.

> Confabulation at venue scale often begins when evidence is treated as truth without provenance, confidence, and freshness. The system starts to speak as if it knows, when it only suspects. That error becomes dangerous when the output influences routing, safety, accessibility instructions, queue decisions, eligibility, or staffing interventions.

> A publish-safe implementation requirement is therefore:

> - Every evidence-bearing signal must carry an authority class (authoritative or advisory),
> - A confidence posture (high, medium, low),
> - A freshness window (how long it remains valid),
> - And a provenance pointer (what produced it, and when).

When confidence or freshness falls below policy threshold, the system must not guess. It must abstain, degrade to safe defaults, or escalate.

A mature implementation begins by classifying sources into two groups:

- Authoritative sources define reality. If the show controller says a scene is running, that is authoritative. If a door sensor says a door is locked, that is authoritative. If a device reports an error state, that is authoritative.
- Advisory sources are evidence that can be wrong. Occupancy estimates, crowd density inference, and many forms of pattern recognition belong here.

The reason to make this distinction early is practical. A venue that treats every sensor as truth will behave confidently and incorrectly. A venue that knows which sources are authoritative can act responsibly under uncertainty.

(2) Ingress and normalization

Venues produce messy signals. Ingress is where those signals become usable and safe.

Ingress performs five duties:

- timestamps events consistently,
- tags source and authority class,
- normalizes payloads into a standard format,
- bounds size and rate so a bad device cannot flood the system, and
- enforces privacy minimization at the boundary.

That last duty is non-negotiable. Privacy posture is not enforced by good intentions. It is enforced by architecture. If a signal arrives with more personal data than is required for the function, ingress is where it is reduced or removed. If you cannot do that, you do not yet have a privacy-forward system.

Evidence Hierarchy, what counts as truth, and what wins when signals conflict

A WorldModel Venue is not only a state store. It is a disciplined relationship with evidence.

In public environments, the most damaging errors are not the dramatic ones. They are quiet conflicts: one surface says a gallery is open, staff tools say it is closed, the app claims it is accessible, and the corridor is blocked. Visitors experience this as incompetence. Operators experience it as fatigue. Systems experience it as drift.

The solution is not more sensors. The solution is an explicit evidence hierarchy that makes conflict resolvable.

This hierarchy should be stated plainly and enforced mechanically.

Level 1: Authoritative state declarations

Some sources define reality. If they disagree with everything else, the correct posture is not to average them. The correct posture is to treat the conflict as a fault condition.

Authoritative sources include:

- show control state, cue state, and hold state,
- operator overrides and closures,
- safety and incident mode state,
- life-safety system state (for example, fire alarm or emergency communications activation),
- emergency stop chain state (E-stop, show-stop, ride-stop, or equivalent),
- asset-reported error state and latching fault state,
- entitlement state when issued by an authoritative system.

When an authoritative source changes, the WorldModel state must reflect it immediately, and downstream systems must be prevented from contradicting it.

Level 2: Verified outcomes

Verification is stronger than belief. A system that can observe whether a surface actually switched, whether audio is actually present, whether motion is actually playing, and whether a zone is actually restored has evidence that is closer to truth than internal declarations.

Verified outcomes exist to prevent optimism.

If a command was issued but the outcome was not verified, the system must treat the action as unconfirmed and behave conservatively.

Level 3: Advisory measurements and inferred signals

Occupancy estimates, crowd density inference, acoustic estimates, and many forms of pattern recognition are useful. They are also uncertain.

Advisory signals should carry:

- a confidence value,
- a provenance trail,
- and a decay rule.

Low-confidence signals are not allowed to drive consequential action without additional corroboration, tighter constraints, or explicit operator review.

Level 4: Derived state and computed convenience fields

Some fields exist because they make decisions easier: traversal cost estimates, predicted queue changes, comfort flags, and recommended routing costs.

These derived fields are never treated as primary truth. They are proposals. They can be recomputed when inputs change, and they must never overwrite authoritative state.

Level 5: Proposals, recommendations, and candidate actions

Orchestration proposes. It does not assert truth.
Proposals include route suggestions, mode switches, staffing prompts, and surface changes. They are evaluated under governance and accepted, modified, or rejected. A proposal that conflicts with authoritative state or fails verification is not "wrong." It is disallowed.

What wins, in one rule

When evidence conflicts, the system should not guess. It should refuse. Refuse means: hold current safe state, alert staff if consequential, and do not create new externalities.

Authoritative declarations override everything. Verified outcomes override unverified commands. Advisory inference can guide, but it cannot overrule authority. Proposals are never truth.

If the venue wants a single operational sentence: Truth is what is authoritative or verified. Everything else is guidance.

Why this belongs in a reference architecture

This hierarchy is what turns a smart venue into an operable venue.

It prevents the system from becoming a debate between dashboards. It gives operators a calm way to resolve conflicts. It also makes audits possible, because a third party can evaluate whether the system obeyed its own evidence rules.

A venue that cannot state its evidence hierarchy cannot claim operational truth.

(3) The truth bus, an event stream

The truth bus is the venue's event stream: durable, ordered enough to reason about, and replayable. It makes the venue operable because the system can answer two questions precisely:

1. what changed, and
2. in what order.

It also makes the venue resilient. If a component restarts, it can rehydrate from the stream rather than waking up blind.

A practical truth bus carries event types such as:

- Zone occupancy change event
- Queue estimate update event
- Route availability change event
- exhibit state change event
- Show cue event
- acoustic state change event
- power state change event
- Asset health change event
- entitlement state change event
- Policy version change event
- Operator override event
- Life-safety state change event
- Emergency stop asserted/cleared event

Each event should include a timestamp, a source identifier, an authority class and confidence, a bounded payload, and a unique ID for traceability.

Add one more field category: freshness semantics.

Events that drive visitor-facing guidance should carry an explicit expiry or time-to-live so stale signals cannot masquerade as present truth. This is a practical defense against both sensor drift and confabulation-like behavior in downstream explanation layers.

This is real and publish-safe because it specifies interface shape rather than private algorithms.

(4) The WorldModel state store

The event stream is history. The state store is the present.

WorldModel is the continuously updated operational "now" of the venue, derived from the truth bus and enriched with provenance so the system knows why it believes what it believes.

At minimum, the state store contains objects like:

- Zones: boundaries, adjacency, occupancy, acoustic state, operational mode
- Routes: segments, accessibility constraints, current availability, traversal cost estimate
- Queues: estimate, capacity, fairness posture, gating rules
- Exhibits and shows: state, schedule, dependencies, integrity constraints
- Assets: displays, projectors, audio zones, sensors, compute nodes, with health and drift indicators
- Policies: objectives, hard constraints, allowed actions, operating modes
- Overrides: operator-imposed holds, closures, special event posture
- Entitlements: minimal right-to-access states, not dossiers

Two rules make the state store credible.

Rule A: State must be versioned. Every update increments a version so downstream systems can avoid inconsistent reads and can reason about freshness.

Rule B: State must carry provenance. Important fields must be traceable to sources and events. This prevents drift into truth-by-assertion.

(5) The orchestration plane

The orchestrator consumes state and proposes actions.

At micro scale, those actions are often content-level: select depth, render language, apply comfort modes, maintain continuity, and suggest optional expansions.

At macro scale, actions become city-like: offer guidance routes, balance load between zones, shape queues, adjust surfaces between narrative and guidance modes, propose schedule interventions, and trigger operational automation such as occupancy-aware behaviors.

The key is that orchestration proposes. It does not execute unboundedly. That separation is what makes governance possible.

(6) The governance gate

The governance gate is the system's decision checkpoint. It is where power becomes responsible.

The governance gate enforces the venue's Constitution: the explicit rule set that defines what the system may and may not do, including privacy, safety, accessibility, fairness, truthfulness, and operator authority. In practice, this is bounded authority: explicit action-space limits, least-privilege permissions, and approval checkpoints for high-stakes actions.

Policy modes (normal, busy, incident) tune thresholds and priorities, but the Constitution defines the non-negotiables. In a reference-grade system, every governance decision records the Constitution version and policy mode used, so later audits can evaluate behavior against declared boundaries.

Every proposed action is evaluated against explicit constraints:

- safety constraints,
- accessibility constraints,
- privacy constraints,
- fairness constraints,
- show integrity constraints,
- truthfulness constraints,
- operational limits, and
- operator overrides.

Truthfulness constraints are what prevent confabulation from becoming operational behavior. If an action requires stating an operational fact or giving guidance that could be acted upon, the claim must be grounded in authoritative state, within its freshness window, or in approved sources. If grounding is missing, stale, or contradictory, the correct behavior is to abstain, fall back to a safer mode, or escalate to an operator. Missing state is not a reason to guess.

The governance gate returns only three outcomes:

- approve,
- modify into an allowed form, or
- reject and request alternatives.

The Governance Gate (Logic Flow)

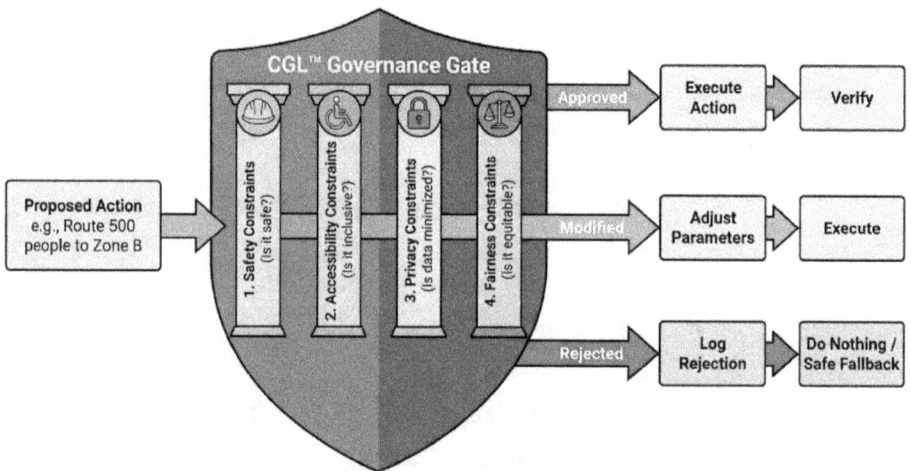

This is the mechanism that makes macro systems deployable. It is also where values become enforceable behavior rather than presentation language.

(7) The action plane

Approved actions must be executed by interfaces that touch the real venue:

- Programmable Canvases and surfaces
- show control systems
- audio zoning and comfort modes
- Personal Channel delivery
- wayfinding guidance outputs
- power management behaviors, including Lifecycle Automation Layer
- staff notification tools

The action plane must report success and failure explicitly. A venue cannot assume that command sent equals outcome achieved.

(8) Outcome Verification Layer and drift detection, defined explicitly below

Verification closes the loop between intention and reality.

It answers two questions: did the action happen in the physical world, and is the system drifting toward failure before a public outage occurs?

Verification includes checks such as:

- this display is showing an image, and it is moving when it should be moving,
- this audio zone is outputting the expected program at the expected level,
- this surface is rendering the correct language mode,
- this projector restored before a visitor could notice after an occupancy-aware transition, and
- drift indicators such as temperature trend, fan escalation trend, and error-rate trend.

Verification is where the Outcome Verification Layer belongs conceptually. Drift detection is the basis of preventive maintenance. Both are required for a system that acts at scale.

The state loop, expressed as venue behavior

A WorldModel Venue runs a continuous loop:

1. Observe
2. Normalize
3. Publish to the truth bus
4. Update WorldModel state
5. Propose candidate actions
6. Evaluate at the governance gate
7. Execute permitted actions
8. Verify outcomes in the physical world
9. Detect drift trends and raise preventive alerts
10. Adjust policy and operations mode as conditions change

WorldModel Implementation Loop: Continuous Adaptive System.

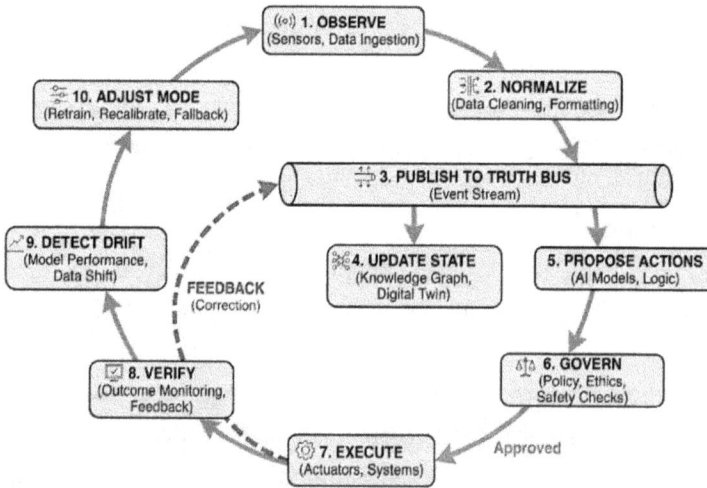

When this loop exists, the venue becomes coherent under stress. When it does not, the venue becomes brittle and unpredictable.

Minimum Viable WorldModel Venue, procurement-grade

A WorldModel Venue is "implemented" only if all items below exist as artifacts and pass basic tests.

State objects (minimum): zones, routes, queues, exhibits and shows, assets (including compute nodes and media endpoints), policies and operating modes, overrides, entitlements (minimal rights states, not dossiers).

Event types (minimum): zone occupancy change, route availability change, queue estimate update, exhibit or show state change, show cue event, acoustic state change, power state change, verification result, operator override event, policy version change event.

Decision artifacts (minimum): decision log that records state version used, constraints checked, outcome (approve, modify, reject), and any human approval or override.

Verification artifacts (minimum): outcome checks proving that critical actions actually happened in the physical world (display state, audio presence, surface mode, restoration after power or mode changes), plus drift signals that detect failures early.

Operator tools (minimum): operating mode switch, override and hold controls, a view of current state by zone, and a "why" view for consequential decisions.

Degraded mode (minimum): critical functions remain coherent with WAN loss, partial subsystem failure produces conservative behavior, and rejoin behavior is defined and tested.

Locality and latency: the three-tier implementation

Venues are physical. Their critical loops must be local. A practical implementation separates compute into three tiers.

Tier 1: Edge zone execution
Local nodes handle time-critical tasks:

- show cue integrity
- local media delivery
- local language switching
- local acoustic behavior
- local fallbacks under degraded connectivity

Tier 2: On-prem operational truth and governance
Venue-local infrastructure holds:

- the truth bus
- the state store
- orchestration
- governance gating
- operator interfaces

This keeps the venue operable even when external connectivity is imperfect.

Tier 3: Cloud async
Cloud is used for non-critical tasks:

- reporting
- long-horizon analytics
- content distribution workflows
- non-critical services that can tolerate delay

This tiering keeps macro systems calm in the real world. It also keeps privacy manageable because local processing reduces unnecessary exposure.

Scenario:

A tactical training space runs adaptive scenarios. The system must personalize to role and language, but remain governed: safety constraints first, bounded variability, clear auditability, and strict data minimization.

WorldModel records safety posture, readiness state, and active segment. Governance constraints tighten allowable actions. Verification confirms the room is delivering what the system believes it is delivering.

Resulting state deltas

Example event list (11 events)

- Session Start:
  training run begins (authoritative).
- Team Recognized Or Registered:
  role and language preferences applied (authoritative).
- Safety Mode Set:
  explicit constraint posture active (authoritative).
- Scenario Branch Proposed:
  next segment chosen based on performance (advisory).
- Media Selection Performed:
  choose approved training media segments (authoritative).
- BoK Retrieved:
  retrieve doctrinal guidance and explanation material (authoritative).
- Governance Check Requested:
  verify constraints before presenting new stimuli (authoritative).
- Playback Committed:
  segment executed (authoritative).
- Verification Result:
  audio and visuals present and synchronized (authoritative).
- After Action Log Written:
  operational evidence stored with bounded retention (authoritative).
- Drift Signal Recorded:
  device health and timing integrity trended (authoritative).

Candidate actions

- Candidate action:
  Select Approved Segment - choose from approved training content sets.
- Candidate action:
  Deliver Role Specific Prompts - different content to different roles where appropriate.
- Candidate action:
  Provide Explanation Layer - doctrine and debrief content, fixed or BoK, per policy.
- Candidate action:
  Age Tuned Delivery - if age_years is provided in a compliant context, adjust vocabulary, otherwise omit.
- Candidate action:
  Log And Forget - record operational evidence and enforce retention bounds.

Governance gate decisions

- Governance gate decision:
  Approve the action only from approved content lists.
- Governance gate decision:
  Approve the action if it does not degrade safety or team coherence.
- Governance gate decision:
  Approve the action with modification - enforce truthfulness and scope limits.
- Governance gate decision:
  Approve the action only if age years is explicitly provided or verified and required for safe delivery.
- Governance gate decision:
  Approve the action - logs are operational evidence with bounded retention.

Verification outcomes

- Verification outcome:
  Verification Result - outputs are present and synchronized.
- Verification outcome:
  Verification Result - safety constraints remained active throughout.
- Verification outcome:
  Drift Alert Acknowledged - preventive maintenance ticket created when needed.

These examples stay publish-safe by describing interface shape and behavior, not proprietary algorithms.

Worked scenarios, the loop in three recognizable moments

Outcome Verification Is a First-Class Layer

Decision-making without verification is opinion, not operation.

A WorldModel system does not end when a decision is approved or an action is issued. It completes the loop only when the system can verify that the intended effect occurred in the physical world, within acceptable bounds, and without unintended side effects.

This is not logging. It is not monitoring. It is **outcome verification**, and it is a distinct architectural responsibility.

Why verification must be explicit

In venue-scale environments, actions have externalities. A display switching modes, an audio zone changing language, a guidance prompt appearing, or a route suggestion being issued all alter human behavior. If the system cannot verify what actually happened, it cannot safely adapt, learn, or recover.

Without verification:

- Errors compound silently.
- Drift goes undetected.
- Operators lose trust.
- Automation becomes brittle.

With verification:

- Actions are accountable.
- Drift is detectable.
- Recovery is possible.
- Governance remains meaningful.

The verification sequence

A verified action passes through four distinct stages:

1. **Command issued**
   An action is proposed, governed, and approved.
2. **Action attempted**
   The system instructs downstream components to change state.
3. **Outcome observed**
   Independent signals confirm what actually happened. This may include display state, audio presence, surface mode, route availability, or environmental conditions.
4. **Outcome evaluated**
   The observed outcome is compared to the intended effect and tolerance bounds. Success, partial success, or failure is recorded.

This distinction matters. Many systems stop at step two and assume success.

Verification as a protection against optimism

Verification exists to counter a predictable failure mode: system optimism.

Optimism assumes that because a command was issued, the world complied. In real venues, that assumption fails constantly. Screens lose sync. Audio zones bleed.

Surfaces fail to switch modes. Staff intervene manually. Visitors behave unpredictably.

A WorldModel system must be pessimistic by design. It must assume that actions fail until verified. Under conflict or missing verification, the system refuses, or falls back conservatively, until operational truth is restored.

Drift detection and recovery

Verification is also how drift is detected.

If outcomes deviate gradually from intent, the system can flag degradation before failure becomes visible. Drift may trigger:

- Re-proposal of corrective actions,
- Downgrading to safer modes,
- Operator notification,
- Temporary suppression of automation.

Recovery is not a fallback. It is part of normal operation.

Why WorldModel Cannot Be Piloted Safely

WorldModel systems do not lend themselves to traditional pilots.

A pilot typically isolates a function, limits scope, and relaxes constraints. WorldModel does the opposite. It integrates functions, expands context, and enforces constraints continuously.

A partial implementation that omits governance, verification, or human authority is not a pilot. It is a distortion.

If a venue wishes to explore WorldModel concepts incrementally, it must do so by narrowing the surface area, not by weakening the architecture. That means:

- Fewer zones, not weaker governance.
- Fewer actions, not missing verification.
- Fewer modalities, not broken consent boundaries.

This distinction matters because a failed pilot can poison organizational trust for years.

WorldModel is not an experiment in intelligence. It is an investment in coherence.

Why this layer cannot be optional

Outcome verification is what allows WorldModel to operate in public space without becoming reckless.

A system that cannot prove what it did cannot justify what it does next.

The scenarios below are worked hypotheticals, written to make the state loop, the governance gate, and verification behavior testable without relying on any existing reference site.

Worked scenario 1,
show release and gentle rebalancing

Context

A theatre show releases approximately 400 visitors into a primary corridor that also serves as the step-free route to Gallery B. Gallery B is already busy, and the corridor's acoustic state has shifted from calm to loud. The venue is in normal mode.

Signals and authoritative sources

- Show cue event:
  show end cue, authoritative.
- Zone occupancy change event:
  corridor and Gallery B occupancy, advisory with confidence.
- Route availability change event:
  step-free route availability, authoritative.
- acoustic state change event:
  comfort and intelligibility flags, advisory with confidence.
- Operator override event:
  temporary closures or holds, authoritative.

Objectives and hard constraints

- Objective:
reduce congestion without breaking discovery or accessibility.
- Constraint:
rarely propose an inaccessible route for a visitor who has opted into step-free routing.
- Constraint:
guidance remains advisory, and visitors can ignore it without penalty.
- Constraint:
privacy remains default-anonymous. No identity is required for flow guidance.

State loop walkthrough

1. Observe:
the truth bus receives Show cue event and occupancy deltas within seconds of show end.
2. Update:
WorldModel updates corridor occupancy, adjacent zone availability, and acoustic comfort flags, including provenance and confidence.
3. Propose:
orchestration generates three candidate actions: switch nearby canvases into guidance mode, propose alternate routes to diffuse the pulse, and notify staff of a predicted bottleneck.
4. Govern:
the governance gate rejects any route proposal that would reduce accessibility, exceed comfort thresholds for "calm" opt-in visitors, or conflict with an active operator hold.
5. Act:
canvases render a short, plain-language guidance prompt. The Personal Channel (when a visitor opts in) can offer a calmer route option and language-specific guidance.
6. Verify:
Outcome Verification Layer confirms that canvases switched modes successfully and that guidance surfaces, Personal Channel guidance, and staff tools are consistent with the same state version.

Example event excerpt (8 events)

- Show cue event:
  Theatre A end-of-show (authoritative).
- Zone occupancy change event:
  Corridor C occupancy rises, advisory, confidence 0.8.
- Acoustic State Changed:
  Corridor C comfort=loud, advisory, confidence 0.7.
- Route availability change event:
  step-free route open (authoritative).
- Candidate Actions Proposed:
  guidance and staff prompts (authoritative).
- Governance Decision:
  approve canvases, approve advisory routing, approve staff alert
  (authoritative).
- Canvas Mode Changed:
  Corridor C surfaces set to guidance mode (authoritative).
- Verification Result, guidance rendered and consistent across planes,
  authoritative.

Visitor and operator outcome

Visitors see clear, non-coercive guidance that reduces decision fatigue at the exact moment confusion would otherwise form. Operators receive a single alert tied to a state version, plus a short-lived trend view that shows whether congestion is resolving or compounding.

Worked scenario 2, archive selection and assembly under constraints

Context

A visitor in a WWII museum asks, "What happened at Dunkirk in 1940," and selects a 30-second summary, Spanish language, and Stroller depth. Gallery state indicates moderate crowding and marginal intelligibility.

Inputs and constraints

- Inputs: the visitor's explicit time budget, language, and depth mode; gallery occupancy and acoustic state; and an indexed archive of approved primary and secondary sources.
- Constraint: responses are assembled only from approved archive material. No un-sourced invention is permitted.
- Constraint: age tuning is used only if an age parameter is explicitly provided in a compliant context, otherwise it is omitted.

Assembly loop

1. Retrieve:
   the system selects a small set of relevant passages from the indexed archive, preserving citations and provenance.
2. Select cut:
   it chooses a 30-second structure aligned to Stroller depth: headline, one context line, and one consequence line.
3. Compose:
   it assembles a short retelling in Spanish, using the venue's approved tone rules, and binds synchronized captions to the same time base.
4. Deliver:
   audio is delivered via the room mix where appropriate, and via the Personal Channel for the visitor's selected language and captions, keeping synchronization within the venue's timing envelope.
5. Verify:
   the system verifies that the correct language track is active, captions are present, and playback is synchronized.

Example event excerpt (9 events)

- Question Asked:
  topic=Dunkirk_1940 (advisory).
- Preference Applied:
  language=es, depth=stroller, time_budget=30s (authoritative).
- Archive Passages Retrieved:
  approved corpus (authoritative).
- Cut Selected:
  30s structure (authoritative).
- Governance Decision:
  approve assembly from approved archive only (authoritative).
- Playback Committed:
  Spanish audio plus captions (authoritative).
- Verification Result:
  audio present, captions present, sync within threshold (authoritative).
- Optional Deep Link Prepared:
  90s and source list available on request (authoritative).
- Operational Log Written:
  state and decisions recorded with bounded retention (authoritative).
  Visitor and operator outcome

The visitor receives a coherent answer at the requested depth and language, without forcing the room into bilingual clutter. Operators gain evidence of what was delivered, with provenance, without collecting a personal dossier.

Worked scenario 3,
capacity shock in themed entertainment
Context

In a theme park land, an attraction goes down during a major show release. Queue estimates spike, and pedestrian density rises across two primary corridors. The venue shifts into busy mode.

Signals and authoritative sources

- Attraction State Changed:
  ride down, authoritative.
- Show cue event:
  show release, authoritative.
- Queue estimate update event:
  predicted waits and shock detection, advisory with confidence.
- Zone occupancy change event:
  land and corridor density, advisory with confidence.
- Operator Staffing State Changed:
  staffing posture, authoritative.

Governed response

1. Mode shift:
   busy mode tightens constraints, prioritizing safety, fairness, and clarity.
2. Propose:
   orchestration generates options: update canvases to truthful queue guidance, suggest alternate attractions with available capacity, stagger guidance prompts to prevent secondary bottlenecks, and issue a staff alert for crowd management.
3. Govern:
   the gate rejects any action that would create accessibility regressions, privilege one group unfairly, or conflict with emergency procedures.

Responsibility boundaries, at a glance

Keep decision responsibilities separate so one failure does not become venue-wide failure.

- WorldModel™ framework:
  operational truth and state reconciliation.
- CGL™:
  constraints, policy gating, and escalation rules.
- MAOL™:
  orchestration of tasks across agents and systems.
- EDE™:
  environment dynamics predictions and flow constraints.
- ICL™:
  continuity state, consent state, and identity boundaries.

Operations:

1. Act: surfaces switch to guidance mode with simple, truthful prompts. The Personal Channel can offer language-specific guidance and a calm-route option when a visitor opts in.
2. Verify: the system verifies that the same state version is reflected across canvases, staff tools, and any Personal Channel guidance, and that the ride-down state is not contradicted by any downstream surface.

Example event excerpt (10 events)

- Attraction State Changed: Ride X down (authoritative).
- Show cue event: Show Y release (authoritative).
- Queue estimate update event: Ride X queue invalidated (authoritative).
- Zone occupancy change event: Land Z density high, advisory, confidence 0.85.
- Operating Mode Set: busy (authoritative).
- Candidate Actions Proposed: route suggestions and staff alerts (authoritative).
- Governance Decision: approve truthful guidance and advisory routing only (authoritative).
- Canvas Mode Changed: land signage set to guidance mode (authoritative).
- Staff Alert Issued: crowd pulse and alternate routing (authoritative).
- Verification Result: guidance consistent and correct (authoritative).

Visitor and operator outcome

Guests experience the system as calm and truthful, not as an optimizer. The land remains navigable, queue shock is handled transparently, accessibility constraints are preserved, and staff receive a single actionable prompt rather than a flood of alarms.

Acceptance tests that define "implemented"

A WorldModel Venue should not ship without acceptance tests that operators can run.

Truth consistency

- Surface guidance, Personal Channel guidance, and staff tools cannot contradict each other under the same state.
- State versions advance monotonically and are observable.

Degraded conditions

- External internet loss does not break core functions.
- Partial subsystem failure yields safe fallback behaviors, not chaos.

Governance constraints

- Accessibility violations are rejected.
- Privacy posture violations are rejected.
- Show integrity violations are rejected.
- Operator overrides are respected immediately.

Verification integrity

- After any change to canvases or screens, verification confirms visible outcome and motion where expected.
- After Lifecycle Automation Layer transitions, restoration is verified before the zone is declared ready.
- Under uncertainty, the system behaves conservatively.

Drift detection

- Thermal drift, fan escalation, and rising error rates trigger preventive alerts with clear operator workflow.

Auditability

- For any action, operators can see what happened, why it was permitted, what state was used, and what constraints were checked.
- Logs remain operational evidence, not visitor biographies.

These are the tests that separate a reference-grade venue system from a demo.

Evidence Artifacts Checklist (operator and procurement ready)

A WorldModel Venue is not considered implemented until the venue can produce the following artifacts on demand. These are not "nice to have." They are the minimum evidence that makes governance defensible, recovery safe, and operations repeatable.

1) Authoritative declarations (Level 1 evidence)

- Operator override log: holds, closures, mode switches, approvals, expiries
- Show control state log: cues, holds, scene state, integrity flags
- Safety and incident mode state log: operating mode changes and triggers
- Asset fault log: authoritative fault states, latched errors, and clears
- Entitlement state log: minimal rights state changes, not dossiers

2) Verified outcomes (Level 2 evidence)

- Outcome verification records for:
- surface mode changes (story vs guidance)
- display state (image present, motion when required)
- audio presence and zone state (program present, within bounds)
- restoration success after occupancy-aware transitions
- Verification failure records with:
  - what failed
  - what fallback executed
  - whether an operator was alerted
  - time-to-restore

3) Advisory measurements and inferred signals (Level 3 evidence)

- Occupancy and flow telemetry with confidence and decay rules
- Queue estimates with provenance and confidence
- Acoustic state telemetry with confidence and update cadence
- Pattern recognition aggregates that remain place-based and non-identifying

4) Derived state (Level 4 evidence)

- WorldModel state snapshots by version, with provenance fields
- Derived fields definition (what is computed, what it depends on, and its decay rules)
- State freshness policy (how stale is too stale for which action classes)

5) Proposals and decisions (Level 5 evidence)

- Candidate action log (what was proposed, by whom or what agent)
- Governance decision log for each consequential action:
- state version used
- constraints checked
- approve, modify, or reject
- Constitution version and policy mode
- human approval or override, if applicable

6) Drift and preventive maintenance

- Drift signals summary (temperature trend, fan escalation, error-rate creep, sync drift)
- Preventive maintenance tickets linked to drift thresholds
- Trend review cadence (weekly and monthly operator evidence)

7) Publishing discipline and rollback

- Content version history for key surfaces and Personal Channel packages
- Configuration version history for policies, modes, and constraints
- Rollback evidence: at least one executed rollback drill with logs

If a venue cannot produce these artifacts, it may have intelligence. It does not yet have operational truth.

Why this is the way forward

The next decade will reward places that behave coherently. Coherence requires one operational truth, governed decisions, executed actions, verified outcomes, and an upgrade path based on provisioning compute rather than rebuilding.

That is what this implementation delivers. It is useful because it is real. It is safe because it specifies structure, interfaces, and discipline, not proprietary recipes.

In the next chapter, we return to the decision architecture layers, CGL, EDE, ICL, and MAOL, now grounded in a concrete implementation loop that makes their roles inevitable rather than abstract.

25

CGL, EDE, ICL, AND MAOL AS DECISION ARCHITECTURE

A WorldModel™ Venue is not impressive because it can display content on command. It becomes impressive when it can decide repeatedly, under pressure, in public space, and still behave like a good host.

That is the threshold between a venue that is merely automated and a venue that is genuinely responsive. Automation plays a script. Responsiveness negotiates reality. Reality includes competing objectives, hard constraints, incomplete information, and the uncomfortable fact that decisions in public space have externalities.

This chapter introduces the decision architecture that makes macro behavior possible without turning the venue into a machine that guesses. It introduces four layers that together form a disciplined pipeline from understanding to action:

- ICL™ (Identity Continuity Layer™)
- EDE™ (Environmental Dynamics Engine™)
- MAOL™ (Multi-Agent Orchestration Layer™)
- CGL™ (Cognitive Governance Layer™)

These are architectural separations of responsibility. The separations make the system operable, auditable, privacy-forward, and capable of improvement over time.

The four jobs of a living venue

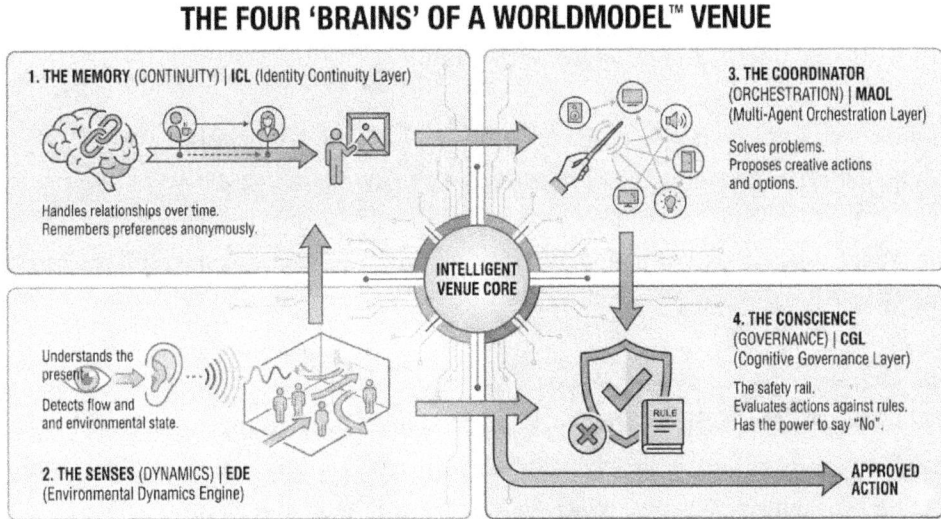

THE FOUR 'BRAINS' OF A WORLDMODEL™ VENUE

1. THE MEMORY (CONTINUITY) | **ICL** (Identity Continuity Layer)

Handles relationships over time.
Remembers preferences anonymously.

3. THE COORDINATOR
(ORCHESTRATION) | **MAOL**
(Multi-Agent Orchestration Layer)

Solves problems.
Proposes creative actions
and options.

INTELLIGENT VENUE CORE

Understands the present.
Detects flow and
and environmental state.

2. THE SENSES (DYNAMICS) | **EDE**
(Environmental Dynamics Engine)

4. THE CONSCIENCE
(GOVERNANCE) | **CGL**
(Cognitive Governance Layer)

The safety rail.
Evaluates actions against rules.
Has the power to say "No".

RULE

APPROVED ACTION

Before assigning technical names to the layers that drive a WorldModel Venue, it helps to name what we are asking the building to do. A single monolithic "AI" that runs a complex environment is difficult to test, difficult to trust, and difficult to upgrade without breaking. A better approach is to divide the system's intelligence into four distinct jobs, four components that work together but have separate responsibilities.

The Memory (Continuity). The venue needs to handle relationships. It needs to know that the visitor in the lobby who asked for Spanish is the same visitor now standing in the gallery, without demanding their name or building a dossier. It needs to remember preferences for an hour or a day, and then reliably forget them.

The Senses (Dynamics). The venue needs to understand the present moment. Sensors are not enough on their own. The system needs to understand flow. It needs to detect that a crowd is forming, that a queue is stalling, or that a room has become too loud for comfortable conversation.

The Coordinator (Orchestration). The venue needs to propose actions. When it sees a bottleneck, it needs to generate options: change signage, reroute the next group, or alert staff. This layer is the problem-solver that produces candidate moves.

The Conscience (Governance). The venue needs a safety rail. It needs a distinct layer that evaluates every proposed action and has the power to refuse. This layer enforces hard constraints: do not route a wheelchair user to a stairwell, do not collect data without consent, do not interrupt the show.

Those constraints are not informal preferences. They are Constitutional rules, written explicitly and enforced at runtime by the governance gate.

Governance should output bounded decisions, with explicit outcomes and reference-based justification.

Confabulation risk does not disappear inside governance. If any AI component is used inside the Cognitive Governance Layer, it must not become an author of explanations. The CGL is an evaluator. Its primary output is a decision object with a bounded vocabulary: approve, reject, modify, require human review, or abstain.

Justification must be reference-based. A good justification trace references the policy clauses and the state inputs used, not a narrative story. If required state is missing, stale, ambiguous, or contradictory, the correct outcome is abstain, followed by safe fallback or operator escalation. Missing state is not a reason to guess.

Preventing confabulation inside the governance layer

A common misconception is that confabulation is only a conversational problem. It is also a governance problem, if governance is implemented with AI components that can generate fluent but ungrounded output.

The governance layer evaluates and constrains, using structured decisions rather than prose.

The Cognitive Governance Layer is an evaluator, not an author. Its output should not be prose. Its output should be a decision object with a bounded vocabulary:

- approve,
- reject,
- modify,
- require human review, or
- abstain.

If the CGL produces narrative explanations, it invites the same failure mode that affects conversational systems: fluent certainty without grounding. Wrong governance is worse than wrong narration because it can authorize actions.

If AI is used inside the CGL, it should be used only for bounded tasks such as classification, constraint matching, conflict detection, and uncertainty estimation. The CGL can use AI to propose which Constitutional clauses may apply, or to detect that an action is safety-adjacent, privacy-adjacent, or accessibility-adjacent. It must not use AI to invent facts, infer visitor attributes, or fabricate justification.

Two enforcement rules keep this safe:

- Rule 1: No new facts.
  The CGL may evaluate only against state that is already present in operational truth, consent state, and declared policies. If a required input is missing, stale, ambiguous, or contradictory, the correct result is abstain, followed by safe fallback or operator escalation. Missing state is not a reason to guess.
- Rule 2: Justification must be reference-based.
  A governance decision should carry a justification trace that references the policy clauses and state inputs used, not a narrative story. A good trace is reconstructable: which clauses applied, which constraints were checked, what the mode was, what the confidence posture was, what outcome was selected, and whether a human override occurred.

Finally, governance must be measurable in operation. Treat governance integrity as an operational signal the same way you treat drift and failure trends. Track abstain rate, override rate, post-hoc reversal rate, and policy conflict rate across updates. A governance layer that becomes more fluent but less conservative has degraded, even if the rest of the system appears functional.

In this chapter, we give these four jobs their architectural names: ICL (Identity Continuity Layer), EDE (Environmental Dynamics Engine), MAOL (Multi-Agent Orchestration Layer), and CGL (Cognitive Governance Layer). They are separated for a practical reason. Keeping the conscience separate from the coordinator helps ensure that no matter how clever the system becomes, it cannot break the rules you set.

Government frameworks are converging on agentic AI requirements

As agentic systems have moved from experimentation to deployment, governments and regulators have begun publishing model governance frameworks specifically addressing autonomous and semi-autonomous systems. These documents are not implementation guides. They are signals that the governance problem has become operational.

Once systems can plan, act, delegate, and adapt, accountability, controllability, and auditability can no longer be treated as policy language added after build. They must be enforced by design.

A recent government-published model governance framework for agentic AI (IMDA, 2026)[xxxvii] organizes this problem into four recurring dimensions:

1. Assess and bound risks upfront
2. Make humans meaningfully accountable
3. Implement technical controls and processes across the lifecycle
4. Enable end-user responsibility

This book treats those dimensions as engineering constraints. Where governance frameworks describe what must be controlled, the WorldModel™ architecture defines how those controls are enforced in real systems through bounded authority, identity continuity, meaningful oversight checkpoints, and continuous lifecycle monitoring.

Crosswalk: governance dimensions to WorldModel™ enforcement mechanisms

Governance dimension (common across emerging frameworks)	WorldModel™ enforcement mechanism in this book
Assess and bound risks up-front	Value System and Constitution define non-negotiables, and bound action-space through least-privilege permissions and explicit authority limits
Make humans meaningfully accountable	CGL™ enforces accountability through approval checkpoints, escalation paths, and auditable decision traces
Implement technical controls and processes	Pre-deployment testing, phased rollout, continuous monitoring, drift detection, and controlled rollback
Enable end-user responsibility	Transparent capability disclosure, consent boundaries, retention limits, and explicit "must forget" rules

Independence note: External frameworks are referenced here only to demonstrate convergence in requirements. They are not dependencies, and they did not drive the architecture.

Why one brain is not enough

When people imagine venue intelligence, they often imagine a single brain: one system that knows everything, decides everything, and pushes commands everywhere. That image is natural and usually wrong in practice.

A large venue is a bundle of different problems that operate on different time scales, have different error costs, and carry different constraints:

- wayfinding and flow
- queue dynamics
- scheduling and show management
- multilingual delivery
- accessibility and comfort modes
- power behavior and equipment life
- staffing, incident response, and operational constraints
- commerce state, entitlements, and service delivery

If you merge these into one undifferentiated decision engine, the system becomes too complex to trust and too opaque to operate.

A layered decision architecture is the opposite. It allows specialized reasoning without surrendering coherence. It makes the system capable while keeping it bounded.

The decision pipeline, in human terms

Before naming the layers in detail, it helps to name the shape of the loop. A macro venue repeats the same cycle all day:

1. observe state
2. update operational truth
3. propose candidate actions
4. evaluate actions under policy constraints
5. execute actions
6. verify outcomes
7. learn what drifted and correct

The key point is that the venue does not merely select content. It selects actions, and actions affect other people. That is why decision architecture matters more than model quality. The system's capability has to be constrained by responsibility, and responsibility has to be enforceable.

Each of the four layers in this chapter exists to make one part of that cycle disciplined.

ICL: the Identity Continuity Layer

Identity continuity sounds like a feature. In a WorldModel Venue it is a layer because continuity is rarely one thing.

A visitor chooses a language, and the system honors it for a session. That is anonymous continuity.

Sometimes the venue needs continuity across multiple exhibits within one visit but still does not need real-world identity. It needs continuity of intent: language, depth mode, interests, and comfort choices. That can be achieved with short-lived, privacy-forward mechanisms.

Sometimes the visitor explicitly asks for continuity across visits, such as membership benefits, saved itineraries, accessibility persistence, or entitlements. That is the moment identity becomes appropriate, because the visitor is requesting a benefit that requires memory.

ICL exists to prevent identity from becoming accidental. It makes continuity explicit, scoped, and consent-bound. It also enforces forgetting as a feature. When the venue defaults to anonymous operation, ICL maintains that default and prevents continuity from drifting into dossiers through convenience.

A useful way to think of ICL is that it manages the shape of the relationship between a person and a place, including the right to remain anonymous.

EDE: the Environmental Dynamics Engine

A venue is not static even when it looks static. Crowds pulse. Queues form, dissolve, and reappear. A show releases a wave of people into a corridor. A gallery becomes calm and then suddenly loud. A closure collapses a route into a bottleneck. A lunch rush shifts the entire rhythm of the day.

If you treat all of this as raw sensor input, you react late. EDE exists to model dynamics, not just snapshots. It takes the living present, occupancy, flow, congestion, acoustic state, and operating modes, and turns them into something usable: what is changing, how fast it is changing, and what is likely to happen next if the venue does nothing.

This is basic venue competence, expressed in system form. Anticipate that a corridor will become crowded when a show ends. Anticipate that a quiet refuge space will become congested if every guidance sign points toward it. Anticipate that a queue will spike after a schedule shift.

EDE is also where comfort becomes operational. Acoustic state, sensory load, and the lived feel of spaces can be treated as dynamics rather than anecdotes. When the venue can see that a zone is becoming overwhelming, it can respond by offering alternate routes, adjusting guidance, or shifting modes. A good venue does this with staff intuition. EDE supports that intuition with state.

MAOL: the Multi-Agent Orchestration Layer

Once you have continuity and dynamics, you still face a hard truth. There is rarely one correct action. There are candidate actions.

A macro venue often has multiple plausible moves at once:

- route some visitors away from a congested area
- hold others in a calmer gallery
- adjust signage to distribute flow
- delay a prompt to avoid sending too many people at once
- open or close an optional experience path
- change a surface from narrative mode to guidance mode
- trigger a staff notification
- adjust comfort modes in a zone
- invoke an occupancy-aware power posture

MAOL exists to generate and coordinate proposals from specialized reasoning processes. It is where the system can run multiple agents or modules in parallel, each concerned with a slice of the venue's needs: flow, accessibility, scheduling, acoustic comfort, retail service, safety posture, and more.

This is where the architecture becomes more humane. A monolithic system tends to optimize what it can measure easily. An orchestrated system can consider multiple objectives without flattening them into a single number. It can generate options, then allow governance to decide.

Governance operating cadence

Governance only works if it has a cadence.

The cadence must include integrity signals, not only outages.

Daily and weekly reviews should include abstain rate, fallback rate, override rate, and policy conflict rate. These are the early warning indicators that the system is drifting toward confabulation-like behavior, either by issuing ungrounded guidance or by treating weak inference as certainty.

- Daily: review incidents, overrides, and safety-adjacent logs.
- Weekly: review drift signals, false alarms, and tuning changes.
- Monthly: review retention compliance, access controls, and evidence pack freshness.
- Quarterly: run rollback drills, audit logs, and commission re-verification in sampled zones.

MAOL is also where redundancy and resilience become natural. When orchestration is distributed, the venue can continue functioning even if some components degrade, because other agents can still propose safe actions and the system can fall back to simpler behaviors.

CGL: the Cognitive Governance Layer™

If MAOL generates candidate actions, CGL decides what is permitted. CGL may be implemented using rules, AI, or hybrid methods, but its defining property is enforceable constraint evaluation with bounded, auditable decision outputs. This is where macro capability becomes deployable.

A venue-scale system must be able to enforce:

- safety constraints
- accessibility constraints
- privacy constraints
- fairness constraints
- operational constraints
- brand and cultural constraints
- the venue's explicit objectives

Without governance, orchestration becomes a clever machine that can do harm accidentally. With governance, orchestration becomes a disciplined system that can be trusted to act in public space.

CGL is not a set of vague values. It evaluates proposals against explicit rules, constraints, and policies and produces three outcomes:

- approve the action
- modify the action into a permitted form
- reject the action and request alternatives

This is the point where the book's covenant becomes enforceable. Privacy-forward becomes a constraint. Inclusion becomes a requirement. Operator override becomes part of the operating loop. Governance is what allows powerful systems to remain kind.

Why the architecture wants multiple machines

Early in macro design, teams often assume there should be one central server that runs the intelligence. Then reality arrives. A venue is physically distributed. So are its failure modes. So are its time constraints. So are its safety requirements.

A resilient architecture provisions compute across nodes, not because centralization is wrong in principle, but because it is fragile in practice. There are three reasons to distribute compute.

1. Latency and stability. Time-critical loops should run close to the experience. A venue cannot wait for distant dependencies when a room cue must land now.
2. Fault containment. If one node fails, the whole venue should not collapse. Distributed compute supports graceful degradation.
3. Load spreading and redundancy. Macro reasoning, analytics, verification, orchestration, and interfaces all consume compute. A venue that expects growth should provision compute the way it provisions capacity: with headroom and clear upgrade paths.

This is where "provisioning sufficient compute" becomes strategic. Replacement becomes about capacity and role rather than brand and bespoke wiring. Upgrades become about adding compute where needed rather than rebuilding the venue.

Future compute without rewriting the venue

New accelerators emerge. New architectures appear. New classes of compute may become useful for certain simulation and optimization tasks. The important point is not to forecast timelines. The important point is architectural.

If your system is built as a set of layers that consume compute rather than as a fixed stack of single-purpose devices, you can incorporate new compute capabilities without redefining what the venue is. A WorldModel Venue designed around modular nodes, distributed execution, and clear interfaces between layers can adopt better compute as it becomes available. That is how the architecture stays alive for a decade rather than aging into a museum exhibit of its own.

Key takeaways

The WorldModel architecture gives the venue shared operational truth. These layers turn truth into action responsibly.

- ICL defines continuity, consent, and forgetting.
- EDE models dynamics so the system can anticipate rather than react late.
- MAOL generates options across competing objectives.
- CGL enforces governance so power stays bounded, auditable, and inclusive.

This is the decision architecture that makes Venue Concierge Interface credible. It is also the reason macro hyper-personalization can remain positive. The venue becomes more capable without becoming invasive, more adaptive without becoming chaotic, and more intelligent without becoming unaccountable.

In the next chapter we move from architecture to practice: how a venue writes objectives, constraints, audit posture, operator overrides, and decision logs in a way that is operational rather than ceremonial. That is where governance becomes a daily discipline.

Governance as a runtime capability

In public environments, personalization is a governance problem because decisions are made continuously about what a person sees, hears, is told, and is guided to do. This book treats governance as an operational capability, not as a policy binder.

In this book, governance means three practical things: defining what the system is allowed to do and not do, enforcing those limits at runtime, and producing evidence of enforcement so decisions can be audited, investigated, improved, and trusted.

Role clarity is part of governance. A WorldModel deployment typically spans at least three roles: provider, deployer, and operator. The system remains trustworthy when these boundaries are explicit in documentation, operator tools, escalation paths, and evidence logs.

Architecture alone is not compliance. Compliance depends on implementation, operational procedures, and evidence. The publish-safe claim is simpler: the architecture is designed to support compliance, and can be mapped to common governance expectations when the required controls, processes, and evidence are present.

A procurement-ready system should produce five evidence categories: decision, data, oversight, monitoring, and improvement. The back matter templates exist to make this measurable and repeatable. For named framework references, see the updateable "Assurance and Compliance Snapshot" appendices, which can change between editions without changing the core architecture.

26

GOVERNANCE IN PRACTICE: SAFETY OVERRIDES, MODES, AND OPERATOR AUTHORITY

Governance fails most often for a simple reason. It is treated as paperwork. A committee writes a policy. A system goes live. The policy sits in a folder. The venue operates under pressure, and the system makes choices anyway, because the building does not pause while you look for a PDF.

In a WorldModel Venue, governance cannot be paperwork. It has to be behavior. It has to show up in three places at once: in the system's constraints, in the operator's tools, and in the venue's operating culture. When those three disagree, governance becomes theatre. When they align, governance becomes the quiet reason a powerful system can be trusted in public space.

This chapter is about alignment. It is about writing objectives, hard constraints, audit posture, and override in a way that survives real life, including busy days, bad days, and emergency days.

The venue is a bundle of competing goods

Every serious venue balances competing goods. Experience quality competes with throughput. Discovery competes with clarity. Comfort competes with spectacle. Inclusivity competes with the temptation to standardize. Privacy competes with the temptation to measure everything. Staff time competes with ambition. Energy usage competes with equipment life.

These are not design failures. They are the reality of public space. Governance begins when a venue stops pretending these goods can be optimized into a single score. Instead, the venue states what it wants, what it will not do, and what it will do when those collide.

That is the difference between a system that behaves like an optimizer and a system that behaves like a host.

The Constitution: the boundary that makes governance enforceable

Venues often publish values statements. Values statements are not governance.

It is the venue's explicit, testable rule set that constrains what the system may and may not do, say, recommend, infer, retain, or optimize. It is the boundary layer that turns "we care about privacy, inclusion, and fairness" into enforceable behavior at runtime.

Truthfulness belongs in the same category.

In governed environments, "do not present guesses as facts" is not a stylistic preference. It is a constraint class. Confabulation is predictable under uncertainty, and that is why truthfulness must be enforceable at runtime: either the system can ground a claim in approved sources or authoritative state, or it must abstain, fall back, or escalate.

A Constitution that cannot be evaluated, enforced, and audited in operation is not governance. It is aspiration. In practice, constitutions exist to bound authority, require justification for consequential actions, define escalation paths, and constrain lifecycle behavior long after deployment. Treating constitutional rules as testable, enforceable system inputs is what allows autonomy without loss of accountability.

In this book, the Constitution is upstream of every other governance artifact. It is what the governance gate enforces. It is also what makes audits meaningful, because auditors cannot evaluate enforcement if the boundary is not written.

How the Constitution differs from objectives and constraints

It helps to separate three layers:

1. Value System
   The Value System describes what the venue is trying to protect and promote, such as dignity, safety, inclusion, truthfulness, and calm. It is the "why" and the prioritization stance when goals collide.
2. Constitution
   The Constitution is the "must and must not." It converts values into rules that can be checked. It defines prohibitions, requirements, escalation triggers, and what counts as acceptable evidence.
3. Policy and operating mode
   Policy is the tunable layer. It is where thresholds, operating modes (normal, busy, incident), and local program choices live. Policy can change more frequently. The Constitution changes rarely.

A practical Constitution for a governed venue

A venue Constitution should be short enough to be used, and strict enough to matter. The most reliable format is a numbered list of clauses, each written so it can be enforced and tested.

Scope and variability

A Constitution is not one-size-fits-all. Every venue should share a small core of non-negotiables, such as safety, accessibility, privacy minimization, truthful communication, operator override, and auditability.

Beyond that core, Constitutional clauses should reflect the venue's category and risk profile (museum, theme park, airport, hospital, retail, cruise ship), its jurisdictional obligations, and its explicit brand promises.

A children's museum, an airport, and a luxury resort can all be governed, but they will not choose identical boundaries. The Value System sets the priorities that reflect identity and intent. The Constitution turns those priorities into enforceable rules. Policy and operating modes tune thresholds and behavior within those rules.

Minimum clause categories

Privacy and identity

- The venue is available without identity by default.
- Identity may be requested only as an explicit benefit exchange, with clear purpose and retention.
- Minimization is mandatory: collect only what is needed for the current purpose.
- Retention is bounded by declared limits and enforced by deletion as a normal system behavior.
- Covert profiling is prohibited.

Safety and accessibility

- Safety constraints are hard constraints and override all other objectives.
- Accessibility constraints are hard constraints and are never treated as "best effort."
- The system must fail conservatively under uncertainty in safety-adjacent and access-adjacent decisions.

Fairness and non-discrimination

- The system must not produce outcomes that systematically disadvantage a protected class or a declared accessibility need.
- When fairness tradeoffs exist, they must be represented explicitly as constraints or as governed policy choices, not as hidden optimization.

Truthfulness and transparency

- The system must not present guesses as facts.
- Confabulation control: when grounding is insufficient, the system must refuse to invent and must switch to safe defaults or operator escalation.
- If guidance is based on uncertain state, the system must communicate uncertainty or fall back to safe defaults.
- The system must prefer informing and offering options over steering, except where safety or procedure requires directive action.

Content integrity and provenance

- When content is generated or adapted, the system must preserve source attribution where feasible and must not invent citations or authorities.
- Where the system summarizes or paraphrases, it must be able to point to the source material used, or clearly indicate that the output is a best-effort synthesis.
- Safety-critical and compliance-critical instructions must come from approved sources and must not be generated from unverified context.

Agency and non-manipulation

- Guidance is advisory by default. The visitor can ignore it without penalty.
- The system must not use personalization to exploit vulnerability, apply hidden pressure, or shape behavior through undisclosed optimization.
- If a system is optimizing a metric, the metric and the constraints must be declared at the governance level.

Auditability and operator authority

- Every consequential decision must be attributable to: the state used, the rule checks applied, and the outcome permitted or denied.
- Operator override is always available, immediate, and logged.
- The system must be able to produce a plain-language explanation of why an action was permitted or refused when the venue needs to defend behavior.[xxxviii]

Vulnerable populations and safeguarding

- The system must not target, profile, or personalize in ways that exploit vulnerability.
- Where a venue serves vulnerable groups, additional constraints may apply, including stricter consent requirements, stricter data minimization, and stricter content and interaction boundaries.

How the Constitution is enforced

The Constitution is enforced at the governance gate. Proposals from orchestration are evaluated against Constitutional clauses and the current policy mode. The gate returns only three outcomes: approve, modify into an allowed form, or reject and request alternatives.

Every governance decision should record the Constitution version, the active policy mode, and the relevant clause identifiers that were checked. This makes auditability real while keeping logs as operational evidence rather than visitor biography.

How to write a Constitution without making it performative

Write clauses that can be tested. If a clause cannot be checked by a system or an operator, it is not yet a clause. It is a slogan.

A practical working format for each clause is:

- Clause statement (must or must not)
- Scope (which decisions or outputs it applies to)
- Trigger (when it is evaluated)
- Evidence (what the system must log to prove compliance)
- Escalation (what happens when a clause blocks an action)

Once a venue has a Constitution in this form, the rest of governance becomes implementable: outcomes can be pursued without violating boundaries, policies can change without changing ethics, and audits can evaluate behavior rather than paperwork.

Governance has four artifacts, and the order matters

The order matters because it prevents you from writing lofty values that rarely touch the machinery.

Governance Artifacts and Decision Gate Process.

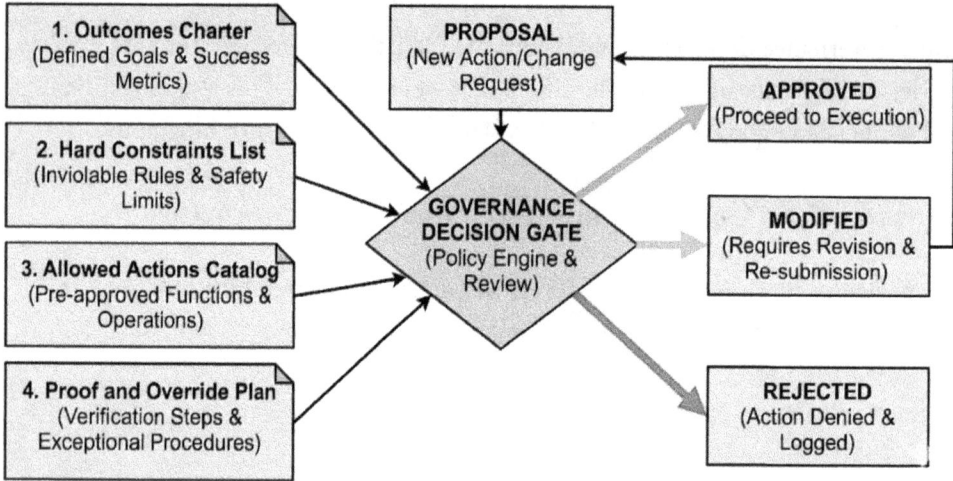

1) The Outcomes Charter

This is the venue's statement of what it is trying to produce, written in language operators and designers can both use. The charter should be operational, not aspirational. It belongs on one page. If it cannot fit on one page, it cannot be used under pressure.

Examples of outcomes:

- Reduce confusion and decision fatigue.
- Protect discovery while improving flow.
- Maintain language dignity across key visitor journeys.
- Expand inclusion by making comfort modes normal.
- Keep queues fair, predictable, and calm.
- Preserve show integrity under load.
- Reduce sound bleed between zones during unoccupied periods.
- Extend equipment life through occupancy-aware behavior without visible degradation.
- Maintain privacy-forward operation as default.

2) The Hard Constraints List

Outcomes can be negotiated. Constraints cannot. A hard constraint is a line the system must not cross even if doing so would improve an appealing metric.

Hard constraints in a WorldModel Venue typically include:

- Safety constraints: do not route into unsafe areas, do not create crowd hazards, do not conflict with emergency procedures.
- Accessibility constraints: do not propose routes that violate declared accessibility needs, do not make basic participation dependent on personal devices.
- Privacy constraints: available without identity, do not retain beyond defined limits, do not expand collection scope silently.
- Fairness constraints: do not systematically disadvantage a class of visitors in routing, queuing, or access to time-bound experiences.
- Experience integrity constraints: do not interrupt certain show beats, do not degrade intelligibility below defined thresholds.
- Operational constraints: do not propose actions that require staff intervention unless staff capacity is available.

Constraints must be written so they can be enforced. If a constraint cannot be expressed as something the system can check, it is not a constraint yet. It is an aspiration.

3) *The Allowed Actions Catalog*

A system can only be governed if the venue is explicit about what the system is allowed to do. The catalog becomes a safety rail. It prevents scope creep from becoming behavior creep.

Examples of allowed actions:

- Provide guidance suggestions to visitors, with clear opt-out.
- Change programmable surfaces between narrative mode and guidance mode under defined conditions.
- Offer alternate languages and depth modes through the Personal Channel.
- Adjust zone volume within defined bounds, especially during unoccupied periods, to reduce sound bleed.
- Power down or dim defined equipment classes when zones are unoccupied, and restore before visitors can notice, under verified restoration rules.
- Notify staff when thresholds are exceeded: congestion, faults, or impending maintenance.
- Propose schedule adjustments within defined limits and approval requirements.

This is also where you decide what requires human approval. Some actions can be autonomous. Some should be advisory. Some should be blocked unless an operator accepts them. A mature venue chooses deliberately.

4) The Proof and Override Plan

The venue must be able to answer, calmly, when asked:

- What did the system do?
- Why did it do it?
- What did it believe to be true at the time?
- Did the outcome match intent?
- Who could override it, and how?

This plan includes audit logs, operator dashboards, and a defined override posture. It is also where you prevent a common operational failure: a system nobody trusts because nobody can show what it did.

Governance is also modes

A venue does not operate under one condition. It operates under modes. A serious governance design defines at least three:

1. **Normal mode**
   The venue is calm. The system prioritizes experience quality, discovery, inclusion, and gentle flow shaping.

2. **Busy mode**
   The venue is at high occupancy. The system shifts emphasis toward clarity, queue fairness, and load balancing while preserving dignity and comfort.

3. **Incident or emergency mode**
   Allowed actions narrow to those that support safe operations and clear guidance.

Emergency overrides and the Emergency Constraint Set

In incident mode, the system's primary job is to stop making the situation worse.

Priority order is not negotiable:

1. Life-safety activation and emergency communications.
2. Emergency stop chains (E-stop, show-stop, ride-stop).
3. Operator holds, closures, and responder-only restrictions.
4. Safety-adjacent constraints derived from state (crowd crush risk, blocked egress, unsafe density, hazardous conditions).

Emergency Constraint Set (minimum required behaviors)

- Suppress automation: any action that could move people, concentrate crowds, or contradict staff instructions is disallowed.
- Remove hazard zones: routing graphs must exclude emergency and responder-only areas immediately.
- Refuse unsafe requests: if a requested destination is inside a restricted zone, the system must refuse and escalate, rather than "finding a clever path."
- Use pre-authored safety language only: do not generate evacuation or hazard instructions. Render approved messages tied to the current authoritative state.
- No auto-restart: recovery from emergency stop or life-safety events requires explicit operator action and must be logged as an authoritative transition.

Audit requirement: every transition into, and out of, incident mode must record which authoritative trigger caused it (life-safety, E-stop, operator hold), when it began, when it ended, and what automation was suppressed.

Modes prevent a clever system from behaving the same way under radically different conditions. A place should not sound, route, and signal the same way during a calm morning and during a surge after a show release. Modes also make the acoustic layer operational. In busy mode, noise floor rises and sound bleed increases. A governed system responds by shifting outputs and guidance rather than by turning everything up.

Logging is necessary, and logs must not become dossiers

A venue-scale system must log decisions. Without logs, you cannot audit. Without audit, you cannot improve. Without improvement, the system becomes brittle or distrusted.

Logging has a trap. Logs can become an accidental archive of personal behavior. A privacy-forward posture treats logs as operational evidence, not biographies.

That means:

- log actions and reasons at the level of system state rather than named individuals,
- use minimal identifiers and prefer session-scoped identifiers that expire,
- separate diagnostic logs from any identity continuity records and minimize the latter, and
- apply retention limits aggressively, because deletion is part of governance.

Revocation and rollback are governance, not customer service

Retention limits matter only if they can be enforced under pressure.

Public venues live in the messy middle. A visitor changes their mind. A parent withdraws consent on behalf of a child. A staff member realizes a setting was applied too broadly. A subsystem begins behaving strangely after an update. A vendor discovers that a cache was "temporary" for longer than anyone intended. In these moments, governance is not a philosophy. It is the system's ability to unwind.

A venue that can adapt must also be able to reverse.

This is a practical requirement, not a legal flourish. If a system can personalize, it can also mis-personalize. If it can remember, it can remember too much. If it can correlate signals, it can create linkability accidentally. The only way to keep capability deployable is to treat revocation and rollback as first-class system behaviors.

Consent is a runtime state

Consent is not a form the venue collects. It is a live condition that gates what the system is permitted to do right now.

If a visitor withdraws consent, the system must not merely stop collecting new signals. It must stop using the consented state immediately, and it must unwind any personalization state that was derived from that consent where policy requires it.

That unwinding should be visible in the same way other consequential decisions are visible: as operational evidence, not as a private promise.

A practical posture is to treat revocation as an event that propagates across the stack:

- It invalidates the current session linkage.
- It disables any personalization behaviors that depended on the revoked scope.
- It triggers deletion and purge actions for data categories that were retained under that scope.
- It produces a decision record showing what was revoked, what changed in behavior, what was deleted, and what remains retained under other lawful or operational bases.

This does not require the venue to log personal biography. It requires the venue to log that a change in permission state occurred, and that the system complied.

Rollback is how systems stay trustworthy after change

Most failures of governance do not arrive as dramatic breaches. They arrive after updates.

A content package changes. A policy file changes. A new model version is deployed. A logging configuration is modified. A vendor component is swapped. The venue remains "up," but behavior subtly shifts.

This is why rollback discipline is not only an IT practice. It is a public-space trust practice.

A venue should be able to revert:

- content versions that altered messaging, tone, or factual claims,
- policy and constraint versions that altered what actions are permitted,
- configuration that altered retention, logging, or identity handling, and
- any update that increased confabulation incidents or reduced abstain behavior.

Rollback is also part of privacy posture. If a change accidentally expanded retention, the venue must be able to roll back both the configuration and the unintended data created during the interval, then prove that it did so.

Deletion must include the side channels

In modern systems, retention creep rarely happens only in the primary database. It happens in the side channels.

A venue that claims deletion must treat the following as retention surfaces and govern them explicitly:

- caches and temporary stores,
- logs and telemetry streams,
- analytics warehouses and derived aggregates,
- backups, snapshots, and replicated copies,
- debugging artifacts retained "for later," and
- training or evaluation datasets for any model component.

If any of these retain sensitive state beyond declared limits, the venue has not implemented deletion. It has implemented delayed surprise.

A privacy-forward posture is not that these systems never exist. It is that they are bounded, access-controlled, retention-limited, and purgeable by design, with proofs that purge occurred.

Evidence of enforcement is part of governance maturity

A venue should be able to answer, with artifacts:

- When consent was withdrawn, what changed immediately?
- What data categories were deleted?
- Which subsystems confirmed deletion, and when?
- What remained, if anything, and under what declared basis?

These answers should be obtainable without reconstructing personal stories. They are about system behavior, not visitor biography.

This is why the evidence pack matters. Logging minimums exist so revocation and rollback are not "trust us" narratives, but observable, testable behaviors that can be commissioned and audited. See Appendix G for the logging and evidence posture required to make this practical.

If a venue needs tamper-evidence, it can anchor logs without making blockchain a prerequisite. A periodic hash commitment to an external ledger can support integrity checks later. The tool is not the point. The posture is the point: demonstrate integrity without collecting more than you need.

Operator override must be normal

The most dangerous systems are the ones that assume operators will rarely need to intervene. Operators intervene because reality is messy. A door sticks. A show is delayed. A VIP tour arrives. A maintenance crew is in a corridor. A staff member needs the system to stop rebalancing a queue for five minutes so they can solve something physically.

> *Override is not failure. Override is coordination.*

A well-governed system gives operators a clear view of what the system is proposing and why, a way to pause or constrain certain action classes, a way to approve higher-impact actions, and a way to return to normal mode without confusion. The goal is not to keep humans out. The goal is to keep humans responsible while the system reduces burden.

Verification is governance's quiet partner

A system can be perfectly governed and still fail if it cannot confirm outcomes. If it powers down a zone and fails to restore it, the visitor does not care that the policy was correct. The visitor sees a broken moment.

Operational compatibility

Treat identity tooling as a capability, not a dependency.

- What works fully offline: session continuity, core guidance, captions, and language selection.
- What can be cloud-assisted: advanced retrieval, long-form content, and model improvement.
- What must degrade gracefully: any feature that touches safety, access, or critical show operation.

Implementation pointers: see Appendix K, Appendix G, and Appendix J.

This is why the governance loop must include proof that actions took effect, and proof that the environment remains in the state the model believes it is in. Verification and trend analytics close the loop between intent and reality. They allow the venue to catch drift early and prevent failures rather than apologize for them.

That is efficient. It is also humane.

A practical governance template for a project meeting

If you want to make governance real quickly, use this structure in a working session. It fits on a whiteboard.

1. **Outcomes**
   List five to eight outcomes that describe what a good day looks like.
2. **Hard constraints**
   List ten to fifteen hard constraints across safety, accessibility, privacy, fairness, and show integrity.
3. **Allowed actions**
   List the actions the system is allowed to take, and mark each as autonomous, advisory, or requires approval.
4. **Modes**
   Define normal, busy, and incident modes and specify how priorities and allowed actions change.
5. **Proof and override**
   Define what gets logged, how long it is retained, how operators view it, and how override works.

If you can complete that exercise with the people who will operate the venue, you have begun real governance. If you cannot complete it, the system is not ready to behave at macro scale, no matter how impressive the prototype looks.

Governance is how a place becomes more capable without becoming less human

The promise of a WorldModel Venue is not that it will be clever. The promise is that it will be coherent. Coherence in public space requires power, restraint, and accountability at the same time. Governance is the mechanism that makes that combination possible.

In the next chapter, we move from governance as daily discipline to the identity stack that can support trust without oversharing. When a system offers entitlements, age gates, membership benefits, and cross-venue continuity, it needs to answer questions with "yes" or "no" without demanding full documents. That is where privacy becomes a technical capability as well as a posture.

Part VI

Identity, Consent, and Agent Layers

This part deals with identity and agency. The goal is not hype. The goal is to give visitors a way to carry context across moments and places, while keeping consent real and minimizing what the venue must retain.

Chapter 27 covers Portable Identity (Web3 in industry terminology) and the practical problem of proving without oversharing.

Chapter 28 explores Agent Coordination (Web5) and AI Negotiation (Web6), where the visitor's device becomes an active participant rather than a passive screen.

Chapter 29 covers Consented Context (Web4), emphasizing consent-driven prosody and the "comfort dial", the idea that experience intensity, guidance style, and social posture should be adjustable by the visitor.

Chapter 30 focuses on spatial AR interfaces: what they can do well, what they do badly, and where they fit in a governed venue stack.

If your near-term plan is strictly on-premise experiences with minimal identity, you can skim this part and return when you want cross-visit continuity. If you do read it, read it as an options map: which mechanisms support consent, portability, and reversibility, without turning identity into a new surveillance surface.

This book uses four layer names - Portable Identity, Consented Context, Agent Coordination, and AI Negotiation - with Web3/4/5/6 shown parenthetically where these terms appear in industry RFPs. These are not claims about official Web versioning. See section 0.2 for full definitions.

In summary:

Portable Identity (Web3): Portable, cryptographically verifiable identity and claims, focused on DIDs and verifiable credentials - not cryptocurrency narratives. W3C DID Core defines DIDs and DID documents.[xxxix] W3C Verifiable Credentials Data Model v2.0 defines the core credential model and exchange roles.[xl]

Consented Context (Web4): Consent-driven human context and negotiated experience, where the visitor's device mediates posture, preferences, and disclosure boundaries in real time. This is an author-defined layer; Web4 is not a standards body term.

Agent Coordination (Web5): An identity and data ownership layer that builds on DIDs and verifiable credentials and adds decentralized data storage and message relay via decentralized web nodes.

AI Negotiation (Web6): A forward-looking layer for higher-order agent coordination across systems and venues, where governance, identity continuity, and outcome verification must remain enforceable. This is an author-defined layer; Web6 is not a standards body term.

What these labels are not:

- They are not claims that the industry has formally standardized "Web3", "Web4", "Web5", or "Web6."
- They are not claims that a venue must adopt a particular blockchain, token model, or ledger to implement consent-driven identity.

Implementation anchor: Regardless of labels, identity assurance and risk selection should follow established digital identity guidance, such as NIST SP 800-63-4[xli].

27

PORTABLE IDENTITY, AND THE ART OF PROVING WITHOUT OVERSHARING

There is a small ritual that happens every night in bars, restaurants, concert venues, cruise ships, stadiums, and theme parks. A person reaches into a wallet and hands over a driver's license. The bartender looks at it, nods, and hands it back. Sometimes the license is scanned. Who knows where that information goes.

In that exchange, the venue receives far more than it needs: a full name, a home address, a date of birth, a license number, and sometimes height, weight, and other details that belong nowhere near a casual transaction. The staff member does not want this information, and the patron did not intend to donate it. The ritual persists because the world has lacked a widely deployable mechanism for trust.

In a governed venue, the governance gate should treat identity claims as bounded inputs, not as permission to infer and retain.

In this chapter, "Web3" and "Web5" are used as shorthand for identity and credential patterns grounded in W3C DID Core and W3C Verifiable Credentials Data Model v2.0, while "Web4" and "Web6" are author-defined labels used only to organize consent and agent-mediated interaction concepts.

The useful promise of this stack is simple: demonstrate what matters and reveal nothing else.

This posture also prevents a common failure mode in intelligent environments: invented attributes.

A system that is allowed to guess will eventually behave as if it knows things about a visitor that were never proven. In identity and eligibility contexts, that is unacceptable. Portable Identity exists to replace narrative with proof: the system receives a bounded claim, tied to consent and scope, and it does not fill gaps with plausible stories.

When a fact matters, the system should prefer proof or explicit choice. When proof is not available, it should fall back to conservative defaults rather than inventing.

The "adult beverages" example is the point

A venue does not need your biography to serve you a drink. It needs one fact: you are above the legal age threshold. If the venue can receive that fact as a reliable yes or no, the interaction becomes safer, faster, and less vulnerable to identity theft.

The "Yes/No" Identity Handshake

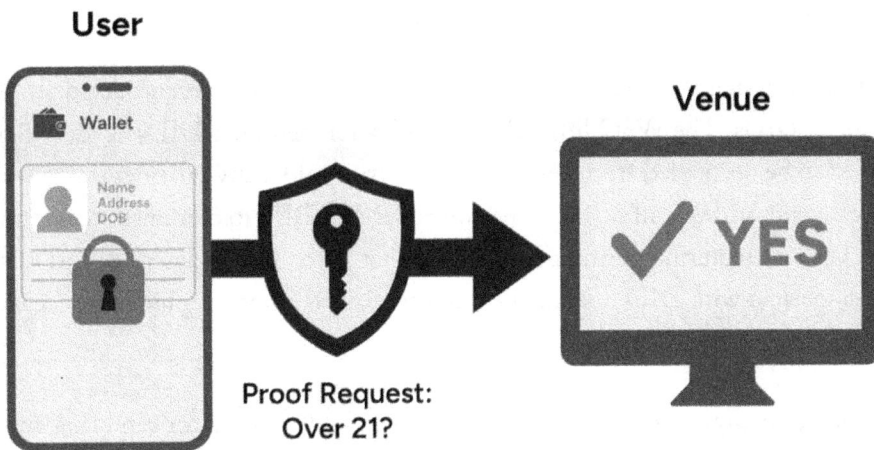

User

Wallet

Name
Address
DOB

Proof Request:
Over 21?

Venue

✓ YES

The venue receives the answer, not the document

W3C's Verifiable Credentials Use Cases describes exactly this scenario.[xlii] A customer can demonstrate they are over 21 without revealing their date of birth, home address, or state ID number.

That single example is worth dwelling on because it captures what changes for physical venues. Identity shifts from a document you surrender to a claim you present selectively.

The shift from documents to claims

A driver's license is a document designed for human inspection. A verifiable credential is a cryptographically secured way to express claims issued by a trusted issuer, then presented by a holder to a verifier. The important point is not the file format. The important point is the relationship.

W3C's Verifiable Credentials Data Model v2.0 describes an ecosystem with roles that map directly to venue reality: issuer, holder, and verifier[xliii]. It also explains that holders can generate verifiable presentations and share them with verifiers to demonstrate they possess credentials with specific characteristics.

That is the architectural move that makes selective disclosure plausible in public space.[xliv] The presentation can be smaller than the credential.

DIDs: identifiers that do not require permission

Underneath verifiable credentials sits a second building block: decentralized identifiers, DIDs. The W3C DID Core specification describes DIDs as identifiers designed to be decoupled from centralized registries[xlv], identity providers, and certificate authorities, and to allow the controller of a DID to demonstrate control over it without requiring permission from another party. It also describes DIDs as URIs associated with DID documents that can express cryptographic material and services.

For venues, the practical implication is straightforward. The visitor can bring an identifier they control, rather than being forced into a venue-issued username and password relationship for every interaction. A visitor can also use more than one DID across contexts, which reduces the pressure to carry one universal identifier everywhere.

Verifiable presentations and the "yes or no" world

The most important capability in this stack is the presentation. W3C's Verifiable Credentials Data Model v2.0 defines a verifiable presentation as a tamper-evident presentation of information, and notes that certain presentations can contain data synthesized from credentials, such as zero-knowledge proofs.

This is the doorway to a better public life. The bar interaction becomes what it should have been: a check, and a yes or no to the question of whether service is permitted. No home address exchanged. No birthdate revealed. No license number copied. No temptation to glance at details that are none of anyone's business.

This pattern applies to many venue interactions where eligibility is binary and documents are dangerously rich:

- age-gated attractions, and age-gated retail
- restricted areas, such as back-of-house tours, VIP access, or safety-controlled zones
- membership entitlements
- staff credentials, and contractor access
- jurisdictional requirements that can be satisfied by a limited proof rather than full identity disclosure

Once you design around yes or no proofs, over-collection becomes harder to justify.

Blockchain is optional

At this point, many people assume blockchain is required. W3C's DID Core specification is explicit that it does not presuppose any particular technology or cryptography underpinning the generation, persistence, resolution, or interpretation of DIDs. It also notes that many, but not all, DID methods use distributed ledger technology or other decentralized networks.

This matters because it frees venues from ideology. A venue can be pragmatic. If the venue wants tamper-evident auditability, an external ledger can be useful as a logging substrate. For example, a venue can anchor hashes of decision log batches periodically to demonstrate integrity later, without storing sensitive visitor information on a public chain. The identity exchange itself does not require a public ledger.

This posture aligns with the covenant of this book: capability without unnecessary risk.

Why this belongs in a WorldModel Venue

It is easy to treat identity as separate from experience, as if it belongs only to checkout or access control. In a WorldModel Venue, identity becomes part of operational truth only in the minimal ways justified by service.

The venue needs to know what the visitor asked for, such as language, depth, modality, and comfort. It may need to know what the visitor is entitled to, if the visitor opted in. It generally needs to know what constraints apply, including safety, accessibility, and policy. It does not need to know more than that.

DIDs and verifiable credentials help keep that boundary clean. They allow a visitor to carry credentials in a wallet and present proofs only when needed, in the moment, with minimal disclosure. They also allow the venue to avoid becoming a warehouse of sensitive data it rarely wanted to store and cannot defend indefinitely.

The phone becomes a trust device

Chapter 1 argued that phones are now a default interface and that venues should design for phones without requiring phones. Minimal-disclosure credentials extend that argument in a more specific direction. The phone becomes a trust device. It is where credentials live, where consent is expressed, and where a visitor can see what they are sharing.

For physical venues, this creates a rare alignment of incentives. The visitor reduces exposure and risk. The venue gets stronger verification with less liability. Staff get simpler point-of-service interactions and fewer arguments.

The result is safer and calmer, and calm is a currency of good public space.

A standard for dignity

Identity systems fail when they feel like control. They succeed when they feel like dignity. A visitor should not feel that a venue's first instinct is to demand documents. A venue should feel as if it is asking only what it needs, and only when it has earned the right to ask.

That is the standard held throughout this book. Privacy is not a compliance layer added at the end. Privacy is a design requirement that determines whether venue-scale capability can exist in public.

In the next chapter we move from credential proofs to the next step in the trajectory: a world in which the visitor's phone, and the visitor's on-device agent, negotiates with the venue on the visitor's behalf, carrying preferences, credentials, and consent forward without turning public spaces into document-checking factories.

28

AGENT COORDINATION: FROM PHONE AGENTS TO AI-TO-AI NEGOTIATION

The phone began as a device you carried. It became a platform you lived through. The next shift is quieter, and it reshapes public space at a deeper level. The phone becomes an agent.

Agent coordination introduces a new class of confabulation risk.

When an agent represents a person, it can be tempted to fill gaps with plausible assumptions about intent, comfort, and priorities. Confabulation shows up here as confident representations that are not grounded: inferred preferences treated as facts, or inferred reasons treated as truth.

The discipline in this book is consistent: preferences are chosen, not invented. Inference may be used to propose options, but it must be labeled as uncertain, it must be correctable by the visitor, and it must never become an authority that overrides declared preferences or accessibility constraints.

Agent here means something practical. A piece of software on your device can hold your preferences, hold your credentials, negotiate what you share, and answer venue requests on your behalf with consent and minimal disclosure. This is not a replacement for human judgment. It is a way to make routine public-space interactions calmer, faster, and less exposed.

This chapter describes that shift, and why "Web5" is useful language for it.

The labels in this chapter are convenient shorthand, not a claim of standards authority.

The patterns that matter are stable: minimal disclosure, consent, bounded claims, and refusal to infer when it is not required. If the shorthand distracts, ignore the label and keep the posture. The substance is a cleaner social contract between visitor and venue.

Web5 overviews frame the ambition as user-controlled identity and user-controlled data, built around decentralized identifiers (DIDs) and decentralized web nodes (DWNs). For venues, the value is operational dignity. A visitor should not need to surrender documents, create throwaway accounts, or install yet another app to be served, guided, included, and respected.

A note on terminology

In this chapter, "Web3", "Web5" are used as shorthand for identity and credential patterns grounded in W3C DID Core and W3C Verifiable Credentials Data Model v2.0. "Web4" and "Web6" are author-defined labels used only to organize consent and agent-mediated interaction concepts.

From apps everywhere to negotiated interactions

An app can be excellent and still be too much to ask of a casual visitor. It adds friction at the threshold of entry, when a person is deciding whether the place feels easy or difficult. It also creates fragmentation: every venue becomes a separate account, a separate privacy policy, a separate set of permissions, and a separate set of settings that the visitor rebuilds again and again.

A negotiated posture flips the relationship. The phone becomes the stable layer, and venues become interoperable requesters rather than permanent custodians. In practical terms:

- The visitor carries a wallet, a preference store, and an agent on their own device.
- The venue requests a specific, minimal item for the moment.
- The visitor approves, or declines.
- The venue receives a minimal proof, or minimal data.
- The interaction proceeds without turning the venue into a long-term custodian of personal life.

One additional rule makes negotiated interactions defensible: bounded claims, not stories.

The phone agent should respond with minimal proofs, minimal attributes, or explicit choices. It should not narrate a person. The venue agent should request only what it needs, and it should not treat negotiated signals as permission to infer and retain.

This is how agent coordination stays humane: it avoids both surveillance and the quieter failure mode where systems become confident about people without evidence.

This is pro-visitor, and it tends to increase adoption because it reduces friction and reduces regret.

DWNs: the personal data store that makes the phone a boundary

Phygital continuity depends on a visitor-controlled boundary. This is where DWNs become enabling infrastructure rather than ideology.

DWN, Decentralized Web Node[xlvi], is defined by the Decentralized Identity Foundation (DIF) as a mesh-like datastore construction that enables an entity to operate multiple nodes that sync state across one another, allowing the owner to secure, manage, and transact their data without reliance on provider-specific infrastructure. DIF's DWN user guide also describes a DWN as a decentralized personal and application data storage and message relay node, and notes that users may have multiple nodes that replicate their data between them.

If the visitor's DWN is on-device, or controlled by keys on-device, the visitor has a practical control point. Venues and apps do not automatically own the profile. They request permissioned access, scoped in time and scope, and the visitor can revoke it.

Alternative approaches can reach similar goals. Solid Pods use user-granted permissions to read and write personal data stores. Encrypted Data Vaults store encrypted data so the storage provider cannot read it. Centralized profiles, such as SSO with a vendor cloud, can be simpler, and they tend to keep the visitor-side of the bargain inside provider silos rather than under visitor-controlled infrastructure. The structural requirement is the same. A system cannot negotiate responsibly if it has nowhere to keep the visitor's side of the bargain.

A DWN can hold language preferences, depth modes (Streaker, Stroller, Student), interest vectors by vertical, accessibility and comfort modes, and a record of what the visitor chose to share with which venue, for how long, and why. The venue can still operate anonymously for most visitors. The DWN becomes relevant when a visitor wants continuity without surrender.

DIDs: the handshake that does not require an account

Negotiation requires a way for both sides to identify a counterparty without creating venue-specific usernames and passwords. DIDs are the standard vocabulary for that.

W3C DID Core defines decentralized identifiers as a new type of identifier designed to be decoupled from centralized registries, identity providers, and certificate authorities, and to allow the controller to demonstrate control over the DID. In a venue context, this enables a clean handshake:

- The visitor's agent presents a DID under the visitor's control.
- The venue presents a DID representing the venue service.
- Trust can be established, and credentials can be presented, without a new account silo.

The details of DID methods do not belong in this book. The design implication does. A venue can request proofs and receive them without becoming the identity provider for the visitor's life.

Verifiable Credentials: the "yes or no" posture

Chapter 27 established the core moral of modern identity for public space: a venue should receive answers, not dossiers. The W3C Verifiable Credentials Data Model v2.0 is a W3C Recommendation (15 May 2025). It defines verifiable presentations and supports selective disclosure patterns, including approaches that can synthesize proofs from credentials.

For venues, the operational pattern is simple. The venue asks a question and receives a minimal proof. The visitor does not surrender a document. The "adult beverages" example from W3C use cases captures the essence: demonstrate age eligibility without revealing date of birth, address, or ID number.

This applies to any scenario where the venue needs a binary decision, and documents are dangerously rich:

- age checks,
- entitlements,
- membership status,
- accessibility accommodations that a visitor chooses to persist, and
- any case where a yes or no is sufficient.

Venue AI and phone AI: a division of labor

Once you accept the phone as the visitor's agent layer, the venue's intelligence takes a healthier shape. The venue side becomes environment intelligence, responsible for operational state, proposing actions, and maintaining coherence under constraints. The phone side becomes personal intelligence, responsible for preferences, credentials, consent, and comfort boundaries.

Each side does what it is suited to do. The venue is good at state: occupancy, schedules, closures, acoustic conditions, and operational constraints. The phone is good at the personal layer: language preferences, sensory comfort, and proofs of entitlement.

When the two negotiate, the experience becomes more helpful and less invasive. The venue can ask, "Do you want guidance to the next gallery in your preferred language?" The phone can respond with the chosen mode without requiring the venue to store the preference permanently. The venue can ask, "Are you entitled to this timed entry?" The phone can answer yes or no without sharing address, birthdate, or the rest of the document that invites identity theft.

This is operational elegance as well as privacy. It reduces the number of systems a venue must defend, and it reduces the number of disclosures the visitor must regret.

A practical pattern for negotiated experiences

The pattern is simple enough to state, and it scales across verticals.

1. Request
   The venue requests a specific, minimal item: a preference, a proof, or a mode.
2. Negotiate
   The phone agent presents what is needed and nothing else. The visitor approves or declines.
3. Act
   The venue executes the outcome through the right plane: surface, room mix, Personal Channel, or staff prompt.
4. Forget
   The venue retains only what operations require, for as short a time as possible, and treats deletion as a normal behavior.
5. Audit without dossiers
   Where audit integrity matters, the venue logs decisions at the level of state and actions, not personal biography, and can make integrity tamper-evident if desired.

This is the practical form of privacy-forward macro architecture. It is also why Web5 is relevant as a direction: interoperable, user-controlled negotiation rather than venue-controlled profiles.

Offline-first, because venues are real places

Negotiated experiences must survive physical reality. Venues have dead zones, interference, and moments when the network degrades. A system that assumes perfect connectivity fails precisely when the venue is busiest.

This is where the DWN concept is useful. DIF describes DWNs as enabling multiple nodes that can sync state across one another. That implies a resilience posture in which the visitor's preferences and credentials can remain accessible through a personal store, and the venue can execute locally. The chapter does not prescribe a single implementation. The posture is the point. The experience should remain dignified under imperfect conditions.

Cross-vertical use cases without changing the pattern

Once you have the pattern, it travels.

Museums and cultural venues
A visitor's phone agent holds language, depth mode, and comfort mode. The venue requests only what it needs to deliver the next moment. The visitor does not need an account for basic participation.

Cruise ships and resorts
Entitlements and reservations can be proven as yes or no, and guidance can be delivered without turning the ship into a document-checking factory.

Retail and supermarkets
A visitor can opt into assistance, such as preferred language and accessibility modes, without creating a permanent marketing identity. If loyalty is desired, it can be a credential the visitor holds rather than a record the store owns.

Smart cities and civic spaces
Public infrastructure can deliver service while avoiding a default posture of collecting personal life. Eligibility and entitlements can be proven with minimal disclosure, and guidance can be offered without turning movement into surveillance.

The more public the environment, the more valuable this posture becomes.

The boundary that keeps it humane

A phone agent negotiating with a venue agent is powerful, and it must remain voluntary. The covenant of this book applies here with special force:

- The venue works beautifully without opt-in.
- Opt-in is a benefit exchange, not a toll.
- A visitor can say no and still participate.
- Privacy is not punished with worse service.

A system that violates these principles will not be trusted and should not be deployed. A negotiated experience that follows them increases trust because it signals that the venue is not hungry. It is simply helpful.

Acceptance tests to require for negotiated experiences

1. Opt-out does not punish. The venue works well without opt-in, and refusal does not degrade inclusion.
2. Minimal disclosure is real. The venue requests only what it needs for the moment, and requests are time-bounded.
3. No invented attributes. The phone agent does not generate new claims about the visitor as facts. Inference is treated as advisory, correctable, and subordinate to explicit preference.
4. Confabulation containment. When state is missing or ambiguous, both sides abstain from confident claims, fall back to conservative defaults, or escalate to human assistance rather than inventing.

Web6, Beyond Scripted

Web5 agents are powerful but limited. They coordinate based on explicit preferences and declared capabilities. They cannot infer what a visitor might want but hasn't stated. They cannot adapt to patterns that emerge during a visit. They execute protocols; they do not reason.

Web6 introduces AI systems that reason. The visitor's AI observes patterns, infers preferences, and represents visitor intent - not just stated preferences, but also understood goals. The venue's AI dynamically understands its own capabilities and can propose experiences the visitor hasn't explicitly requested.

This is the emerging frontier.

Today, this direction is speculative, but the direction is clear. Inference is also where confabulation risk rises.

The moment an agent begins to infer intent, it can produce confident explanations that are not grounded. That is why AI-to-AI negotiation must remain governed by the same constraints as the rest of the architecture: bounded outputs, explicit uncertainty, and refusal to invent when state is missing.

In practice, the AI layer can propose, but governance still decides. A proposal that cannot be justified against declared preferences, consent, and operational truth should not be executed.

As AI capabilities on personal devices increase, the negotiation between visitor and venue will increasingly occur at the inference layer.

The Visitor's AI

A visitor's AI runs on their personal device - phone, AR glasses, or future wearables. It builds a model of the visitor through observation and feedback:

Behavioral Patterns

The AI observes (with consent) how the visitor engages with experiences. Do they linger at interactive exhibits? Do they skip text-heavy panels? Do they prefer social history over military history? Do they engage more deeply in the morning than afternoon?

Implicit Feedback

The AI interprets signals that the visitor doesn't explicitly provide. Dwell time indicates interest. Quick exits indicate disengagement. Return visits indicate satisfaction. Skipped recommendations indicate preference mismatch.

Explicit Refinement

The visitor can correct the AI's model. "I skipped that exhibit because I was hungry, not because I wasn't interested." The AI incorporates corrections and improves its model.

Over time, the visitor's AI develops a rich representation of their interests, preferences, and patterns - one that far exceeds what any preference form could capture.

The Venue's AI

The venue's AI maintains a dynamic model of venue capabilities and conditions:

Real-Time State

Current occupancy, queue lengths, show timings, exhibit availability, and environmental conditions.

Experience Matching

Understanding of which experiences serve which visitor interests, beyond simple category tagging.

Capacity Optimization

Awareness of how routing and timing decisions affect overall venue flow.

Emerging Patterns

Recognition of visitor engagement patterns that suggest unmet needs or opportunities.

The venue's AI doesn't just respond to requests. It anticipates needs, identifies opportunities, and proposes experiences that visitors might not have known to request.

The AI-to-AI Conversation

When a visitor with an AI-equipped device enters a venue with AI systems, a conversation begins. Consider a visitor wearing AR glasses entering a history museum. Their AI has learned their patterns: they prefer interactive exhibits, respond well to narrative framing, engage deeply with social-history themes, and typically sustain high engagement for about two hours before energy flags.

The visitor's AI initiates contact with the museum's AI. The conversation might proceed:

Visitor AI: "This visitor prefers interactive, narrative-focused experiences. Social history themes resonate strongly. They have shown interest in daily-life topics rather than military or political topics. Current engagement energy suggests approximately 90 minutes of focused attention remain."

Venue AI: "Understood. Current conditions: the social-history wing has moderate occupancy and low wait times at interactive stations. I can queue up a narrative-driven path through three interactive exhibits and a culminating experience, timed to complete in 80 minutes. The path avoids the military gallery and prioritizes daily-life themes. Does this match the intent?"

Visitor AI: "Confirmed. Additional context: this visitor responds well to unexpected discoveries. Consider including an off-pattern recommendation if confidence is high."

Venue AI: "Noted. I'll include the hidden gems corner of the textile gallery: it has interactive elements and strong narrative content, though it's categorized as craft history. Confidence: 0.7 that this visitor will appreciate it."

Neither AI has shared the visitor's identity, nor has it exposed raw behavioral data. Coordination occurs through intent and inference.

Privacy in AI-to-AI Communication

Web6 amplifies both the opportunity and the risk. AI systems that infer intent can deliver remarkable personalization. They can also enable remarkable surveillance if poorly architected.

The privacy requirements for Web6 are strict:

- Visitor AI processes raw behavioral data locally; only derived intent is communicated.
- Venue AI receives intent signals, not behavioral traces.
- Neither AI stores the conversation beyond the current session.
- Visitors can inspect what their AI has communicated.
- Visitors can constrain what categories of inference their AI may share.
- Venue AI does not attempt to reverse-engineer visitor identity from intent signals.
- AI-to-AI communication is encrypted and authenticated.

The principle: the visitor's AI is their advocate, not the venue's spy. It shares what serves the visitor, in the way the visitor authorizes.

Integration Requirements

For spatial AR and AI-to-AI coordination to be coherent with venue experience, they must integrate with venue systems:

Spatial Registration

AR content must be accurately anchored to physical locations. This requires spatial mapping that aligns with venue coordinate systems.

Content Coordination

AR content should complement rather than conflict with physical content. This requires coordination between AR content management and venue content operations.

State Synchronization

AR layers should reflect the venue's current state. If an experience is closed, the AR layer should be aware. If content changes, the AR layer should update.

Governance for AR and AI

AR content and AI recommendations displayed in the venue should be subject to venue governance:

- Content approval:
  what AR content is permitted in venue spaces
- Experience coherence:
  ensuring AR and AI enhance rather than conflict with venue experience
- Safety:
  preventing AR content that could create hazards
- Privacy:
  ensuring AR devices do not capture other visitors without consent

Current Limitations

Web6 is not production-ready today. Current limitations include the following:

- Device capability:
  On-device AI inference is improving rapidly but remains limited compared to cloud systems. True Web6 requires AI capable of nuanced intent inference, running on personal devices.
- Protocol maturity:
  No standard protocols exist for AI-to-AI intent communication. Early implementations are proprietary and non-interoperable.
- Trust frameworks:
  No established methods exist to verify that a venue's AI will respect privacy commitments. Visitors must trust venue claims.

These limitations are temporary. Device capabilities are improving at an accelerating pace. Protocol standardization will follow deployment experience. Trust frameworks will emerge as the industry matures.

Preparing for Web6

Venues building systems today should prepare for Web6 even while implementing Web3, Web4, and Web5:

- Architect systems with AI integration points, even if current AI capabilities are limited.
- Design privacy frameworks that can accommodate AI-to-AI communication.[xlvii]
- Build visitor agent interfaces that can evolve from scripted protocols to AI reasoning.
- Collect the operational data that will train venue AI systems when capabilities mature.
- Monitor emerging standards for intent-based communication protocols.

The chapter's conclusion

Web5 is best understood as a banner for a direction: user-controlled identity and user-controlled data, with DIDs and DWNs as core building blocks. For venues, the value is a cleaner social contract. A visitor should not need to surrender a driver's license to buy a drink. A visitor should not need to create a new account to receive captions. A visitor should not need to hand over a personal life to be treated with respect in public space.

The phone agent is one of the most credible ways for those conditions to become normal.

In the next chapter, we take negotiated experiences out of the abstract and put them into space. If Web5 makes the phone the visitor's agent, the next question is how that agent expresses outcomes in the real world without turning the venue into signage clutter or forcing every visitor into an app. That is where Spatial AR Interface comes in: an opt-in spatial lens that renders governed WorldModel truth as guidance, meaning, accessibility cues, and layered interpretation, in the visitor's language and preferred depth, without baking those layers into permanent walls.

This pattern is privacy-forward and operationally useful. It reduces sign-in friction, reduces liability, and makes cross-venue continuity possible without creating a single database that someone will eventually breach.

Example: a museum issues a consent credential at entry that includes language preference, accessibility needs, and an age parameter. Exhibits request only what they need. The visitor's device responds with proofs, and nothing else. If the visitor revokes consent, the proofs stop.

In venues, this avoids central data hoarding. A kiosk does not need a profile database to learn that a visitor prefers Spanish or that they require step-free routes. It needs a verifiable proof for that moment. The proof can be short-lived, consented, and logged as operational evidence rather than a biography.

In the issuer - holder - verifier model, an issuer signs claims, the holder keeps them in a wallet, and a verifier checks proofs when needed. The key implementation detail is selective disclosure: the holder can present only the minimum necessary claim without revealing the full credential.

DWN gives the visitor a boundary. Verifiable Credentials give the boundary a practical trust mechanism.

29

CONSENTED CONTEXT: PROSODY, EMOTION, AND THE COMFORT DIAL

A venue can be immersive without being overwhelming. That sentence sounds obvious, and in practice it is difficult. Many modern experiences are built on intensity, and intensity is a blunt instrument. It creates peaks quickly. It also excludes quietly. The visitor who loves thrill stays. The visitor who becomes overloaded leaves, often without complaint, and often without anyone realizing the experience failed them.

Web4, as the term is used in this book, is not a claim that machines can read souls. It is a practical posture: a venue can adapt to human state only when the human grants permission, and only when the system treats that state as uncertain, bounded, and subordinate to safety, inclusion, and agency. Web4 is the extension of good design into an explicit comfort layer. It does not replace good design.

The key word is explicit. This book does not advocate systems that silently infer emotions and then act as if they know the truth of a person. It advocates systems that give people control over intensity, loudness, pacing, and social demand, and then use consent-driven feedback to improve hospitality. That is the Web4 posture: consent, control, and restraint.

Confabulation risk is especially sharp in human-state systems.

A system can easily produce confident interpretations of stress, engagement, fatigue, or intent that are not warranted by the signals. That is confabulation expressed as diagnosis. In public environments, it is unacceptable to treat inferred emotional state as truth about a person.

The defensible posture is: treat state inference as uncertain, bounded, and reversible. The visitor can override it. The system can use it only inside explicit constraints. When uncertainty is high, the system should abstain from intense adaptations and default to conservative, visitor-controlled comfort settings.

Inference and feedback are different tools

A system can respond to human state in two broad ways.

One is inference. The system observes signals such as voice tone, facial expression, and movement, then guesses what the person feels.

The other is feedback. In governed venues, feedback is the preferred tool because it replaces guesswork with agency. If a slider can solve it, use a slider.

In public spaces, feedback is the default for ethical and operational reasons. Inference is uncertain, and uncertainty is risky when actions have consequences. Feedback is explicit, and explicit signals can be governed, audited, and constrained. Feedback also preserves dignity. A visitor is not being analyzed. They are being offered control.

A simple rule follows. If an experience can be improved with a slider, a button, or a clear opt-in choice, implement that first. If inference is used at all, treat it as weak evidence, not truth, and keep it local and ephemeral.

The comfort dial: a new kind of interface

The most useful Web4 capability is not an emotion detector. It is a comfort dial.

A comfort dial allows a visitor to choose the intensity of an experience and change it midstream without embarrassment. It can be implemented through a personal device, a wearable, a kiosk, a staff tool, or a small set of physical controls, depending on the venue.

A comfort dial can include:

- Intensity: low, medium, high
- Audio load: room mix, narration-forward, quiet mode
- Visual load: reduced strobes, reduced motion density, calmer transitions
- Social load: more guidance, less surprise, more autonomy
- Language and modality: personal audio, captions, sign language video, and other modes described earlier in this book

This is access, not coddling. It expands participation without flattening the core experience. Thrill seekers still get thrill. Visitors who need a calmer path get a calmer path. A family can keep a day coherent even when one member has a different tolerance. The venue stops forcing one nervous system on everybody.

Haunted houses and Halloween

Intensity is the product. That is the point. Haunted experiences also reveal a truth that applies everywhere: the same intensity can be delight for one person and distress for another.

A Web4-ready haunted experience does not need to guess fear. It can allow the visitor to set a scare level. A scare level system can be as simple as:

- a slider at entry, and
- a mid-experience control that allows lowering intensity instantly.

The system can then adapt within guardrails:

- actor proximity rules,
- audio levels,
- lighting intensity,
- frequency of jump cues,
- routing toward calmer paths, if available, and
- content selection, such as which scenes are triggered.

The visitor remains in control. The venue remains responsible. This is humane, and it reduces risk. A venue that gives visitors agency is more likely to produce excitement without crossing into regret.

Prosody, voice, and the Personal Channel

Prosody, the rhythm, pace, and tone of speech, is often associated with emotion. In venues, prosody matters in two ways: as an output and as an input.

The output side is the simpler win. A venue can offer a conversational layer, such as a docent interface, that adapts its tone and pacing to the visitor's selected mode: concise for Streakers, more context for Strollers, deeper detail for Students. It can also offer calmer delivery for visitors who choose a comfort mode, and more energetic delivery for visitors who choose a thrill mode. This is not emotional manipulation. It is a hospitality choice.

The input side, prosody as signal, requires restraint. If the visitor opts in, the system can treat prosodic cues as a weak, temporary indicator that the visitor might be overwhelmed, confused, or excited. It should not treat that as proof, and it should not retain it beyond what the moment requires.

A privacy-forward pattern is straightforward:

- process prosodic signals locally, preferably on the visitor's device,
- convert the result into a minimal suggestion such as "calm mode suggested," and
- require an explicit accept or decline.

Vertical deployment quick-start template

Use this template to make the vertical chapters commissioning-ready.

- Outcomes:
  list the top five outcomes you are optimizing.

- Minimum architecture:
  which layers are mandatory for this site.

- Acceptance tests:
  what must be proven at handover, and what is monitored thereafter.

The visitor remains the authority on themselves.

The governance boundary

Web4 inputs do not override safety and policy. They can inform choices inside a safe envelope. This is where CGL becomes practical.

The system can propose an adaptation, shift to narration-only, offer captions, reduce intensity, or suggest a calmer route, but actions still pass through explicit constraints: safety constraints, accessibility constraints, privacy constraints, fairness constraints, and show integrity constraints.

In plain language, the venue is allowed to be kind, and it is required to remain predictable. Governance prevents comfort systems from turning into private exceptions that quietly break the shared experience. It also prevents comfort systems from being used as levers for manipulation. A visitor should be able to tell whether the system is helping or pushing. In a governed system, help is the default.

A preference-first toolkit across verticals

The haunted house makes the pattern vivid, and the pattern generalizes.

Museums and cultural venues
Comfort modes reduce sensory fatigue, especially in spaces with high audio density or visual complexity. A visitor can choose quiet narration, captions, or calmer pacing without asking staff.

Theme parks and immersive attractions
Guests can choose an intensity profile for a ride, a walkthrough, or a show. The system can route them toward experiences that match intent without segregating them.

Retail and supermarkets
Visitors can choose language, guidance level, and a calmer assistance mode. Staff tools can respect these preferences without requiring personal questions in public.

Cruise ships
Guests can set comfort and language profiles once, then use the Personal Channel to receive consistent guidance, quieter mixes, and accessibility support as they move between venues and decks.

Cities and civic spaces
Comfort mode becomes a public service: quieter routes, clearer wayfinding, and reduced sensory overload, offered as a choice rather than inferred surveillance.

The vertical changes. The posture stays the same: consent-driven, preference-first, and privacy-forward.

Commitments that make Web4 safe and worth doing

A venue that wants Web4 capabilities should commit to these rules early. They are design discipline.

1. **Preference first**
    If a slider can solve it, use a slider.

2. **Explicit consent**
    No hidden effect sensing. No surprise collection. No silent analysis.

3. **Local processing by default**
    Treat the visitor's device as the privacy boundary whenever possible.

4. **Minimize and forget**
    Comfort signals are ephemeral. The system does not build biographies.

5. **Governed actions only**
    Adaptations stay inside explicit constraints, with auditability and operator override.

6. **No confabulated diagnosis**
    The system does not claim to know what a visitor feels. It treats inferred state as uncertain, allows override, and defaults to conservative settings when confidence is low.

7. **Visible agency**
    The visitor can see, change, and disable comfort modes instantly.

If you follow these commitments, Web4 becomes what it should be: a way to make experiences more human, not more invasive.

30

SPATIAL AR INTERFACES

This chapter introduces a fourth delivery plane. Earlier chapters described three: the room plane for shared experience, the surface plane for programmable canvases and signage, and the Personal Channel for private, synchronized delivery. Spatial AR adds the ability to overlay guidance, meaning, and accessibility cues directly onto the visitor's view of the physical environment - without baking those layers into permanent signage and without requiring the visitor to look away from the space to consult a separate screen.

But AR carries a prerequisite. It is useful only when it is correct, and correctness at scale requires the governed operational truth established in Part V. It is trustworthy only when the visitor's privacy is protected, which requires the consent and negotiation architecture established in Chapters 27–29. Spatial AR is therefore not an independent layer. It is a delivery surface that becomes possible - and responsible - only when WorldModel truth and phone-agent consent are already in place.

Spatial AR is also a confabulation amplifier.

A wrong statement on a screen is a nuisance. A wrong statement anchored into the world can become an instruction. If a floating arrow points the wrong way, or an overlay asserts an exit is open when it is not, the system has not merely misspoken. It has misled.

That is why AR must be a rendering of governed operational truth, not a storytelling layer. It must not invent. When state is uncertain or stale, AR should communicate uncertainty, reduce specificity, and fall back to safer planes such as surfaces and staff tools.

The most persistent fantasy in venue design is that the perfect sign will solve everything. It will not. Signs are static. Venues are dynamic. Signs are scarce in physical space. Language needs are not. Signs force compromises between readability and density, and visitor attention does not cooperate.

The result is familiar. A building becomes cluttered with guidance, disclaimers, and bilingual compromise, and visitors still feel lost. They still miss context. They still fail to find the experience they would have loved if the venue had met them clearly, in their language, at their pace.

Spatial AR Interface exists for a different premise: a spatial interface can make the venue legible without making it visually noisy.

It is the AR-facing interface layer of the architecture, a way to overlay guidance, meaning, accessibility cues, and personalized layers onto the real environment without baking those layers into permanent physical signage. It does not replace good physical wayfinding. It sits above it as an opt-in tool that makes the environment more intelligible and more inclusive.

This chapter places Spatial AR Interface where it belongs: as a client of governed operational truth. AR is useful only when it is correct. Correctness at scale requires WorldModel state, and behavior at scale requires governance. Spatial AR Interface is not an independent storyteller. It is a spatial rendering of what the venue knows and what the venue is permitted to do.

Operational claims in AR must be grounded in authoritative state within its freshness window, or withheld.

The AR posture that stays humane

AR can easily become a gimmick: a floating label, a dancing arrow, a novelty layer used once and then abandoned. AR becomes valuable when it reduces real friction without creating new friction.

The humane posture has four rules:

1. **Opt-in, always**
   Spatial AR Interface is a benefit, not a gate. The venue works beautifully without it.
2. **Privacy-forward, by construction**
   AR does not require identity to be useful. It should default to local processing and minimal retention.
3. **Governed truth only**
   Spatial AR Interface renders what governance allows from WorldModel operational truth. It does not invent.
4. **Assist first, decorate second**
   The primary job is clarity, accessibility, and comfort. Spectacle is optional and must not undermine trust.

These rules protect the visitor and the venue. AR can increase trust or destroy it.

A spatial query interface, not a map

A Spatial AR Interface is not a map. It is a spatial query interface.

Traditional navigation treats the visitor as a dot on a map. That model works until it does not. In real spaces, visitors experience themselves as bodies facing a direction, interpreting crowds and sound, and trying to decide what is safe, what is open, what is worth their time, and what will not exhaust them.

A spatial interface can answer questions in the form visitors have:

- Where is the nearest restroom that is accessible?
- How do I reach the next gallery without stairs?
- Which route is calmer right now?
- What is that object, and why does it matter?
- What is the one-minute story here?
- What is the deep story here?
- Where is the next show that fits my time and comfort preferences?

A Spatial AR Interface is therefore best understood as a way to query WorldModel state through the physical environment itself. The visitor points the lens, and the venue answers in context. This chapter follows Web5 because once the phone becomes an agent that can negotiate preferences, AR becomes an output plane for that agent.

The three planes, now with a fourth

Earlier chapters described three planes of delivery:

- the room plane, shared experience
- the surface plane, Programmable Canvases and signage
- the Personal Channel plane, language and modality delivered privately

Spatial AR Interface adds a fourth plane:

- the spatial plane, AR overlays aligned to the environment

This plane has a clear virtue. It can deliver clarity without occupying physical wall space. That matters in multilingual and accessibility-heavy environments where printed solutions become clutter and compromise.

It also carries a clear risk. If the overlay is wrong, it is worse than useless because it actively misleads. That is why Spatial AR Interface must be bound to operational truth.

What Spatial AR Interface should render

A reference chapter needs scope. Spatial AR Interface is strongest when it renders four categories of information, derived from WorldModel state and bounded by governance.

1) Guidance and routing

- arrows and paths that reflect current route availability
- accessibility-aware routes as a normal option
- load-balanced suggestions that remain advisory rather than coercive
- time-to-arrive estimates that reflect current congestion

2) Context layers

- headline, context, deep layers aligned to the Streaker, Stroller, Student model
- object annotations that respect curatorial intent
- multilingual delivery as runtime rather than fixed bilingual compromise

3) Comfort and accessibility cues

- quiet route suggestions when acoustic state is high
- captions and personal audio availability cues
- sign language availability cues
- comfort mode toggles that are visible and reversible

4) Operational truth messaging

- closures and reroutes
- showtime and capacity truth
- safety notices and incident-mode guidance where governance allows

What it should not do by default

Spatial AR Interface should not default to behaviors that break trust:

- persistent personal tracking as a requirement for basic guidance
- covert inference of emotion or identity
- manipulative prompts disguised as guidance
- overlays so dense that cognitive load increases rather than decreases

The rule is practical. If the AR layer adds cognitive load without adding clarity, it should not exist.

Spatial AR Interface as a governed client

Spatial AR Interface is not the brain. It is a client.

It consumes:

- WorldModel state in a minimal form appropriate to the user's request
- governance decisions that determine what is allowed to be shown or suggested
- the visitor's declared preferences, language, depth, and comfort modes, held locally and consent-driven

It produces:

- spatial overlays that reflect truth
- Personal Channel triggers for audio, captions, and other modalities
- voluntary feedback signals such as "this was helpful" or "this was too much"

This is publish-safe and operationally real. It specifies interfaces and behavior without publishing proprietary mapping or tracking recipes.

The privacy posture: AR without creepiness

AR carries an obvious fear: the camera. People assume AR implies recording, analysis, or persistent tracking. A privacy-forward posture makes the design explicit.

A responsible implementation follows these principles:

- local-first processing wherever possible, especially for spatial alignment and preference application
- ephemeral session state by default
- no identity required for basic AR guidance and interpretation
- explicit consent for any data-sharing beyond local rendering
- a clear off switch and a clear degradation path, with the venue remaining usable without AR

This aligns with the larger thesis of the book: the phone should be a privacy boundary, not a data funnel. In a Web5 posture, the phone agent negotiates what is shared, and Spatial AR Interface becomes one of the agent's output modalities rather than a separate tracking system.

In the next section, we move from consent-driven human context back into deployment patterns, because even the best ideas matter only when they can be implemented in real venues by real teams, with operability, maintainability, and long-term trust built in from the outset.

AR and Programmable Canvases: complementary planes

A common mistake is to treat AR as a replacement for venue surfaces. It is not.

Programmable Canvases are shared truth. They serve groups. They remain visible without requiring a device. AR is personal truth. It serves the individual. It can carry long-tail language without clutter, offer depth without occupying physical space, and provide accessibility cues privately.

The two should work together. A Programmable Canvas can set the shared posture of a space. Spatial AR Interface can provide optional overlays, language and depth expansions, and comfort controls that do not need to be printed into the room. When these planes are complementary, the venue gains elegance and inclusion.

AR and the acoustic layer: seeing the quiet path

One of the most practical uses of Spatial AR Interface is comfort-aware navigation. Visitors do not generally know why a space feels uncomfortable. They simply feel it and they leave.

If the WorldModel framework represents acoustic state, the venue can offer calm routes as a normal option. Spatial AR Interface can render those routes spatially, turning a comfort mode into a visible path that reduces fatigue without turning the visitor into a special case. This is inclusion that feels like hospitality. It also reduces operational stress. When the venue can distribute visitors gently toward calmer spaces, the entire environment becomes more stable.

AR and governance: why overlays must be constrained

A spatial interface can influence behavior strongly. An arrow is a command-like suggestion even when it is intended as help. That is why AR must be governed.

Governance must constrain:

- Which routes can be suggested
- How load balancing is expressed
- How fairness is preserved
- What is shown during incident mode

What privacy posture is applied to data associated with the AR session

Spatial AR Interface should default to advisory guidance. The visitor remains in control. The system remains explainable. The venue remains defensible.
A governed AR layer can be ambitious because it is constrained. An ungoverned AR layer should not be deployed.

Cross-vertical use cases

Museums
Interpretive overlays with layered depth, multilingual delivery without signage clutter, and accessibility cues that normalize captions and sign language availability.

Theme parks
Load-aware guidance that remains optional, showtime overlays that reflect real capacity, calm routes during peak congestion, and location-based narrative that supports discovery without forcing phone usage.

Cruise ships
Deck navigation in complex vertical circulation, multilingual guidance, and safety mode overlays that reflect ship operating mode without confusing guests.

Retail and supermarkets
Aisle-level guidance, product context in the shopper's language, accessibility-first routes, and queue visibility that reduces stress.

Cities
Event precinct guidance, detour overlays, accessibility routes that update in real time, and public service messaging that remains truthful during disruptions.

The pattern remains consistent. AR is a client of governed truth and can remain helpful without becoming invasive.

Acceptance tests for Spatial AR Interface

A venue should not ship an AR interface without acceptance tests that protect trust.

1. Truth correctness
   - Overlays reflect current closures, routes, and schedules.
   - Under stale state, the interface fails conservatively rather than misleading.
   - Confabulation control: the interface does not display invented route availability, schedules, eligibility, or instructions when grounding is insufficient.
2. Governance compliance
   - Accessibility constraints are enforced in route suggestions.
   - Incident-mode constraints are enforced immediately.
   - Advice remains advisory, with clear user agency.
3. Privacy posture
   - The AR experience works without identity.
   - Session data is ephemeral by default.
   - Opt-in continuity is explicit and reversible.
4. Comfort and cognitive load
   - Overlays do not clutter the field of view excessively.
   - Quiet routes and comfort controls are accessible and simple.
5. Graceful degradation
   - If AR alignment degrades, the venue still provides usable guidance through surfaces and staff tools.
   - The AR layer communicates uncertainty rather than pretending.

Uncertainty is explicit: overlays visibly indicate confidence posture when guidance depends on state that may change (closures, congestion, incident mode).

These tests prevent Spatial AR Interface from becoming the kind of AR novelty visitors try once and then distrust.

The chapter's conclusion

Spatial AR Interface is valuable because it solves a structural problem. Physical space is scarce. Language needs are not. Guidance has to change faster than fabrication can.

A spatial interface becomes useful when it is opt-in and privacy-forward, derived from WorldModel operational truth, constrained by governance, and integrated with the other planes of delivery rather than competing with them. That is how AR becomes a humane interface to an environment that can finally be intelligent without becoming invasive.

In the next chapter we move into consent-driven human context, including prosody, feedback, and comfort controls, with the same posture: opt-in, minimal, local-first where possible, and generally designed so the visitor remains in control.

Part VII

DEPLOYMENT PATTERNS BY VERTICAL

The same architecture behaves differently in different worlds. Museums optimize for learning and dignity. Theme parks optimize for throughput and spectacle. Cruise ships combine hospitality with logistics in a bounded footprint. Retail is driven by friction reduction. Corporate centers want narrative control. Smart cities must withstand politics, heterogeneity, and long lifecycles.

This part is a set of applied patterns. Chapters 31 - 36 translate the earlier concepts into concrete deployment postures by vertical, with the goal of preventing category errors, for example, importing retail identity habits into museums, or importing theme-park audio habits into civic spaces.

Use these chapters as design review checklists. Pick the chapter describing the vertical that matches the one you are interested in. Ask, "What is the outcome, what is the failure mode, and what must be governed?" You will also see where the same building blocks, such as programmable canvases, non-visual guidance, and verification loops, reappear with different constraints. The point is not to prescribe one stack. The point is to give you a way to reason, compare, and decide with operational truth instead of wishful thinking.

For Museums and Cultural Institutions, Chapter 31.

For Theme Parks and Attractions, Chapter 32.

For Cruise Ships, Chapter 33.

For Retail, Supermarkets and Service First venues, Chapter 34.

For Corporate Experience Centers, Chapter 35.

For Smart Cities and Public Infrastructure, Chapter 36

31

MUSEUMS AND CULTURAL INSTITUTIONS

A museum is not only a building that contains objects. It is a public promise.

It promises that the past will be handled carefully, that the present will be interpreted honestly, and that visitors will be treated with dignity regardless of language, ability, age, education, or temperament. It promises that a place can still exist where attention is not harvested, where meaning is not reduced to speed, and where a person can leave with more than a photo.

That promise is why museums are uniquely suited to hyper-personalization and uniquely vulnerable to doing it poorly. When it personalizes poorly, it feels like surveillance, manipulation, or mission drift. The difference is not technology. The difference is posture.

Confabulation containment as an operational requirement

Confabulation is any confident output that is not grounded in approved sources or authoritative operational truth. In public environments, this is not a minor content defect. It is a trust and safety risk.

Therefore, any visitor-facing claim that can influence behavior must be grounded, time-bounded by freshness, and verifiable, or the system must abstain, fall back to a safer mode, or escalate to staff. "Unknown" is an acceptable output state. Confident guessing is not.

This requirement is enforced through delivery-mode discipline, truthfulness constraints at the governance gate, and operator-visible evidence of abstain, fallback, and escalation events.

Museums must remain the most privacy-forward, values-explicit, and trustworthy of the verticals discussed in this book because their relationship with the public is civic rather than transactional. This chapter translates that responsibility into practical guidance: how museums adopt micro hyper-personalization across multiple exhibits without breaking the room, and how larger museums move toward macro behavior, when ready, without becoming the kind of place people stop trusting.

Goals

Museums are trying to achieve several outcomes at once, and a useful system must honor all of them.

- Preserve meaning, scholarship, and curatorial intent.
- Increase visitor clarity without turning the building into signage clutter.
- Expand inclusion, language dignity, accessibility, and comfort as default behavior.
- Reward different depths of engagement without shaming any visitor mode.
- Improve repeat visitation by making the venue feel alive and responsive.
- Maintain operational calm, reliability, and maintainability over years.

Ladder stage fit

Museums benefit from all three stages, but most should climb deliberately.

- **Stage 0: Exhibit-scale localization**
  Language selection and accessibility modes at a single exhibit, available without identity. This is foundational and often the first practical win.
- **Stage 1: Micro hyper-personalization across multiple exhibits**
  Coherence across a cluster or wing: language, depth mode, and comfort choices persist. This is where the museum begins to feel like a system rather than islands.
- **Stage 2: Macro hyper-personalization in WorldModel Venues**
  Large museums, multi-building campuses, and high-throughput institutions may need governed guidance, load balancing, and schedule-aware flow shaping. Governance becomes mandatory at this stage because decisions have externalities.

Key layers

Museums should treat these layers as the minimum stack for modern, inclusive operations.

- Edge Compute Nodes and provisioned compute: modular nodes that can be replaced and upgraded by provisioning sufficient compute, allowing the venue to evolve without rebuild cycles.
- Programmable Canvases: adaptable interpretation and guidance without fixed bilingual compromise.
- Personal Channel: synchronized, privacy-forward delivery of language, captions, sign language video, personal audio, quiet mixes, and other comfort modes.
- Virtual Docent Layer and Embodied Docent Interfaces where appropriate: curated, governed conversational guidance and interpretive depth without generic internet tone.
- Modular interaction infrastructure: consistent, serviceable exhibit affordances that scale.
- Acoustic layer: intelligibility, sound bleed management, and comfort modes as access.
- Lifecycle Automation Layer and Outcome Verification Layer: occupancy-aware calm and equipment life protection, coupled with verified restoration and drift detection.
- Venue Concierge Interface, WorldModel, and governance: only for macro museums where venue-scale truth, orchestration, and constraints are required.

MUSEUMS AND CULTURAL INSTITUTIONS | 455

Privacy posture

Museums must default to trust.

A museum-grade privacy posture is:

- available without identity operation for core participation, language selection, and interpretive access.
- Opt-in identity only when the visitor receives a clear benefit, such as saved preferences across visits, membership continuity, or explicit entitlements.
- Prefer anonymous recognition and short-lived session continuity over persistent identifiers.
- Process locally where possible and minimize what leaves the venue.
- Treat logs as operational evidence, not visitor biographies, with bounded retention and deletion as normal behavior.

A museum is not a data business. It is a trust business.

Ops posture

Museums are often judged by the visible. They are sustained by the invisible.

A museum operations posture that matches the architecture in this book includes:

- Commissioning as proof, not only as turn-on, with acceptance tests operators can run.
- Monitoring that covers health, integrity, and drift so failures are detected before guests do.
- Lifecycle Automation Layer behaviors that reduce sound bleed and extend equipment life in unoccupied zones while restoring before perception.
- Outcome Verification Layer checks that confirm images are present and moving and that zones are truly ready after automated transitions.
- Content operations with versioning, staging, rollback, and post-publish verification so the museum can evolve safely.
- A support model that is boring by design: clear roles, response ladders, and predictable maintenance rituals.

Reliability is not a backstage metric. It is part of the museum's promise.

What to measure

Measurement in museums should improve clarity and inclusion without turning visitors into inventory. Measure the place first.

- Flow and distribution: where visitors cluster, hesitate, and reroute.
- Friction points: repeated confusion at signs, transitions, and decision nodes.
- Engagement patterns: where Streakers, Strollers, and Students consistently disengage or deepen.
- Mode usage: language selection, caption usage, personal audio usage, quiet mode usage, sign language availability usage.
- Acoustic state correlations: where noise and bleed correlate with abandonment or reduced engagement.
- Operational integrity: uptime of critical nodes, synchronization integrity, restoration success after Lifecycle Automation Layer.
- Drift indicators: thermal trends, fan escalation, increasing fault rates, and repeated verification anomalies.

Use identity-based measures only as an explicit opt-in benefit exchange, and rarely as a default.

The museum's real competition is the living room

Museum strategy is often framed as competition between institutions: who has the better exhibition, the better building, the better traveling show. In practice, the harder competition is the living room.

A visitor who leaves home is making a choice about time, energy, comfort, and friction. Museums win that choice when they offer what the living room cannot: embodied scale, shared presence, real objects, and authentic context, while removing the modern frictions that make people hesitate to go out.

Hyper-personalization, done correctly, is not spectacle. It is friction removal, clarity, and inclusion expressed through a system. A museum does not need to become louder to be modern. It needs to become more legible, more accommodating, and more coherent across a day.

Begin where museums already succeed: meaning

Museums have one advantage no algorithm can counterfeit: curatorial intent. Museums know how to say, "this matters," and then earn the claim through evidence, narrative, and restraint. Personalization should extend curatorial intent rather than compete with it.

Confabulation is the failure mode that most directly violates curatorial intent. In a museum context, a fluent but ungrounded statement is not a harmless error. It becomes interpretive misinformation delivered with institutional authority.

A museum-grade system therefore treats all generative behavior as bounded delivery, not open-ended invention. If a claim cannot be grounded in the museum's approved Body of Knowledge or in an explicitly approved source set, the correct behavior is to abstain, ask a clarifying question, route to a human-curated explanation, or present a clearly labeled uncertainty statement. In other words: the system must remain helpful without pretending.

This is why the Streaker, Stroller, Student model fits museums so well. Museums already write in layers even when those layers are scattered across physical artifacts: an intro panel, a label, a brochure, a tour, a lecture, a catalogue. Hyper-personalization brings those layers into a coherent system that can deliver the right layer at the right moment.

A Streaker wants the headline, the essential meaning, and then they move. A Stroller wants the headline and a little more, enough to connect and commit. A Student wants depth, adjacent sources, and technical detail. A museum does not betray scholarship by serving the Streaker. It betrays scholarship when it serves only the Streaker. The correct posture is layered depth, with the visitor in control.

Language dignity is interpretive dignity

Museums often try to solve multilingual support through printed compromise. It is well intentioned, and it ages poorly. Printed multilingual layouts force hierarchy, and hierarchy is noticed. They also force scarcity. A wall has finite space. The public does not have finite language needs.

A museum-grade solution treats language as runtime rather than architecture.

Runtime language introduces a specific risk: translation confabulation. A system can produce text that sounds plausible in a target language while subtly changing meaning, adding details, or collapsing nuance. In museums, this is an interpretive integrity issue, not merely a quality issue.

The operational rule is simple: the museum must be able to point to what the system is grounded in. For any factual or attributional statement, the system must either

(a) map the statement to an approved source reference,
(b) present the statement as interpretation clearly labeled as such, or
(c) refuse to guess.

This does not mean replacing every label with a screen. It means using programmable surfaces where adaptability is valuable and using the Personal Channel where long-tail languages and accessibility modes are best delivered privately and cleanly.

The most important multilingual rule is not the number of languages supported. It is parity of presence. If the museum offers multiple languages, it should do so without implying that one language is real and the others are add-ons. Runtime language selection delivered through canvases and Personal Channels is one of the simplest ways to achieve parity without turning galleries into typographic clutter.

The Personal Channel is the museum's most scalable inclusion tool

Museums are asked to serve a public that is multilingual, diverse in how it reads and processes information, and diverse in sensory tolerance. Many visitors are excluded by factors that rarely appear in an accessibility checklist: noise, cognitive overload, fatigue, anxiety, and the exhaustion of trying to keep up.

Vertical deployment quick-start template

- Outcomes:
  list the top five outcomes you are optimizing.
- Minimum architecture:
  which layers are mandatory for this site.
- Acceptance tests:
  what must be proven at handover, and what is monitored thereafter.

The Personal Channel is how museums offer inclusion without turning shared space into compromise. A visitor who wants captions can have captions without forcing captions onto every screen. A visitor who wants sign language video can receive it without special seating and special hardware. A visitor who wants a quieter narration-forward mix can receive it through earbuds without demanding that the museum lower the room for everyone else. A visitor who needs long-tail language support can receive it privately, synchronized to the exhibit rhythm.

This is inclusion from the outset rather than inclusion by exception. It also aligns with a museum's trust posture. The Personal Channel can be offered without identity. Visitors can receive value anonymously and opt in only when the benefit is clear, such as continuity across exhibits or saved preferences for a return visit. A museum should rarely feel hungry. The Personal Channel helps the museum stay helpful without over-collecting.

Programmable Canvases as capability, not aesthetic

Museums have long treated walls as permanent interpretive surfaces. That worked when refresh cycles were slow. The world is no longer slow. Exhibits change. Community conversations shift. Sponsors rotate. Programming evolves. Language needs change by season.

Programmable Canvases give the institution the ability to change exhibits, change languages, and change emphasis quickly without construction trauma. In a museum context, the most powerful use of a canvas is often clarity rather than motion. A canvas can switch between narrative and guidance when a gallery becomes crowded. It can show the headline layer for Streakers and invite deeper layers for Strollers and Students. It can present language with dignity, including typographic and layout choices that respect scripts rather than forcing everything into an English-shaped box.

A museum should be visually restrained. A canvas makes restraint easier because it allows specificity in the moment instead of printing everything forever.

The acoustic layer is museum policy, whether you name it or not

Museums often underestimate sound because sound is invisible. Visitors rarely file formal complaints about sound bleed. They simply leave sooner. They stop reading. They stop listening. They become tired. They exit, and they may not return.

A museum that wants to be inclusive treats acoustic comfort as part of the interpretive contract. This is true at design time through materials, zoning, and intelligibility, and it is true at runtime through operational behavior.

Lifecycle Automation Layer belongs in this chapter because it is a museum behavior, not only an energy behavior. When a gallery is empty, there is no reason for it to broadcast at full volume. Lowering volume in unoccupied zones reduces sound bleed, preserves calm, and makes neighboring experiences more intelligible. It also supports visitors who are sensitive to overstimulation by keeping the building from becoming a constant chorus.

In a museum, calm is not a luxury. Calm is access.

Operations is the hidden half of visitor experience

Museums often separate front of house and back of house as if one is about beauty and the other is about mechanics. In practice, mechanics decide whether beauty is delivered reliably.

A projector failure does not feel like a technical issue to a visitor. It feels like the museum did not keep its promise. This is why operational verification and preventive maintenance belong in a reference book about hyper-personalization. The more adaptable the museum becomes, the more it must verify that adaptation is actually occurring.

Outcome Verification Layer is the discipline of proof. If the system powers equipment down to extend life and restores it when visitors return, the museum should be able to verify that the image is present, motion is playing when it should be playing, and a fault is detected before the public becomes the diagnostic tool.

Trend analytics is the museum's long-term ally. Museums are full of devices that drift toward failure rather than failing suddenly. Rising temperature and rising fan speed are the system telling you a maintenance story. A museum that catches drift early preserves uptime, reduces emergency repair cost, and prevents public failure. Reliability is a form of respect.

Analytics that improves the museum without turning visitors into inventory

Museums need analytics for the same reason they need curators: to learn what is happening, and to improve. Museum analytics begins with a strict boundary. Measure the place, not the person.

A second boundary matters just as much: measurement is not meaning. Analytics can tell you where people hesitate, cluster, abandon, or linger. It cannot, by itself, tell you why, and it must not be allowed to generate interpretive stories about visitors.

The discipline is: measure first, interpret second, and keep interpretation bounded to what evidence can actually support. This protects the museum from turning operational metrics into pseudo-psychology, and it protects visitors from being treated as inferred identities.

Anonymous analytics can show where visitors cluster, where they hesitate, where they abandon, where they linger, and where confusion keeps repeating. Those signals are interpretive gold. They reveal where labels fail, where guidance fails, where acoustics repel, and where a narrative is too dense or too thin. They allow improvement without collecting identity.

More precise "unique visitor" analytics can be tempting. If a museum considers it, it must be consent-driven, purpose-bound, and minimal, with retention short enough to remain defensible. The default should remain anonymous. Opt-in should be a benefit exchange, not an expectation.

A museum is not a data business. It is a trust business.

When a museum becomes a WorldModel Venue

Many museums will rarely need macro behavior, and they should not feel pressured to adopt it. Some museums are already operating at city scale. Multi-building campuses, large galleries, high-throughput tourism, timed ticketing, school groups, special events, and complex circulation patterns create the same class of problems as theme parks, with a different tone.

A museum crosses into macro territory when the system begins to manage externalities: routing and load balancing across galleries, queue shaping, schedule coordination, staff deployment prompts, and venue-scale comfort management. At that point, governance is not optional. It keeps the museum aligned with its civic promise. The system must be constrained by explicit objectives and hard constraints: safety, accessibility, privacy, fairness, and interpretive integrity.

Venue Concierge Interface becomes relevant as the interface surface that translates visitor intent into governed outcomes: guide me, reduce friction, keep the day coherent, respect my comfort, and do it without requiring me to surrender privacy. WorldModel state becomes relevant as operational truth: what is open, what is full, what is calm, what is loud, what is scheduled, and what is safe.

In macro museums, guidance becomes safety-adjacent. A wrong statement about what is open, what is accessible, or what is safe can produce real harm, not just confusion. This is where confabulation must be treated as an operational risk, not a content risk.

Therefore, any visitor-facing guidance that affects routing, scheduling, accessibility, or crowd movement must be derived from operational truth sources, and it must fail conservatively. If the system cannot verify current state with sufficient confidence, it should present "unknown" and route to staff-facing tools or static safe guidance, rather than inventing a plausible answer.

A museum should adopt macro behavior only when it can adopt it with restraint. A museum that governs its system well becomes calmer and more inclusive. A museum that does not govern becomes unpredictable, and unpredictability erodes trust.

The museum deployment posture in one page of commitments

A museum can keep itself aligned by holding a small set of commitments and repeating them consistently.

- Treat language as runtime, not printed hierarchy.
- Deliver layered depth: headline, context, deep, with the visitor in control.
- Use the Personal Channel to make inclusion normal, not exceptional.
- Treat acoustic comfort as access and manage sound bleed intentionally.
- Use Programmable Canvases to stay adaptable without becoming visually noisy.
- Measure the place first and stay available without identity.
- Design operations as part of experience quality: Lifecycle Automation Layer for calm and equipment life, Outcome Verification Layer for verification, and trend analytics for preventive maintenance.
- If the museum becomes macro, govern it explicitly, and treat privacy as the product.

Museums do not need to become something else to become more capable. They need to become more coherent versions of what they already are: places that handle truth carefully, and people respectfully.

In the next chapter we move to a vertical that shares many of the same scale problems but with a different emotional tempo: themed entertainment, where throughput and spectacle increase pressure and governance becomes the difference between wonder and chaos.

32

THEME PARKS AND ATTRACTIONS

A theme park is a city that pretends to be a story. It has streets, districts, signage, schedules, power systems, queue lines, crowd pulses, staffing constraints, food and beverage throughput, retail transactions, incident response, and the daily reality that a small change in one corner can ripple across the entire map. Yet it must feel effortless. The day has to feel as if it is unfolding naturally even when the underlying machine is working hard to keep it from unraveling.

This is why theme parks are the clearest case for macro hyper-personalization. Museums often begin with meaning and then add systems. Parks begin with motion and then demand coherence. The experience is built on rhythm, and rhythm is a systems problem.

Visitors do not evaluate a park by architecture diagrams. They evaluate it by how the day feels. Did the park feel joyful or exhausting? Did it feel magical or managed? Did the queues feel fair or punishing? Did the signage tell the truth? Did staff feel present where needed? Did the soundscape feel immersive or noisy? Did the app and the physical reality agree?

Hyper-personalization in this environment is not primarily about telling a different story to each person. It is about helping each person have a better day while keeping the place stable for everyone.

Confabulation containment is an operational safety requirement

In attractions, a confident wrong statement can move crowds, trigger conflict, misroute guests, or undermine show credibility. Any guidance that influences guest behavior must be grounded in approved facts and current operational state, carry a freshness posture, and expose an evidence pointer or source trace. If grounding is missing, stale, or ambiguous, the system must abstain, fall back to a stricter delivery mode, or escalate to staff.

Enforce this through delivery-mode discipline, a governance gate that can refuse or constrain outputs, and operator-visible logging of abstain, fallback, and escalation events.

Goals

Theme parks are trying to produce emotional outcomes under extreme operational constraints.

The goals are:

- Preserve wonder, delight, and pace without making guests feel managed.
- Improve day coherence so choices feel easy rather than exhausting.
- Increase fairness and predictability in queues and capacity-limited experiences.
- Expand multilingual support and inclusion as default behavior rather than a service-desk ritual.
- Reduce sensory fatigue by treating acoustic comfort as access.
- Maintain reliability because public failures are reputation events.
- Improve operational calm so staff are supported rather than overloaded.
- Enable continuous evolution through modular compute and publishing discipline rather than rebuild cycles.

Ladder stage fit

Parks use every stage of the ladder, often simultaneously.

- Stage 0: Exhibit-scale localization
  Language selection and accessibility modes at a single attraction, preshow, kiosk, or interactive queue moment, available without identity.
- Stage 1: Micro hyper-personalization across multiple experiences
  A guest's preferences persist across a land or cluster: language, depth mode, comfort modes, and interest vectors, without demanding identity.
- Stage 2: Macro hyper-personalization, the park as a governed environment
  Flow control, load balancing, queue shaping, schedule coordination, staff prompts, and comfort-aware routing across lands and districts. Governance becomes mandatory because actions have externalities.

Key layers

A park-grade implementation is a full-stack problem. The key layers are:

- Edge Compute Nodes and provisioned compute: modular nodes at the edge so time-critical loops stay local and upgradeable.
- Modular interaction infrastructure: consistent, serviceable physical affordances at scale.
- Programmable Canvases: operational truth, guidance, and dynamic narrative surfaces without permanent signage compromise.
- Personal Channel: multilingual delivery, captions, personal audio, quiet mixes, and comfort modes without cluttering shared spaces.
- Virtual Docent Layer and Embodied Docent Interfaces where appropriate: governed, curated conversational guidance and role-based guest assistance.
- Acoustic layer: sound ecology management, intelligibility, theming boundaries, and comfort as access.
- Lifecycle Automation Layer and Outcome Verification Layer: occupancy-aware calm and equipment life protection coupled with verified restoration and drift detection.
- Venue Concierge Interface: the interface layer for guest intent and staff intent at macro scale. WorldModel plus CGL, EDE, ICL, and MAOL: operational truth, dynamics modeling, continuity, orchestration, and governance for park-scale behavior.
- Analytics and observability: measure the place first, prevent failures through drift detection, and keep guidance truthful.

Privacy posture

Parks must remain helpful without becoming creepy. A park-grade privacy posture is:

- available without identity participation for core guidance, language, and comfort modes.
- Opt-in identity only when the guest receives a clear benefit, such as entitlements, continuity across visits, or membership features.
- Prefer anonymous recognition and short-lived session continuity over persistent identifiers.
- Local-first processing where possible, especially for time-critical and privacy-sensitive loops.
- Logs as operational evidence, not guest biographies, with bounded retention and deletion as a normal behavior.
- Clear opt-out and clear degradation, the park still works beautifully when a guest declines.

Ops posture

Parks are unforgiving. Operations must be engineered rather than improvised.

- Commissioning as proof, including timing integrity, queue and guidance correctness, and degradation behavior under load.
- Monitoring for health, integrity, and drift, not only uptime.
- Verified restoration for occupancy-aware automation, including audio reduction, display and projection power states, and signage state.
- Modes as first-class operations: normal, busy, and incident, with different priorities and allowed actions.
- Operator override as normal coordination rather than a panic button.
- Content operations with staging, rollback, and post-publish verification so the park can update without risking show integrity.

Reliability is guest experience.

What to measure

A park should measure outcomes that improve the day, not vanity metrics. Measure the place first.

- Flow and distribution across lands, including congestion pulses after show releases.
- Queue fairness, predicted waits versus actual experience, and repeated bottlenecks.
- Guidance integrity, whether signage, apps, and staff tools agree with reality.
- Acoustic state and fatigue signals, including where sound bleed degrades theming boundaries.
- Comfort mode uptake: captions, personal audio, quiet mixes, and multilingual selections.
- Show integrity: cue success rates, synchronization stability, and public failure frequency.
- Operational stability: time-to-recovery, manual reset frequency, and verified restoration success after Lifecycle Automation Layer actions.
- Drift indicators: thermal trend, fan escalation, error-rate creep, and repeated verification anomalies.

Use identity-based measurement only as explicit opt-in benefit exchange.

The park's competition is time

It is tempting to think parks compete mainly on new rides, new lands, and new intellectual property. They do. The harder competition is the visitor's patience.

Modern visitors arrive trained by a world of immediate feedback and personalized media. They accept waiting when waiting feels purposeful, predictable, and fair. They reject waiting when it feels arbitrary, confusing, or avoidable. They also arrive with phones. They photograph, share, translate, coordinate, and document. The park can treat this as distraction, or treat it as a default interface for navigation, guidance, and optional personalization.

Design for phone-era expectations without requiring a phone, and the park gains a Personal Channel that can deliver language, captions, accessibility support, comfort modes, and individualized guidance without forcing the physical environment to carry every variation publicly.

A park day is a sequence of decisions. Every decision is a moment where friction accumulates or is removed.

In parks, confabulation is not a philosophical AI issue. It is a guest-trust failure. A fluent but wrong statement about wait times, closures, show schedules, accessibility routes, weather impact, or capacity will be interpreted as deception, even when it was merely an error.

A park-grade guidance system must therefore be designed to avoid confident guesses. Where the system is not reading authoritative operational truth, it must label uncertainty, provide ranges, defer to verified signage, or route to staff tools. Guests can tolerate "we do not know yet." They will not tolerate "we are sure" when it is wrong.

Parks that remove friction feel larger than they are. Parks that accumulate friction feel smaller than they are.

Flow is the story you cannot ignore

Attractions are built on flow. Flow is an emotional pacing problem as well as a logistics problem. A guest who moves smoothly from a great moment to the next great moment experiences the park as magical. A guest who repeatedly collides with congestion experiences the park as tiring even if the attractions are world-class.

Macro systems exist to protect flow without making guests feel herded.

Flow guidance becomes defensible only when it is grounded in current state and remains advisory. The system can propose a calmer route, a shorter queue, or a nearby option, but it must not fabricate justifications or claim certainty it does not have.

This is where confabulation containment becomes a governance requirement: guidance must be traceable to current park state, show integrity constraints, and explicitly allowed action catalogs. If those inputs are missing, degraded, or contradictory, the correct behavior is conservative fallback, not improvisation.

Suggestions have to feel like help rather than control. Guidance has to feel like clarity rather than manipulation.

This is where Venue Concierge Interface becomes a turning point. A concierge translates intent into outcomes under constraints. In a park, intent is rarely "optimize my day." It is human: "We have two hours," "My child is tired," "We need shade," "We want thrills," "We want calm," "We are hungry," "We do not want another long wait."

The concierge layer is where the park can respond with governed proposals: a calmer route, a shorter queue, a show that is starting soon with capacity, a restaurant that can serve quickly, or a nearby experience that fits the visitor's chosen comfort and language modes. Guidance remains an offer. Agency remains with the guest. When that boundary is held, distribution improves and the park feels better for everyone, including guests who rarely use the guidance layer at all.

Queues are experience

A queue is the most common place where a park earns trust or loses it. A guest can tolerate a long wait if it feels fair, predictable, shaded, and purposeful. A guest becomes angry when the wait feels deceptive, chaotic, or physically uncomfortable.

Queue management at macro scale is not simply posting wait times. It is shaping demand and release. Because these outputs influence behavior, any wait-time statement must be grounded in authoritative operational truth or explicitly presented as an estimate with bounded uncertainty, and the system must not guess when signals are missing or stale:

- predicting when a queue will spike and why,
- coordinating show releases so one crowd does not crash into another crowd,
- balancing capacity across attractions so the park does not starve one land and overload another, and
- offering choices that feel like options rather than directives.

Truthfulness is part of guest trust.

A precise number delivered with confidence is not automatically better than a range. Under uncertainty, the correct behavior is to show a range, disclose that it is an estimate, or present "unknown" with safe alternatives. Confabulation in this context looks like confident numbers that are not warranted by measurements. Avoid that failure mode by requiring evidence pointers and freshness windows for any published wait-time claim.

This is where EDE, the Environmental Dynamics Engine, matters in plain terms. A park has pulses. A parade ends. A show lets out. A land becomes the fashionable place to be at 2 p. m. EDE models these dynamics so the system can anticipate rather than react late.

Anticipation still needs governance. The temptation in parks is to chase throughput at the expense of wonder. Governance, expressed through CGL, prevents that drift. It protects fairness, accessibility, privacy, and show integrity.

Vertical deployment quick-start template

- Outcomes:
  list the top five outcomes you are optimizing.
- Minimum architecture:
  which layers are mandatory for this site.
- Acceptance tests:
  what must be proven at handover, and what is monitored thereafter.

Show control, show integrity, and the macro system

Parks contain two truths that must coexist.

One truth is show integrity. Certain moments must land exactly, and their timing must be deterministic. This is classic show control. It is the score.

The other truth is operational reality. Guests are late, crowds form, weather shifts, an attraction goes down, and the park adapts. This is orchestration. It is the conductor.

A macro system must not corrupt the score. It can propose changes around the score. It must protect show integrity as a hard constraint.

This is also a confabulation boundary. When timing, status, or availability is tied to show control, the system must not guess. If show state is unknown, stale, or contradictory, the system must not narrate certainty.

Operational rule: for any guest-facing statement that references show state, ride availability, safety mode, or incident mode, the system must either cite a current authoritative state source or abstain and provide safe fallback guidance.

This is also where operating modes matter. A park does not operate the same way at rope drop, midday peak, parade release, and evening close. Governance includes modes that shift priorities and allowed actions without making the system unpredictable. Normal mode can prioritize discovery and delight. Busy mode can prioritize clarity, queue fairness, and calm routing. Incident mode narrows actions to safety, guidance, and operational coordination. Visitors may not see these modes explicitly. They feel them as coherence.

The acoustic layer: parks as sound ecologies

Theme parks often treat sound as spectacle. That is understandable, and it is also where fatigue is born. A park is a sound ecology. Music, effects, announcements, character moments, crowd noise, and mechanical noise combine into an environment that can feel immersive or exhausting. Sound bleed between lands can destroy theming. Excessive volume can exclude visitors who are easily overstimulated. Intelligibility can collapse, which reduces perceived quality even when content is strong.

This is why the acoustic layer must be treated as state rather than set-and-forget design. At micro scale, a Personal Channel can offer alternatives: narration-forward mixes, captions, language tracks, and quieter modes. At macro scale, occupancy-aware behaviors become acoustic hygiene. If a zone is empty, there is no reason for it to be loud. Lifecycle Automation Layer becomes sound management as well as energy management. Lower volume in unoccupied zones reduces bleed, preserves theming boundaries, and keeps the park calmer.

A calm park is not a boring park. It is a park where intensity is intentional and recovery is available. Visitors need rhythm, not constant intensity.

Power, equipment life, and keeping the day reliable

A park is a machine that performs all day, in public, at scale. Reliability is experience quality. A screen that goes dark, a projection that does not start, or a show moment that fails to trigger is not interpreted as maintenance. It is interpreted as a broken promise.

Occupancy-aware power behavior extends equipment life, reduces heat, and reduces wear without visible downside. If projectors can be powered down when unoccupied and restored before guests notice, runtime drops, brightness is preserved, and heat-related failures are reduced. Dynamic behavior adds a requirement: proof.

Outcome Verification Layer becomes essential. Commands are not outcomes. If the system powers equipment down and restores it, the park must verify that content is playing, the image is present, motion is correct, and faults are detected early. A macro system without verification humiliates the venue. A macro system with verification can operate at scale with calm confidence.

Analytics that improves the day

Parks need analytics because parks are complex, not because parks want to be invasive. The healthiest posture remains the same as in museums: measure the place first.

Occupancy, flow, dwell, congestion, and repeated confusion points can improve operations dramatically without requiring identity. Trend analytics on equipment health can prevent failures by detecting drift: rising temperature, rising fan speed, and other slow indicators.

If a park considers more precise uniqueness analytics, the discipline remains unchanged: consent, purpose limitation, minimal retention, and clear benefit exchange. Most park operations can be improved without collecting biographies. A park that feels hungry will lose trust, and trust is a revenue driver in disguise.

Food, beverage, retail, and the small-city truth

Parks are not rides plus sidewalks. They are economies. Food and beverage throughput, retail capacity, reservation timing, staffing, and service friction shape the day. Guests do not experience these as separate categories. They experience one day.

Macro hyper-personalization can help here under governance:

- suggest dining options based on capacity, proximity, and guest intent without pressure,
- coordinate reservation timing and guidance so guests are not forced into sprints,
- use programmable surfaces to reduce confusion rather than shout offers, and
- link entitlements to service state without oversharing and without demanding identity for basic participation.

This is where the "yes or no" identity posture becomes practical. The park verifies entitlements without copying documents. The phone becomes a trust device rather than a data funnel. Tone matters. Parks are there for joy. A system that nags or pushes will be resented even when it is correct.

Smart parks, and why governance protects wonder

When parks become macro, system choices begin to shape the environment's character. A system can optimize throughput and destroy discovery. It can reduce congestion and create a feeling of being managed. It can increase revenue per guest and cheapen the space.

Governance prevents these failure modes. In practice, governance means the park writes down its outcomes, including joy, clarity, and fairness, not only throughput. It writes hard constraints, including safety, accessibility, privacy, and show integrity. It defines allowed actions and decides what is autonomous versus advisory versus approval-required. It maintains a proof and override plan, so operators remain in control and the system remains accountable.

The best macro systems are not visible as systems. They are visible as calm. The park feels easier. It feels more legible. It feels more inclusive. It feels as if the day is cooperating.

A deployment posture for attractions

A park can adopt the architecture in this book in stages:

1. Start with micro wins visitors feel immediately: multilingual dignity, captions, personal audio, and comfort modes across a cluster of experiences.
2. Build programmable surfaces where adaptability improves clarity, queue experience, and operational truth.
3. Treat the acoustic layer as state and use occupancy-aware behaviors to reduce bleed and fatigue.
4. Build analytics that measures the place and trend analytics that prevents failures.
5. Introduce Venue Concierge Interface when the park is ready to treat guidance as a governed service.
6. Formalize WorldModel operational truth so decisions are grounded in reality.
7. Add governance explicitly before expanding macro action classes. rarely after.

A park does not need to become more complicated to become more capable. It needs to become more coherent.

The chapter's conclusion

Theme parks and attractions are where the architecture of this book becomes unavoidable because parks already behave like living systems. The only question is whether that behavior remains manual and reactive or becomes supported by an architecture that is privacy-forward, inclusive, and governed.

The goal is not to build a park that feels automated. The goal is to build a park that feels hospitable under scale.

In the next chapter we move to the most bounded form of city-scale experience: the cruise ship, where language diversity, scheduling density, safety posture, and personal comfort collide inside a finite footprint, and where the Personal Channel becomes rather than only inclusion, sanity.

33

CRUISE SHIPS AS BOUNDED CITIES

A cruise ship is the most honest kind of venue. It cannot pretend that logistics is optional, because logistics is the ocean. It cannot pretend that people will simply leave if they are unhappy, because leaving is not possible. It cannot pretend that language diversity is an edge case, because the guest list changes every week. It cannot treat safety as background, because safety on a ship is a daily operating posture.

And yet the ship must feel like leisure. It must feel like ease, abundance, and choice delivered in a finite footprint, under constant schedule pressure, with thousands of people sharing the same corridors, elevators, restaurants, theaters, pools, and decks.

That combination makes cruise ships one of the clearest demonstrations of a WorldModel Venue. A ship is a small city with an enforced boundary, a single operator, and a dense, constantly shifting program. It is a city you can instrument, a city you can govern, and a city where trust is not philosophical. Trust is the difference between delight and chaos.

Confabulation containment is a ship-safety requirement

On a ship, errors about schedules, deck locations, muster guidance, venue hours, accessibility routes, or policy can produce immediate operational harm. Visitor-facing statements that influence movement or decisions must be grounded in approved sources and live operational truth, with freshness and evidence pointers. When the system cannot verify, it must abstain, fall back to a safer mode, or escalate to crew.

Enforce this through delivery-mode discipline, governance-gated outputs, and operator-visible audit traces of abstain, fallback, and escalation.

Goals

A cruise ship must deliver hospitality under hard constraints that most land venues rarely face. The goals are:

- Preserve a sense of ease in a bounded environment with dense schedules.
- Provide multilingual, accessible service as default behavior, not as an exception.
- Reduce friction in navigation, reservations, and day planning so guests feel calm rather than managed.
- Maintain fairness in access to capacity-limited venues and time-bound programming.
- Treat comfort as access, including acoustic comfort in dense, reflective environments.
- Maintain reliability, because failures are amplified when guests cannot go elsewhere.
- Support safety posture as an always-present constraint layer.
- Enable continuous evolution through modular compute and publishing discipline, not refits.

Ladder stage fit

Ships use the same ladder, and macro behavior becomes relevant earlier because the environment is bounded and schedule-driven.

- Stage 0: Localized clarity
  Language and accessibility choices at a single venue, theater entry, kiosk, or deck station, available without identity.
- Stage 1: Micro coherence across venues
  Preferences persist across multiple ship spaces: language, depth mode, comfort modes, and interests, without demanding identity.
- Stage 2: Macro orchestration, the ship as a governed environment
  Flow and routing across decks, elevator load shaping, scheduling coordination, reservation guidance, staffing prompts, and incident-mode behavior. Governance becomes mandatory because actions have externalities, and safety posture is generally present.

Key layers

A ship-grade implementation is a full stack with strong emphasis on resilience, safety, and multilingual delivery.

- Edge Compute Nodes and provisioned compute: distributed local nodes so critical loops stay on ship, upgradeable by provisioning sufficient compute.
- Modular interaction infrastructure: consistent, serviceable station I/O and triggers in high-use environments.
- Programmable Canvases: programmable surfaces for guidance, programming, and multilingual delivery without permanent signage compromise.
- Personal Channel: synchronized multilingual delivery, captions, personal audio, quiet mixes, and other comfort modes without cluttering shared spaces.
- Virtual Docent Layer and Embodied Docent Interfaces where appropriate: governed, curated conversational assistance for guest questions, orientation, and program discovery.
- Acoustic layer: sound ecology management in dense footprints, with bleed control and comfort-aware routing.
- Lifecycle Automation Layer and Outcome Verification Layer: occupancy-aware calm and equipment life protection with verified restoration and drift detection.
- Venue Concierge Interface: the interface layer for guest intent and staff intent across the ship.
- WorldModel plus CGL, EDE, ICL, and MAOL: operational truth, dynamics modeling, continuity, orchestration, and governance for ship-scale behavior.
- Analytics and observability: measure ship state, detect drift early, and keep service truthful.

Privacy posture

Ships must be helpful, and they must rarely feel hungry. A ship-grade privacy posture is:

- available without identity participation for wayfinding, language, and comfort modes.
- Opt-in identity only when the guest receives a clear benefit, such as entitlements, reservation continuity, and membership-like services.
- Prefer anonymous session continuity over persistent identifiers for day-of-cruise coherence.
- Use minimal entitlement states, demonstrate access without oversharing documents.
- Local-first processing where possible, because ships are constrained and because privacy is easier to defend locally.
- Logs as operational evidence, not biographies, with bounded retention and deletion as normal behavior.

Ops posture

On a ship, operations is not backstage. It is the guest's lived reality. A ship operations posture that matches this architecture includes:

- Commissioning as proof under dense schedule conditions, not only in quiet hours.
- Monitoring for health, integrity, and drift, because drift is the silent prelude to guest-visible failure.
- Verified restoration for occupancy-aware automation: audio reduction, display and projection states, and venue readiness before doors open.
- Strong degraded-mode behavior: core functions remain operable under imperfect connectivity and interference.
- Modes as first-class: normal, busy, and incident, with safety posture tightening allowed actions automatically.
- Operator override as routine coordination, because ship reality includes constant exceptions.
- Content operations with staging and rollback, because programming changes and sailing demographics change.

Reliability is hospitality at sea.

What to measure

Measure what improves the day and protects safety, without turning guests into inventory. Measure the place first:

- Flow and distribution across decks and vertical circulation, including elevator and stair pulse patterns.
- Congestion and bottlenecks around theaters, dining venues, and excursion departure points.
- Guidance integrity: whether signage, Personal Channel prompts, and staff tools agree with operational truth.
- Schedule adherence and pulse impacts: show releases, dining windows, excursion departures.
- Acoustic state correlations: where noise and bleed correlate with avoidance or fatigue.
- Comfort mode uptake: language selections, captions, personal audio, quiet mixes, and accessibility routing requests.
- Operational integrity: readiness verification success, restoration success after Lifecycle Automation Layer actions, time-to-recovery.
- Drift indicators: thermal trends, fan escalation, error-rate creep, and repeated verification anomalies.

Use identity-based measurement only as explicit opt-in benefit exchange, and keep retention bounded.

Why ships demand a Personal Channel

On a ship, the visitor's phone is not a distraction. It is a lifeline. It is where itineraries live, where reservations appear, where excursions are coordinated, where deck plans are consulted, where groups coordinate, and where language becomes a practical requirement rather than a cultural preference.

The Personal Channel is therefore not an extra. It is the ship's most scalable inclusion layer. A guest should receive guidance in a preferred language without forcing every corridor sign to carry five translations. A guest should receive captions, personal audio, and comfort modes without being routed to a special seating section or a service desk ritual. A guest who becomes overstimulated should be able to choose calmer modes without explaining themselves publicly.

This is the difference between a ship that feels modern and a ship that feels like a maze. It is also why micro hyper-personalization scales quickly on ships. Once language, depth, and comfort modes are available on the Personal Channel, the ship can serve diverse guests without turning shared spaces into compromise.

The ship's core problem is schedule density

A museum visitor can drift. A park visitor can improvise. A cruise ship guest lives inside a dense schedule whether they want to or not. Shows start on time. Dining windows matter. Excursions depart. Safety drills exist. Port calls impose hard boundaries. Some venues are reservation-based. Others first come, first served. Elevators and corridors pulse as entire decks move.

The ship is therefore a scheduling system wearing a vacation costume.

In a schedule-dense, bounded environment, confabulation becomes safety-adjacent. Wrong guidance about time windows, muster procedures, deck routing, elevator access, venue capacity, or port-call constraints can cascade into crowding, missed safety actions, and guest distress.

Ship-grade guidance must therefore be truth-first: it should only state what it can verify from current ship state, and it should fail conservatively when verification is unavailable. The guest experience depends on calm certainty, which must be earned by evidence rather than performed by tone.

This is why macro orchestration becomes more than a nice idea. It becomes the mechanism by which the ship stays calm. Guests do rather than need to be managed, they do need guidance that matches reality. Reality changes quickly and confusion is amplified in a bounded environment.

A ship that guides well feels effortless. A ship that guides poorly feels cramped.

Multilingual is operational truth

Cruise lines already know this in practice. Announcements are made in multiple languages. Staff are multilingual. Printed materials often include translations. Static multilingual delivery reaches its limit quickly because ship audiences are not only diverse, they are variable.

A sailing dominated by one set of languages can be followed by a sailing dominated by another. A single ship can have different mixes depending on itinerary, season, and port.

This is why language must be treated as runtime. The ship needs the same three language planes described earlier:

- Room plane:
  a coherent shared language in theaters and shows when necessary.
- Surface plane:
  programmable signage and canvases that adapt to the sailing, deck state, and current conditions.
- Personal Channel plane:
  long-tail language support and accessibility modes delivered privately, in sync, on the guest's device.

When language is runtime, the ship adapts without reprinting and without permanently privileging one language over others. Runtime language interacts with confabulation risk most sharply during safety or incident messaging. In those contexts, the system must not improvise phrasing, add interpretation, or compress instructions in ways that change meaning.

Practical posture: safety-adjacent messages should use approved phrase sets per language, with controlled variants for clarity and accessibility. If a requested language is not available at the required fidelity, the system should fall back to an approved alternative delivery mode rather than generating an unverified translation.

It can also respond to safety needs with more agility because critical messaging can be delivered clearly and consistently through multiple channels.

Flow is comfort

Cruise ships have a distinctive flow problem. The same corridors and vertical circulation carry guests to everything. On a ship, there are fewer alternate paths, congestion is more shared, and the subjective experience of congestion is more intense because the environment is enclosed.

That makes flow a comfort issue. When a corridor is crowded, the ship feels smaller. When elevators are unpredictable, the ship feels stressful. When guests arrive late to a show because vertical circulation failed, the ship feels unfair.

Vertical deployment quick-start template

- Outcomes:
  list the top five outcomes you are optimizing.
- Minimum architecture:
  which layers are mandatory for this site.
- Acceptance tests:
  what must be proven at handover, and what is monitored thereafter.

Macro hyper-personalization on ships is therefore about calm routes, distributed demand, and smoothed pulses:

- staggering prompts to reduce simultaneous surges,
- suggesting alternate venues that fit a guest's intent and comfort,
- balancing restaurant capacity by timing and proximity,
- guiding guests to quieter routes when they choose calm modes, and
- coordinating show releases so one crowd does not crash into the next.

This is precisely the kind of work EDE is suited to: modeling dynamics, not just snapshots. A ship needs to anticipate pulses because pulses are built into the schedule.

The acoustic layer is amplified at sea

Sound on a ship behaves differently than sound in many land venues because the environment is dense, layered, and often reflective. Multiple venues operate simultaneously within a limited footprint, and sound bleed can destroy clarity and mood. A theater's exit should not flood a corridor with a soundtrack that fights the adjacent lounge. A quiet dining area should not be punctured by an energetic deck program nearby. A guest seeking calm should not be forced into sonic crossfire.

This is why the acoustic layer must be treated as state. A macro ship system should know where the ship is loud, where it is calm, and how that interacts with guest comfort choices. It should also be able to manage sound bleed operationally, not only through design-time construction.

Occupancy-aware behaviors become both comfort and efficiency. If a venue is unoccupied, there is no reason for it to broadcast at full volume. Lowering volume in empty zones reduces bleed into active zones, preserves mood boundaries, and makes the ship feel calmer. It also aligns with energy management and equipment life because systems run for long continuous hours. Lifecycle Automation Layer is the simplest expression of that posture.

Reliability is a psychological contract

On a ship, a failure is not merely a broken exhibit. It is a breach of trust. Guests cannot go to a competitor across the street. Reliability is amplified. When systems fail, guests notice, talk, and generalize. A broken screen becomes a narrative about how the ship is run. A failed queue system becomes a narrative about fairness. A missing language option becomes a narrative about who the ship was designed for.

This is why verification and preventive maintenance become central. If the ship uses occupancy-aware power behavior, it must verify restoration. If it relies on digital signage, it must verify signage truth. If it relies on synchronized Personal Channel delivery, it must verify timing integrity. Commands are not outcomes. The ship needs proof.

Outcome Verification Layer is the posture that verifies displays are functioning, guidance is accurate, and failures are detected early rather than discovered by the public. Trend analytics provides the preventive posture: detect drift before failure. On ships, equipment runs hard. Thermal drift, fan escalation, and environmental stress are normal. Systems that watch trends and schedule maintenance prevent the most damaging failures; the ones guests experience directly.

Reliability is hospitality at sea.

Identity, entitlements, and the "yes or no" ship

Cruise ships are entitlement systems. Cabin access, beverage packages, dining reservations, excursion tickets, age-gated service, VIP zones, and event entries all involve eligibility. Historically, these systems required guests to carry cards, wristbands, or identities in ways that can be clumsy and occasionally invasive.

The posture advocated in this book becomes unusually compelling on ships:

- default to anonymous service where possible,
- use opt-in identity for benefits guests want, and
- use yes or no proofs where eligibility is all the ship needs.

A bartender should not require a guest to display a full identity document with home address and birthdate. The ship should be able to ask, "Is this guest eligible for this service?" and receive a minimal answer.

This is where Web3 credentials and the phone as trust device become practical in a way guests immediately understand because the alternative is constant friction at points of service. A ship that makes eligibility interactions calm feels modern. A ship that makes them invasive feels risky.

The ship as a governance environment

If any venue makes governance obvious, it is a ship. Ships are already governed.

A ship-grade governance boundary should explicitly include confabulation containment, because the ship is an environment where guests rely on guidance as part of daily functioning. The system must be constrained to speak truthfully about state, schedule, and safety posture.

Operationally, this means: claims about ship state must be traceable to operational truth; claims outside that scope must be labeled as optional, informational, or uncertain; and the system must be designed to abstain under missing evidence rather than filling gaps with plausible text.

Safety procedures are explicit. Operating modes exist. Emergency behaviors are rehearsed. A WorldModel Venue approach applies the same discipline to guest-experience automation.

A macro ship system must have:

- explicit outcomes, comfort, fairness, clarity, delight, and inclusion,
- hard constraints, safety, accessibility, privacy, and show integrity,
- allowed action catalogs,
- mode-aware behavior, normal, busy, and incident modes, and
- auditability, what happened, why it happened, and how operators can override.

Governance does not reduce the ship's magic. It protects it. It helps ensure the ship can be more adaptive without becoming unpredictable.

A deployment path that works on ships

Cruise lines do not need to rebuild ships to adopt the architecture in this book. They can stage it.

1. Begin with the Personal Channel: language, captions, personal audio, and comfort modes.
2. Add programmable surfaces where adaptability improves clarity and reduces confusion.
3. Treat the acoustic layer as state and use occupancy-aware behaviors to reduce bleed and fatigue.
4. Build analytics that measures the place: flow, dwell, congestion, and repeated friction points.
5. Add verification and trend analytics so reliability improves over time rather than degrading.
6. Introduce Venue Concierge Interface when the ship is ready to treat guidance as a governed service.
7. Formalize WorldModel operational truth so decisions remain grounded in lived ship state.
8. Use credential-based yes or no interactions for entitlements where they reduce friction and reduce risk.

A ship that follows this path becomes calmer, more inclusive, and more resilient without becoming invasive.

Acceptance tests for confabulation containment on ships:

1. Schedule truth: showtimes, dining windows, and excursion departure guidance must match authoritative schedule state, or present "unknown."
2. Deck and route truth: accessibility routes, closed corridors, and elevator advisories must match current operational truth, or fail conservatively.
3. Safety-mode truth: under drill or incident posture, the system must switch to approved safety messaging sets and suppress non-essential narrative output.
4. Language integrity: translations for safety-adjacent content must come from approved phrase sets, not free generation.
5. Evidence logging: every safety-adjacent guidance statement must be logged as an evidence event so post-incident reconstruction is possible.

The chapter's conclusion

Cruise ships are bounded cities, and bounded cities reveal what matters. They reveal that inclusion is a system, language is runtime delivery, sound is comfort and access, reliability is trust, and governance is the mechanism that keeps power humane.

A ship that adopts these ideas becomes a blueprint for other destination-scale venues. It shows that a place can be more adaptive without becoming creepier, more efficient without becoming colder, and more intelligent without becoming less human.

In the next chapter we move into retail and supermarkets, where the phygital layer is often treated as marketing and where privacy-forward, service-first personalization can transform customer experience without turning commerce into surveillance.

34

RETAIL, SUPERMARKETS, AND THE SERVICE-FIRST PHYGITAL FUTURE

Retail is where the future arrives first and forgives least. A museum visitor will tolerate a little mystery. A theme park guest will tolerate some friction when the payoff is wonder. A shopper, especially in a supermarket, is not there to solve puzzles. They are there to complete a task, often under time pressure, often with children, often with a phone in one hand and a list in the other. When the environment feels confusing, slow, or indifferent, most people do not file complaints. They shorten the visit, then shorten the relationship.

That makes retail a demanding laboratory for the ideas in this book. If hyper-personalization can be made helpful, privacy-forward, inclusive, and operable in a supermarket on a busy weekend, it can be made helpful anywhere. If it cannot, then the architecture is not ready to be public infrastructure.

The thesis of this chapter is straightforward. Retail personalization should not begin with advertising. It should begin with service. Service is what makes the store feel modern, trustworthy, and worth returning to. Advertising becomes appropriate only when it behaves like assistance, and the shopper remains in control.

Confabulation containment is a trust and liability requirement

In retail, wrong answers about availability, price, substitutions, returns, ingredients, allergens, promotions, or policy damage trust instantly and create legal exposure. Any claim that influences purchase behavior must be grounded in approved content and verified operational state, with freshness posture and evidence pointers. If the system cannot support the claim, it must abstain, fall back to a stricter mode, or route the customer to staff or a proof artifact.

Enforce this through delivery-mode discipline, governance constraints on claims, and operator-visible evidence of abstain, fallback, and escalation events.

Goals

Retail environments succeed by reducing friction. The goals are:

- Make the store easy to navigate and easy to understand without signage clutter.
- Provide multilingual support as runtime behavior, not printed compromise.
- Improve accessibility and comfort beyond minimum compliance as default service.
- Reduce queue stress and improve perceived fairness at checkout.
- Support staff with operational truth, not dashboards they cannot use.
- Increase reliability and reduce downtime of digital surfaces and interactive assistance.
- Enable continuous change through content operations, not constant reprinting and rebuilding.
- Maintain privacy-forward behavior by default so trust remains intact.

Ladder stage fit

Retail uses the ladder cleanly, and most value is captured before identity is ever needed.

- Stage 0: Point-of-experience localization
  Language selection, readability, and accessible product clarity at kiosks, service stations, and key surfaces, available without identity.
- Stage 1: Micro hyper-personalization across zones
  Preferences persist across entry, discovery, service counters, and checkout: language, depth mode, and comfort modes, without demanding identity.
- Stage 2: Macro retail, the store as a governed environment
  Queue balancing, staffing prompts, capacity-aware routing, pickup-lane coordination, and incident-mode behavior. Governance becomes mandatory because actions affect other shoppers and staff workload.

Key layers

Retail is operationally harsh and requires a disciplined stack.

- Edge Compute Nodes and provisioned compute: local, replaceable nodes for guidance, media, and assistance, upgraded by provisioning compute rather than rebuild.
- Modular Interaction Infrastructure: consistent readers, buttons, sensors, and service triggers.
- Programmable Canvases: aisle guidance, policy messaging, and product education without permanent signage compromise.
- Personal Channel: multilingual delivery, captions, personal audio, and quiet guidance for shoppers who want it, without requiring it.
- Virtual Docent Layer and Embodied Docent Interfaces where appropriate: governed conversational assistance at service points, in role-appropriate tone, without salesy behavior.
- Acoustic layer: comfort and intelligibility, especially in noisy checkout zones and high-density aisles.
- Lifecycle Automation Layer and Outcome Verification Layer: occupancy-aware calm and equipment life extension with verified restoration and drift detection.
- WorldModel and governance: for larger formats where routing, queues, and staffing are orchestrated across the store.
- Analytics and observability: measure friction points, prevent failures, and tune operations without turning shoppers into inventory.

Boundaries

These are non-negotiable. They exist to keep personalization service-first, privacy-forward, and publicly defensible.

No confabulation about operational facts. Retail systems must never invent prices, discounts, stock status, store hours, return policy details, or allergen and ingredient information. These are operational truth claims with direct consumer harm potential.

If the system cannot verify a retail fact from an authoritative source, it must abstain and route to a verified lookup or staff assistance. "Helpful" does not mean "confident." In retail, confident wrongness is liability.

- Consent requirements: any identity-linked personalization, any biometric processing, and any long-horizon continuity must be explicit opt-in with clear benefit exchange, clear retention limits, and a clear off switch.
- No covert profiling: do not infer sensitive attributes and do not build hidden profiles. If a capability cannot be explained in one plain paragraph, it does not belong in public space.
- Data minimization: collect only what is required for the moment and purpose. Prefer local processing. Delete by default. Logs are operational evidence, not biographies.
- No identity requirement for basic service: core functions, wayfinding, language selection, accessibility modes, and baseline assistance must work anonymously. Identity is invited only for benefits that truly require it.

Ops posture

Retail operations are time-pressured and unforgiving. The system must be boringly reliable.

- Commissioning as proof: guidance correctness, language rendering, and checkout-zone integrity under load.
- Add a truthfulness proof requirement: under peak conditions, the system must either
  (a) display and speak verified operational facts, or
  (b) visibly downgrade to "unknown" with safe fallback.
  It must not fill gaps with plausible-sounding content.
  This is confabulation containment expressed as commissioning discipline, not as an AI aspiration.
- Monitoring for health, integrity, and drift, not only whether screens are reachable.
- Verified restoration for occupancy-aware behaviors: audio reduction, dimming, and powering down.
- Clear fallback: if digital guidance fails, the store remains legible and does not become chaotic.
- Operator override and hold modes for staffing changes and floor resets.
- Content operations with staging, rollback, and verification, because retail changes constantly.

What to measure

Measure what improves service and reduces friction without measuring people by default. Measure the place first.

- Navigation friction: hesitation points, repeated wrong turns, repeated staff questions.
- Queue dynamics: wait variability, fairness perception proxies, and peak-lane stress points.
- Mode usage: language selections, caption and personal-audio use, quiet guidance uptake.
- Accessibility outcomes: whether accessible routes are used and whether barriers correlate with abandonment.
- Acoustic state correlations: noise and stress zones, especially near checkout and service counters.
- Operational integrity: uptime of surfaces, correctness of messaging, and restoration success after Lifecycle Automation Layer actions.
- Drift indicators: thermal trends, fan escalation, error-rate creep, repeated verification anomalies.

Use identity-based measures only as explicit opt-in benefit exchange.

The modern retail bargain is time

Retail used to rely on persuasion: layout psychology, promotions, impulse triggers, and the assumption that a shopper would accept a little wandering. The modern bargain is different. It is time.

A shopper wants to find what they came for quickly, discover a few things that feel relevant rather than noisy, get assistance in a language they understand, avoid stress at checkout, and leave feeling the store respected their day.

This is why the phone changed retail so completely. The phone trained people to expect immediate lookup, immediate translation, and immediate comparison. The store can fight that expectation, or it can use it.

A service-first store uses the phone as a Personal Channel without making it a requirement.

It offers guidance, language, captions, and accessibility support as benefits rather than gates. That stance is how the store stays modern without becoming invasive.

The Personalization Ladder in retail terms

Retail fits the same ladder as museums and parks, but the outcomes are sharper and the tolerance for friction is lower.

- Stage 0: Point-of-experience localization
  This is the simplest improvement, and it is underused. A kiosk, a product education display, or a service station can offer language selection, readability modes, and clear accessible information at the point of use. A shopper does not need an account. They need clarity. Programmable surfaces can change language and layout without the "big English, small Spanish" compromise and can update quickly when inventory, promotions, or policies change.
- Stage 1: Micro hyper-personalization across zones
  This is where retail begins to feel like a system rather than a building. A shopper chooses language, interests, and assistance level once, and the store honors it across entry signage, wayfinding prompts, service kiosks, and education content. Depth modes can be offered without calling them Streaker, Stroller, Student. "Just the basics," "a little more," and "tell me everything" is often enough. This stage works best when it is available without identity.
- Stage 2: Macro retail, the store as a small city
  At peak hours, a supermarket is a small city. Queues form and dissolve. Checkout lanes become bottlenecks. Staff load shifts dynamically. Aisles become congested. Curbside pickup creates a parallel flow system. In some formats, fuel, pharmacy, and returns add their own scheduling dynamics.

This is where WorldModel becomes relevant. The store needs operational truth: what is congested, what is open, what is staffed, what is running slow, and how to guide people without making them feel managed. Governance becomes mandatory because macro actions have externalities. If the system routes people, it can misroute them. If it changes queue prompts, it can create unfair outcomes. If it shifts staffing recommendations, it can overload staff. The same rules apply: explicit objectives, hard constraints, auditability, and operator override.

Recognition in retail and the discipline of not being creepy

Retail is where recognition is most easily abused. It is also where recognition can be most useful when constrained properly.

A shopper does not want a store to know their biography. They want the store to be helpful. A helpful store recognizes context: which areas are congested, where lines are forming, where people are hesitating repeatedly, where staff are being asked the same question, where signage is not being understood, and where a calmer route would reduce fatigue.

These are place truths, not identity truths. Anonymous recognition is often sufficient. A short-lived session token can preserve language and assistance choices across the visit without tying the visit to a real person. Environmental pattern recognition can reveal friction points without tracking individuals across time.

If a store offers deeper personalization, such as loyalty benefits, it should be opt-in and benefit-driven, with minimal retention and clear controls. The shopper must be able to shop without being enrolled into surveillance by default.

Cars, colors, patterns, and the arrival handoff

Retail has a distinctive boundary condition: arrival. In a parking lot, a drive-through lane, a pickup zone, or a fuel forecourt, the journey begins before a person walks through the door.

Correlation between people and cars can be useful under a strict privacy posture. A vehicle can be used as a context signal without being treated as identity.

- A returning car entering a pickup lane can indicate "someone is here for curbside" without requiring a face, a name, or a license plate database.
- Color and pattern recognition can support operational cues, such as distinguishing a pickup vehicle in a crowded lane, without turning the system into personal tracking.
- In drive-through environments, recognizing a vehicle class can support lane management and service timing without identifying the occupant.

The rule is restraint. Recognize context, not identity. Store only what is needed for the moment and forget it on purpose.

Quick-service and drive-through restaurants

A drive-through is a retail environment with hard constraints and fast feedback. Unlike a store, the guest's path is linear, time-bounded, and physically constrained. Every delay is felt immediately, and the cost of confusion is measurable in abandoned orders and amplified frustration. That makes drive-through and quick-service restaurants one of the cleanest places to apply the same architecture described throughout this book: operational truth, governed decisions, verified outcomes, and drift-driven prevention.

Start with the operational truth model. A drive-through has a small set of state objects that matter: lane occupancy, current service phase, estimated time-to-pickup, menu board mode, and audio intelligibility. Treat these as state with provenance and confidence, not as guesses. Name the prohibited behavior: a drive-through system must not confabulate operational state. If the system cannot verify lane occupancy, service phase, time-to-pickup, menu mode, or intelligibility, it must not pretend. It must fall back, alert staff, and keep the interaction calm.

Drive-through succeeds as a pattern because it forces the discipline: truth, governance, verification, then delivery. A lane camera, an order system, and an audio zone are truth sources. They should not contradict each other. The goal is not to identify a person. The goal is to keep the system truthful about what is happening now.

Personalization is allowed only when it is opt-in. The most useful parameters in drive-through are explicit and simple: language, age, and interest, plus a service posture parameter such as fast versus explore. Language changes what is shown and spoken. Age tunes vocabulary and pacing without needing to group people into bands. Interest selects the cut and framing, for example nutrition, ingredients, sustainability, or value. In practice, these parameters can be provided through a consented credential, a QR-based Personal Channel flow, or a loyalty interaction, but the default service must remain available without identity.

Queue shaping is the action layer. When the lane becomes congested, the system should do what a good manager does: reduce decision load and keep throughput stable. That means fewer simultaneous prompts, shorter messages, and bounded recommendations. In busy mode, it may suppress cross-sell prompts that slow ordering or shift them to the pickup moment. It can also suggest an alternate pickup window, or route a large order to a waiting bay, but only as an advisory option that does not punish the guest.

Drive-through is also a proof environment for verification. A command is not an outcome. A menu board can be on and blank. A speaker can be enabled and unintelligible. A microphone can be live and too noisy for comprehension. The verification plane should confirm that the menu board is showing expected content and that audio is within intelligibility bounds. If verification fails, the system should not pretend. It should fall back to a safe posture and alert staff.

Drift is where operators win. Outdoor displays run hot. Fans clog. Audio intelligibility degrades with weather and wear. If brightness drops over weeks, or fan duty rises, or intelligibility flags shift from good to marginal, that is a maintenance event before it becomes a guest-facing failure. The value is not a dashboard. The value is workflow: what triggers an alert, who gets it, and what the recommended fix is.

Vertical deployment quick-start template

- Outcomes:
  list the top five outcomes you are optimizing.
- Minimum architecture:
  which layers are mandatory for this site.
- Acceptance tests:
  what must be proven at handover, and what is monitored thereafter.

Finally, treat intensity as a parameter when the experience includes motion or sensory load. In quick-service this may be as simple as quiet mode versus normal mode for Personal Channel narration, or lower-sensory prompts for visitors who opt in. In adjacent verticals, such as mobile attractions and robot-arm rides, the same idea becomes explicit: a guest who wants gentle should be offered a softer motion profile, and a guest who wants extreme can opt into higher intensity within safety constraints. The underlying principle is the same: do not force a binary choice when a parameter can widen inclusion.

Drive-through works as a reference pattern because it is unforgiving. If the system is not truthful, not governed, and not verifiable, it will fail loudly. If it is, it becomes calm, fast, and quietly excellent.

POS linkage and why "state" beats "profile"

Point-of-sale is where retail systems often make a catastrophic mistake. They treat every transaction as a biography. A better approach treats POS as state. Treating POS as state also prevents a subtle confabulation failure mode: narrative interpretation of transactions. A purchase does not justify invented inferences about intent, identity, or future behavior. It is a state transition that can enable immediate service improvements without creating a story about the person.

Retail governance should therefore constrain how transactions can influence outputs: allow operational consequences, such as receipts, instructions, fulfillment, and immediate guidance, while forbidding speculative narratives about the shopper. A transaction changes what the store should do next in the same way an exhibit state change changes what a venue should do next.

Examples that remain service-first and privacy-forward:

- A purchase triggers a receipt in the shopper's preferred language, with an option for expanded product information through the Personal Channel.
- A pharmacy pickup triggers clear, calm guidance, and a simple verification step that proves entitlement without oversharing.
- A return triggers a support flow that respects accessibility modes and language without forcing staff to guess.
- A loyalty benefit, if the shopper opted in, triggers an entitlement state, not a dossier.

This is where yes or no proofs become relevant. Eligibility should be proven, not disclosed. Retail does not need a home address to verify an age-gated purchase. It needs a policy answer. When retail uses the phone as a trust device, it reduces identity theft risk, reduces staff friction, and increases trust.

Multilingual retail and the end of the bilingual wall

Retail environments often degrade language dignity through signage compromises because the store is full of words: aisle markers, price labels, safety notices, promotions, policies, and service instructions. Printed bilingual solutions do not scale, and they create hierarchy by design.

Runtime language delivery solves this through three planes:

- surfaces for shared truth: aisle markers, policies, operational guidance,
- the Personal Channel for long-tail language support and private depth, and
- staff tools that respect the shopper's selected language and mode.

The store does not need to shout in five languages. It needs to offer the right language at the right moment without implying that one language is default and the rest are add-ons.

Accessibility beyond minimum as a retail advantage

Retail has a hidden inclusion problem: sensory overload. Bright lights, loud music, crowded aisles, and chaotic checkout zones are not merely unpleasant. They exclude. A shopper navigating in a second language, with hearing difficulty, or with sensory sensitivity experiences the store as work.

A service-first store offers comfort modes: quieter narration-forward assistance through earbuds, captions and readable text on surfaces, calmer route suggestions when the store is crowded, and clear simple language modes when requested. These modes should be opt-in and visible. The shopper should not feel analyzed. They should feel respected.

In retail, comfort is not softness. Comfort is throughput. A calm shopper moves more confidently, asks fewer emergency questions, and returns more readily.

Analytics that improves service and prevents failures

Retail analytics is often treated as marketing analytics. That is narrow. The most valuable analytics in retail is service analytics: where people hesitate, where they abandon, where staff are pulled into the same confusion repeatedly, where queue dynamics become unfair, and where signage is misleading because reality shifted. This is place analytics and it can be done anonymously.

The second analytics layer is operational health. Screens fail, sensors drift, networks degrade, and equipment runs hot. Retail is an uptime business. Trend analytics belongs here for the same reason it belongs in museums and parks. Failures often announce themselves slowly. Rising temperature, rising fan speed, and creeping error rates are signals. Catch drift early and uptime improves.

Lifecycle Automation Layer and Outcome Verification Layer in retail terms

Retail has fewer dark galleries than museums, but it has many zones that do not need full intensity all day. Unoccupied areas, closed service counters, overnight hours, and low-traffic periods are opportunities for occupancy-aware behavior: reduce audio in inactive zones to avoid unnecessary bleed, dim or power down certain displays when not needed, restore before shoppers notice, and verify restoration automatically.

This is Lifecycle Automation Layer applied to retail: the environment behaves differently when it is empty, preserving calm, extending equipment life, and reducing waste.

Outcome Verification Layer is the proof layer. If the system changes power and media states dynamically, it must verify outcomes, not merely send commands. A screen that looks on but is frozen is a retail failure because it erodes trust quickly.

A service-first store does not ask staff to walk the floor as human diagnostics. It builds verification into the operating loop.

A practical deployment path for retail and supermarkets

Retail succeeds when it is incremental. Stores cannot close for long and they cannot tolerate fragile experiments. A staged path aligned with the covenant of this book looks like this:

1. Start with Stage 0 improvements: multilingual, accessible point-of-experience clarity on key surfaces and kiosks.
2. Add the Personal Channel as an opt-in benefit: language, depth, and comfort modes that do not require identity.
3. Introduce micro continuity across key touchpoints: entry, discovery, service counter, checkout, and exit.
4. Measure the place, not the person: flow, friction, and repeat confusion points with anonymous analytics.
5. Add operational verification and drift detection for the digital layer.
6. If the store is large and complex enough, introduce macro guidance as a governed service: queue balancing, calmer routing, staff prompts, and capacity-aware suggestions, with explicit constraints and operator override.
7. Use yes or no proofs for eligibility moments, and treat the phone as a trust device, not a data funnel.

This path makes retail more inclusive, calmer, and more efficient without turning it into surveillance.

The chapter's conclusion

Retail and supermarkets will not win the future by becoming louder, more targeted, or more invasive. They will win by becoming more helpful, more multilingual, and more respectful of how people actually live.

A service-first phygital store:

- treats language as runtime,
- treats comfort as access,
- treats analytics as a tool for clarity and reliability,
- treats identity as opt-in and minimal, and
- treats the phone as a personal boundary that can negotiate rather than surrender.

That is better ethics and better business. Trust creates return.

In the next chapter we move to corporate experience centers, where the visitor's time is short, the narrative is deliberate, and the opportunity is to use the same architecture to make the experience feel intelligent and tailored without ever feeling invasive or salesy.

35

Corporate Experience Centers

A corporate experience center is a promise compressed into an hour. It is not a museum, and it should not borrow the museum's voice. It is not a theme park, and it should not compete on spectacle. It is a place where a company asks a visitor for attention and then has to earn that attention quickly with clarity, credibility, and respect.

The visitor who walks into a corporate center is rarely looking for entertainment. They are looking for confidence. They want to understand what is real, what is differentiated, and what will remain dependable after the excitement of the demo fades. They are also increasingly multilingual, time constrained, and surrounded by other options. If the center is confusing, or if it feels like a sales funnel, it fails.

This is why corporate centers are one of the most practical verticals for the architecture in this book. The objectives are clear. The environment is controlled. The visitor journey is intentional. The value of hyper-personalization is immediate: serve different stakeholders differently without building separate rooms for each one and without forcing staff to improvise under pressure.

Confabulation containment is reputational control

In corporate centers, fluent errors about capability, integrations, timelines, compliance posture, performance, or customer outcomes can destroy trust with technical buyers. Visitor-facing claims must be grounded in approved sources, tied to evidence pointers, and bounded by freshness. When proof is unavailable, the system must abstain, fall back to curated delivery, or route the visitor to staff or a documented artifact.

Enforce this through delivery-mode discipline, governance-gated answers, and operator-visible logging of abstain, fallback, and escalation events.

Done well, a corporate center becomes a tool for trust. In corporate environments, confabulation is reputational risk. A fluent, incorrect statement about capability, compliance posture, performance, integrations, customer outcomes, or timelines can destroy trust instantly, especially with technical buyers.

Therefore, any adaptive or conversational layer must be bounded by approved content, and it must default to evidence-bearing answers. If the system cannot support a claim with an approved source reference, it should abstain or route the visitor to a proof artifact, a document, or a staff-led explanation. It explains, adapts, stays calm, respects privacy, and makes visitors feel understood rather than targeted.

Goals

Corporate experience centers are trying to produce confidence quickly without feeling salesy. The goals are:

- Deliver relevance by role and industry without overloading the visitor.
- Maintain credibility through clarity, operability, and proof rather than marketing flourishes.
- Support multilingual delivery and inclusive modes as default behavior.
- Keep the room coherent and calm, even with multiple concurrent tours.
- Make change fast and safe so the center does not go stale.
- Protect trust through privacy-forward defaults and explicit opt-in only where beneficial.
- Prevent public demo failures through verification and drift-aware operations.

Claims, actions, and promises should be emitted as structured decisions, with traceable inputs and evidence.

A credible system treats consequential claims and system actions as decision objects: structured, reviewable records that contain the inputs used, the constraints checked, the outcome selected, and the evidence pointers that justify the result. Narrative output, what a system says conversationally, is treated as a presentation layer that must never outrun the decision object beneath it.

This principle is how you prevent a polished demo voice from becoming a liability.

Ladder stage fit

Corporate centers can use all three stages of the ladder. Most of the value is captured in Stage 1, with a controlled entry into Stage 2 when the center is large enough.

- Stage 0: Localized clarity at the station
  Language selection, readability modes, and clean content at each station, available without identity.
- Stage 1: Micro hyper-personalization across the entire tour
  Visitors choose language, Streaker, Stroller, or Student depth mode, role-aligned interests, and comfort modes once, and the center stays coherent across multiple rooms and stations without requiring identity.
- Stage 2: Macro behavior, the center as an orchestrated environment
  When multiple groups, timed reveals, and limited-capacity rooms create routing and scheduling externalities, the center needs governed guidance, load management, and operator control.

Key layers

A corporate center needs the same architectural layers as other venues, tuned for credibility, calm, and rapid updates.

- Edge Compute Nodes and provisioned compute: modular nodes sized by role, allowing upgrades by provisioning compute rather than rebuilding the center.
- Programmable Canvases: fast narrative updates, clean multilingual delivery, and role-specific story arcs without wall clutter.
- Personal Channel: captions, personal audio, long-tail languages, and quiet mixes as opt-in benefits without making phones mandatory.
- Virtual Docent Layer and Embodied Docent Interfaces where appropriate: governed, curated conversational guidance that adapts to stakeholder roles and depth modes.
- Modular interaction infrastructure: consistent, serviceable triggers, sensors, and affordances across stations.
- Acoustic layer: intelligibility and bleed control as credibility constraints, not AV details.
- Lifecycle Automation Layer and Outcome Verification Layer: occupancy-aware calm and equipment life protection, coupled with verified restoration and drift detection.
- Venue Concierge Interface and WorldModel: only when the center truly operates at macro scale, with explicit governance and operator override.

Privacy posture

Corporate visitors are cautious, and they should be. A center that feels hungry loses trust immediately.

A privacy-forward posture is:

- Anonymous by default for tour delivery, language selection, and comfort modes.
- Opt-in identity only for clear benefits such as follow-up packets, saved preferences across visits, or entitlements.
- Minimal data by design with bounded retention and deletion as normal behavior.
- Prefer session continuity over persistent identifiers.
- Use yes or no proofs for eligibility moments instead of document exposure when applicable.
- Logs as operational evidence, not visitor biographies.

Ops posture

A corporate center cannot afford public failures, and it cannot rely on heroics. A practical operations posture includes:

- Commissioning as proof that the full tour works under realistic load.
- Monitoring that covers health, integrity, and drift, not only device reachability.
- Verified restoration for occupancy-aware automation so a room rarely looks broken when a visitor enters.
- Content operations with staging, rollback, and post-publish verification because messaging changes constantly.
- Drift-based preventive maintenance to avoid demo-day failures.
- Operator controls that are calm and usable, not control-room theater.

Reliability is part of credibility.

What to measure

Measure what improves credibility and reduces friction without measuring people by default. Measure the place first:

- Where visitors hesitate, ask repeated questions, or disengage, by station and segment.
- Which narrative arcs and depths are being selected, Streaker, Stroller, and Student, and where visitors request more detail.
- Multilingual and mode usage: language selections, captions, personal audio, and quiet mixes.
- Acoustic and comfort friction, including where bleed or noise correlates with reduced attention.
- Operational integrity: public failures, time-to-recovery, and verified restoration success.
- Drift indicators: thermal trends, fan escalation, and repeated anomalies that predict failure.

Use identity-based analytics only as an explicit opt-in benefit exchange.

The corporate center's real product is relevance

A corporate center has two enemies. The first is generic narrative. If the story sounds like it could belong to any company in the category, the visitor tunes out. The second is overload. Centers often try to demonstrate everything in one visit and end up proving nothing. Too many features, too many screens, too many talking points, and no coherent arc. The visitor leaves tired rather than convinced.

Hyper-personalization is useful here because it creates relevance without overload. A CFO does not need the same story as a CTO. A safety officer does not need the same story as a marketing director. A procurement lead does not need the same story as a creative lead. A visitor from a regulated industry needs different proof than a visitor from a leisure industry. A visitor on a one-hour schedule needs a different pacing than a visitor on an afternoon schedule.

A corporate center that adapts to these differences feels intelligent. A center that does not adapt feels like a brochure in a building.

The Personalization Ladder in corporate terms

- Stage 0: Localized clarity at the station
  Language selection, readability modes, and clean content at each station.
  Stage 0 matters because a visitor who cannot parse a station quickly often
  disengages rather than asking for help.

- Stage 1: Micro hyper-personalization across the tour
  This is where corporate centers shine. Visitors can choose, explicitly and
  politely, language preference, Streaker, Stroller, Student depth mode,
  interest vectors aligned to role and industry, and comfort modes such as
  captions, personal audio, quieter mixes, and readable typography. The center
  then behaves coherently across rooms and stations. The visitor does not
  repeat themselves. The story stays consistent. The tour feels as if it was built
  for them without requiring identity.

- This stage produces a strong trust effect because it reduces friction and
  increases relevance at the same time.

- Stage 2: Macro behavior, the center as an orchestrated environment
  Not every corporate center needs this. Macro behavior becomes relevant
  when multiple groups tour simultaneously, rooms have limited capacity,
  scheduled demos and timed reveals create pulses, bottlenecks appear, and
  the center wants to route groups differently based on goals and time.

At that point, guidance and scheduling coordination become part of the visitor
experience. Venue Concierge Interface keeps tours calm and coordinated.
WorldModel provides operational truth, so the center does not overpromise itself
under load. Governance matters because routing and scheduling decisions can create
unfair outcomes unless constraints are explicit. In corporate environments,
reputational risk is immediate. A visitor must rarely feel managed, manipulated, or
profiled.

The strongest personalization tool is the narrative arc

Corporate centers often treat personalization as a technology trick. The best personalization is narrative arc selection. A visitor should be able to experience the company's story through different lenses:

- reliability and support,
- safety and compliance,
- operational efficiency and lifecycle,
- design and experience,
- scalability and upgradeability,
- sustainability and resource management, and
- integration with existing systems.

This is where Streaker, Stroller, Student becomes practical. A Streaker wants the headline: the differentiated promise quickly. A Stroller wants the headline and enough context to believe it. A Student wants the full story: architecture, constraints, trade-offs, commissioning posture, support model, and behavior under stress.

Vertical deployment quick-start template

- Outcomes:
  list the top five outcomes you are optimizing.
- Minimum architecture:
  which layers are mandatory for this site.
- Acceptance tests:
  what must be proven at handover, and what is monitored thereafter.

A corporate center that serves Students well is rare and valuable. Students are often the people who make or break deals.

Programmable Canvases as the agility layer

Corporate centers live in a constant update cycle. Product lines evolve. Regulations change. New case studies appear. New industries become targets. A center that cannot update quickly becomes stale, and staleness is poison in a space designed to signal modern capability.

Programmable Canvases are operational here. They allow the center to swap exhibit modules without construction, support multiple languages without signage clutter, present different narrative arcs for different audiences, and update messaging to reflect current realities without reprinting the building.

They also allow multiple layers on a single surface: a fast executive summary layer, an optional deep layer, and a proof layer, without turning the room into a wall of text. This is how a corporate center stays credible over time. It behaves like a living system, not a static brochure.

The Personal Channel: privacy-forward personalization without awkwardness

Corporate visitors are often alert to manipulation. They will notice if a center behaves as if it is harvesting them.

The Personal Channel is useful here because it offers individualized value without creating an invasive atmosphere. The visitor can choose language and captions, personal audio for clarity, quiet mixes when the room is dense, and deeper technical content delivered privately and optionally. The room stays clean. The visitor stays in control.

Most benefits do not require identity. The center can remain available without identity, and that posture tends to increase trust. If the visitor chooses to opt in, for example to receive a follow-up packet, a personalized recap, or an RFP-ready specification set, the benefit exchange is obvious. Identity is invited, not extracted.

The acoustic layer, because trust is fragile

Corporate centers often overuse sound. Open-plan layouts can look modern while sound bleed makes adjacent stations unintelligible. Immersive audio can undermine comprehension when intelligibility is treated as secondary.

In corporate environments, acoustic clarity is credibility. If a visitor cannot hear, cannot focus, or cannot maintain attention because the space is noisy, the center feels undisciplined.

Treat the acoustic layer as state and manage it intentionally: zoning and intelligibility as design constraints, Personal Channel audio for clarity and language, quieter narration-forward modes for comfort and comprehension, and occupancy-aware behavior to reduce sound bleed when areas are not in use.

Lifecycle Automation Layer becomes professionalism. Empty rooms should not remain loud. Projectors should not run unnecessarily. The center should feel calm and intentional.

Analytics that serves operations, not dossiers

Corporate centers need analytics, and the purpose is operational improvement and proof rather than marketing voyeurism. A center should measure where visitors hesitate, where questions cluster, which parts of the story create confusion, and which stations reliably deliver clarity. This can be done with anonymous place analytics.

The second purpose is reliability and preventive maintenance. If the system drifts, it will fail in front of the buyer. Trend analytics that detects drift belongs in the operational posture. Verification matters as well. If the center uses automated power saving or dynamic mode switching, it must verify outcomes. A black screen in a demo is a trust event.

Acceptance tests for confabulation containment in corporate centers:

1. Capability claims: any claim about features, integrations, or limits must map to an approved source or be refused.
2. Compliance claims: the system must not imply certification, approval, or legal compliance. It may describe posture and evidence artifacts only.
3. Performance claims: no invented numbers. Use measured, documented values, or label as "not available."
4. Evidence routing: the system can surface a proof artifact, document, or diagram when a visitor asks "how do you know."
5. Logging: capture a record of high-stakes answers so post-demo disputes can be reconstructed.

Governance as a credibility guardrail

Corporate centers rarely think they need governance because the environment is controlled. The credibility risk is still high. Two failure modes dominate: personalization that feels like profiling and automation that feels like manipulation.

Governance prevents both. The center should hold explicit constraints: privacy-forward by default, opt-in identity only, no covert personalization, clear explanation of what the system does and does not do, and operator override for routing or scheduling decisions when multiple groups are present.

If the visitor cannot understand the posture, they will assume the worst. Clarity is a trust strategy.

Add one more constraint explicitly: confabulation containment. The system must not claim facts it cannot ground in approved materials, and it must not invent compliance or performance assurances. When evidence is missing, the correct behavior is: abstain, ask, or route to proof.

This turns "trustworthy posture" into an enforceable operating requirement rather than a brand promise.

Web3 proofs and the end of the document ritual

Corporate centers are full of eligibility moments: visitor check-in, access to restricted demo rooms, participation in safety-controlled areas, and sometimes age-gated service in hospitality zones. These moments should not require a visitor to surrender a document that exposes more than necessary.

The yes or no proof posture applies cleanly. A corporate visitor proves eligibility without handing over a dossier that invites identity theft. This is also where Web5 negotiation becomes relevant as a near-future posture. The visitor's phone agent can represent preferences and proofs, and the center can request only what it needs for the moment it needs it.

A center that adopts this posture feels modern, and it signals that the company understands privacy as a first-class design requirement.

A deployment template

A practical deployment pattern:

1. Build Stage 0 clarity: multilingual, readable stations, no identity required.
2. Add Stage 1 coherence: language, depth mode, and interest vectors persist across the tour.
3. Use Programmable Canvases to keep surfaces adaptable and updateable without construction.
4. Offer the Personal Channel as an opt-in benefit: captions, personal audio, quiet mixes, and private deep content.
5. Treat acoustics as credibility: manage sound bleed and intelligibility and use occupancy-aware behavior to keep unused rooms calm.
6. Add anonymous analytics for narrative improvement and operational tuning.
7. Add verification and drift detection to prevent public demo failures.
8. If the center hosts multiple groups or timed sequences, add macro guidance under governance constraints.

This path produces a center that feels tailored without feeling invasive.

The chapter's conclusion

Corporate experience centers exist to create confidence. Confidence is not created by polish. It is created by truth, bounded claims, and evidence. A center that can say "we do not know" and then route to proof is more credible than a center that always has an answer.

That is why confabulation containment belongs in the core posture: it protects the company from its own demo layer. Confidence is created when the visitor feels three things at once: the story is relevant, the system is real, and the posture is trustworthy.

Hyper-personalization, done as architecture rather than gimmick, can deliver all three. It allows the same physical center to speak differently to different audiences without losing coherence. It makes inclusion normal by treating language and accessibility as runtime choices rather than printed compromises. It keeps the center upgradeable, so it stays credible as the company evolves. It stays privacy-forward so trust is earned rather than demanded.

A corporate center that behaves this way does not feel like a showroom. It feels like a competent organization made visible.

In the next chapter we return to the broader operations layer, because the difference between a compelling center and a must-have center is not the first visit. It is whether the system stays reliable, updateable, and calm as months and years pass.

36

SMART CITIES AND PUBLIC INFRASTRUCTURE

A smart city is not a city that collects more data. A smart city is a city that serves better.

That distinction matters because cities are not venues in the commercial sense. They are civic environments. They contain people who did not consent to be part of a product experiment simply by walking outside. They contain obligations that private destinations do not carry: equity, public accountability, procurement transparency, legal due process, and the reality that the most vulnerable residents are often the ones most exposed to "innovations" that quietly become coercive.

This is why the architecture in this book belongs in cities, and why cities demand the strongest form of the covenant: positive, helpful, idea-forward, privacy-forward, and governed.

If a theme park is a small city pretending to be a story, a city is a story pretending to be neutral. A city tells people what it values through what it builds: where it makes walking easy, where it makes it hard, which languages it honors, which disabilities it anticipates, and whether it treats the public as citizens or as inputs.

A WorldModel approach is relevant here because cities are not short on devices. They are short on coherence.

Civic coherence has a hard constraint that commercial venues can sometimes ignore: due process.

A city cannot justify action by saying "the model thinks." If a public system issues guidance, denies access, prioritizes service, or changes routing, it must be able to justify the decision against explicit policy and current operational truth. This is why the governance gate must be able to abstain under uncertainty. In civic space, the cost of confident error is not only user frustration. It is loss of public trust.

Confabulation containment is a civic obligation

In public infrastructure, wrong guidance can change behavior at scale, affect safety, and create inequitable outcomes. Public-facing claims must be grounded in published policy and verified operational truth, carry a freshness posture, and expose evidence pointers suitable for audit and accountability. When verification is unavailable, the system must abstain, fall back to safer delivery, or escalate to a human channel.

Enforce this through delivery-mode discipline, governance gates that can refuse or constrain outputs, and operator-visible audit logs of abstain, fallback, and escalation events.

Goals

Cities should adopt these architectures only for outcomes that are civic, legible, and defensible. The goals are:

- Improve legibility of public space so people can navigate without confusion.
- Honor language dignity and accessibility as default public service, not optional add-ons.
- Maintain privacy-forward public infrastructure where basic service does not require identity.
- Coordinate operational truth across fragmented systems so signage, apps, and staff do not contradict reality.
- Improve safety posture in normal, busy, and incident modes without coercive surveillance.
- Reduce operational waste and public failures through verification and preventive maintenance.
- Preserve public trust through explicit governance, auditability, and bounded retention.
- Enable upgradeability through modular compute and clean interfaces, not constant civil works.
- Prove truthfulness under uncertainty: public-facing systems abstain rather than guess when state is missing, stale, or contradictory.

Ladder stage fit

Cities use the same ladder, and Stage 2 requires unusually strong governance because the environment is public.

- Stage 0: Localized public clarity
  Multilingual, accessible information at civic kiosks, hubs, and public surfaces, available without identity.
- Stage 1: Micro continuity across civic touchpoints
  Preferences such as language and accessibility modes persist across multiple interactions in a corridor or facility without identity.
- Stage 2: Macro orchestration as governed public service
  Event precinct flow, transit ingress balancing, accessibility-aware routing, detour management, and incident-mode guidance. Governance is mandatory because actions affect equity, safety, and public behavior.

Key layers

A city implementation is not one product. It is an integration discipline.

- Edge Compute Nodes and provisioned compute, conceptually: distributed local compute at hubs and facilities, upgraded by provisioning compute rather than rebuilding infrastructure.
- Modular Interaction: modular I/O patterns for kiosks, triggers, sensors, and facility interfaces that remain serviceable and consistent.
- Programmable Canvases: programmable public canvases for dynamic guidance, multilingual parity, and operational truth messaging.
- Spatial AR Interface: opt-in spatial interface overlays for guidance and accessibility cues without signage clutter, constrained by governed truth.
- Personal Channel posture: phone as privacy boundary, multilingual and accessibility modes delivered as benefits, not gates.
- Acoustic layer: comfort-aware navigation in noisy hubs, stations, and civic interiors, with quiet-route options as normal service.
- Lifecycle Automation Layer and Outcome Verification Layer: occupancy-aware power and audio behaviors with verified restoration and drift detection for public infrastructure.
- Venue Concierge Interface, conceptually: the service interface that translates citizen intent into governed outcomes.
- WorldModel plus CGL, EDE, ICL, and MAOL: shared operational truth, dynamics modeling, continuity, orchestration, and governance to prevent fragmented reality.
- Analytics and observability: measure the place first, keep systems truthful, and prevent public failures.

Privacy posture

Cities must default to dignity.

A civic-grade privacy posture is:

- available without identity access to core services, navigation, language, and accessibility.
- Identity only when strictly required for specific entitlements, and even then, as minimal yes or no proofs where possible.
- Prefer place analytics over people analytics, measuring congestion and state rather than individuals.
- Local-first processing wherever possible, because the city must not require a surveillance cloud to function.
- Strict minimization at ingress, short retention, and deletion as normal behavior.
- Auditability that proves integrity without creating dossiers.

A smart city that requires identity for basic navigation is not smart. It is coercive.

Boundaries

These are non-negotiable. They exist to keep personalization service-first, privacy-forward, and publicly defensible.

- Consent requirements: any identity-linked personalization, any biometric processing, and any long-horizon continuity must be explicit opt-in with clear benefit exchange, clear retention limits, and a clear off switch.
- No covert profiling: do not infer sensitive attributes and do not build hidden profiles. If a capability cannot be explained in one plain paragraph, it does not belong in public space.
- Data minimization: collect only what is required for the moment and purpose. Prefer local processing. Delete by default. Logs are operational evidence, not biographies.
- No identity requirement for basic service: core functions, wayfinding, language selection, accessibility modes, and baseline assistance must work anonymously. Identity is invited only for benefits that truly require it.

Ops posture

Public infrastructure is judged by reliability because failure becomes public and political.

A city operations posture that matches this architecture includes:

- Commissioning as proof that guidance is truthful and accessible under real conditions.
- Monitoring for health, integrity, and drift, not only whether a device is reachable.
- Verified restoration for any occupancy-aware behavior, displays and audio must be correct before the public relies on them.
- Degraded-mode resilience: core services operate when external connectivity is imperfect.
- Explicit operating modes: normal, busy, and incident, with tightened constraints during incidents.
- Operator override and controlled publishing, because civic messaging must be accountable.
- Preventive maintenance driven by drift detection, so failures are prevented, not discovered by citizens.

What to measure

Measure what improves service and equity without measuring people by default. Measure the place first:

- Repeated confusion points: where citizens hesitate, backtrack, or need staff help.
- Flow and congestion: hubs, corridors, event precincts, and transit ingress patterns.
- Guidance integrity: whether public surfaces and digital services match live closures and detours.
- Accessibility outcomes: whether accessible routes remain truthful and usable in real conditions.
- Multilingual usage: language selections on public interfaces, and where long-tail language support is needed.
- Acoustic state correlations: noise and stress zones, and demand for calm routes in noisy areas.
- Operational integrity: uptime and correctness of public canvases and kiosks, restoration success after Lifecycle Automation Layer behaviors.
- Drift indicators: thermal trends, fan escalation, error-rate creep, and repeated verification anomalies.

Use identity-based measures only when strictly required, consented, and bounded.

The real problem cities face is coordination

Most cities already have systems for everything. Transit has systems. Parking has systems. Events have systems. Permitting has systems. Public safety has systems. Accessibility has systems. Tourism has systems. Digital signage has systems. Sometimes the systems are good. More often, they are fragmented.

Fragmentation creates a predictable civic failure. Different parts of the city tell different truths. A sign says one thing, an app says another, and the street says a third. A transit platform is crowded but guidance behaves as if it is calm. A detour exists but official navigation does not reflect it. A public event changes flow patterns but surrounding neighborhoods absorb the shock without support.

This is where WorldModel as operational truth becomes relevant. A city needs a shared, living representation of state that can support coherent service across systems. The city does not need a single central brain that controls everything. It needs a shared truth layer that prevents reality from fragmenting.

The city version of the Personalization Ladder

Cities benefit from the same ladder described earlier, and the intent is different.

- Stage 0: Localized public clarity
  At this stage the city does not personalize people. It localizes information. A kiosk in a transit hub, a digital sign in a civic building, an information wall in a tourist corridor, or a wayfinding point at a park can offer language selection, readable typography, and clear accessible information at the point of use. Treat key surfaces as Programmable Canvases and messaging can change quickly, detours can be reflected, languages can rotate without permanent hierarchy, and yesterday's truth is less likely to be printed into permanent materials.
- Stage 1: Micro continuity across civic touchpoints
  At this stage the city honors preferences across multiple interactions. A visitor selects a language once and that choice carries across several kiosks, signs, or services in a corridor. A resident chooses accessibility and comfort modes once and those modes carry through a civic building journey. This stage should remain available without identity. Preference continuity does not require a name. It requires intent.
- Stage 2: Macro, governed city-scale orchestration
  This is where the city begins shaping flow and service outcomes:
    - dynamic wayfinding across districts during events,
    - load balancing across transit entrances and platforms,
    - accessibility-aware routing that updates as reality changes,
    - queue and crowd management around stadium precincts, festivals, and emergencies,
    - coordinated signage and messaging that reflects live closures and hazards,
    - power and acoustic comfort behaviors in public facilities, and
    - staff deployment prompts for field operations.

At this stage, governance becomes mandatory because city-scale actions have externalities. Routing decisions affect neighborhoods. Messaging affects crowd behavior. Prioritization affects who gets service and who gets pushed aside. A city cannot deploy macro intelligence without explicit objectives, hard constraints, auditability, and override.

A city must be smarter than surveillance

The most dangerous error in smart-city thinking is to treat recognition as the path to service. The path to service is state.

Most city improvements do not require identity. They require better awareness of what is happening in public space:

- occupancy and congestion,
- closures and detours,
- schedule changes and event pulses,
- accessibility route availability,
- acoustic state in enclosed areas such as stations, underpasses, and civic lobbies,
- equipment status for signage, lighting, escalators, and elevators, and
- safety posture and incident mode state.

These are measurements of place, not measurements of people. Cities should default to anonymous place analytics and be proud of that default.

A smart city is not a city that knows who you are. A smart city is a city that knows what is happening.

Identity, when truly needed, must be yes or no

Cities do have identity moments. The question is how they are handled.

There are civic interactions that require proof:

- eligibility for a benefit,
- access to a restricted area such as a secure facility,
- age-gated service in a licensed environment,
- resident verification for a local permit, and
- staff credentials for field access.

The old model makes people show documents that expose far too much: full name, address, birthdate, document number. The city becomes a collection point for identity theft risk. The modern model is proof without oversharing. The city asks a question and receives a minimal answer.

The bar example is intentionally blunt because it is universal. You should not have to hand over your home address and full identity details to answer a simple eligibility question. The same principle applies to civic services. A resident should be able to demonstrate what matters and reveal nothing else. This reduces liability and reduces the temptation to retain sensitive information forever.

Web5: the phone agent as the citizen's representative

Cities are where Web5 becomes more than a slogan. A city interacts with millions of people who cannot be expected to download a dedicated app for each service, department, and corridor. Fragmentation becomes exclusion because only the most persistent and digitally confident participate fully.

The phone agent changes the model. Instead of the city becoming the identity provider for everyone's life, the person's device becomes the stable layer:

- preferences live with the person,
- credentials live with the person,
- consent lives with the person, and
- the city requests only what it needs in the moment for a specific purpose.

A DWN-style personal data store allows continuity without centralizing personal life inside a municipal database. The city remains helpful. The citizen remains protected. This is a clean path to smart-city systems that do not require the city to become an identity warehouse.

The acoustic layer: cities as sensory environments

Cities are not only visual. They are acoustic. Noise floors in stations, tunnels, civic buildings, and dense precincts affect intelligibility, stress, and accessibility. Wayfinding in a loud station is not only a signage problem. It is a comfort and comprehension problem.

A WorldModel approach treats acoustic state as operational truth: where the environment is already loud, where speech intelligibility is degraded, where sound bleed between zones creates fatigue, and where calm routes can be suggested as normal options.

This is access, not luxury. A person who is easily overstimulated does not need special handling. They need the city to offer a calmer path when possible and the Personal Channel to provide guidance without adding to the noise. Comfort becomes a public service.

Power behavior as public responsibility

Cities run equipment at scale. Lighting, displays, kiosks, escalators, HVAC, and public media surfaces all have energy cost and lifecycle cost.

The Lifecycle Automation Layer posture travels here cleanly:

- If a public facility zone is unoccupied, reduce unnecessary audio output to avoid sound bleed and fatigue.
- Dim or power down non-critical displays where appropriate to extend life and reduce waste.
- Restore before the public notices so service quality stays high.
- Verify restoration so command sent does not become assumed success.

Public infrastructure is where preventive maintenance is most valuable. Failures erode trust and repairs are expensive, visible, and politically sensitive. Trend analytics becomes civic competence. A city that prevents failures feels calm. A city that constantly fails feels neglected.

Governance is not optional in a city

Private venues can sometimes get away with informal governance. Cities cannot. A city-scale system must be able to answer without drama: what did the system do, why did it do it, what policy constrained it, who could override it, what was logged, what was retained, what was deleted, and how fairness and accessibility were protected.

This is why CGL becomes essential in civic environments. It turns values into enforceable constraints and makes decisions auditable. It is also why blockchain, if used at all, should be treated as optional and primarily as a tamper-evident integrity layer for logs. The city does not need blockchain to have identity. It may choose tamper-evidence for decision logs. These are separate choices and separating them prevents ideology from driving procurement.

Governance must be legible to the public. A city cannot deploy black-box routing and claim neutrality. Neutrality is a policy decision, and the policy has to be explicit.

A cross-city pattern library that stays humane

Civic wayfinding as a governed service

Dynamic guidance that reflects closures and congestion, delivered through surfaces and Personal Channels, with clear opt-out and no identity requirement.

Event precinct orchestration

During stadium events and festivals, the city can balance inflow and outflow, adjust signage, coordinate transit prompts, and keep pedestrian routes safe while treating fairness and accessibility as hard constraints.

Eligibility without documents

Yes or no proofs for benefits and services, so citizens do not surrender full documents for simple validations.

Comfort-aware navigation

Calm routes, reduced sensory load options, and Personal Channel guidance in noisy environments offered as normal choices.

Infrastructure truth and verification

Automated verification that public displays are functioning, guidance is accurate, and failures are detected early rather than discovered by the public.

These patterns make cities more capable without making them creepier.

The chapter's conclusion

Smart cities fail when they treat the public as a data source. Smart cities succeed when they treat the public as people.

WorldModel operational truth, Venue Concierge Interfaces as a service interface, privacy-forward continuity, consent-driven modes, and explicit governance are how a city becomes more coherent without becoming coercive. A city that adopts these ideas does not become a surveillance city. It becomes a legible city. Language is honored at runtime. Accessibility is normal. Eligibility can be proven without oversharing. Systems improve through observability and preventive maintenance rather than public failure.

In the next chapter we turn to the economics and operationalization layer, because even in cities, long-term success depends on unglamorous disciplines: commissioning, verification, maintenance, auditability, and the ability to upgrade by provisioning sufficient compute rather than rebuilding infrastructure every time capability advances.

Part VIII

Operations Automation, Lifecycle Automation Layer, Outcome Verification Layer, and Predictive Maintenance

A venue is not a demo. It is a living system with drift, degradation, staff turnover, seasonal content shifts, and budget constraints. This part is about making the experience behave like an upgrade, not like a fragile installation that slowly breaks.

Chapter 37 covers commissioning, monitoring, and support, including what must be measured so support is diagnostic, not guesswork.

Chapter 38 frames Lifecycle Automation Layer and Outcome Verification Layer as the mechanisms that keep promises verifiable over time.

Chapter 39 links power posture to lifecycle posture, because energy behavior and reliability behavior are coupled.

Chapter 40 addresses content operations and fast change without rebuild, the difference between "we can update" and "we can update safely."

Chapter 41 defines the upgrade path as an architectural commitment, not a procurement event.

Chapter 42 closes with a forward posture: how to design systems so future capability arrives as a controlled expansion, not a replacement.

If you only read one part as an operator, read this one. If you are an owner, this part tells you what to demand so the venue remains adaptable without endless rebuild cycles.

37

COMMISSIONING, MONITORING, AND SUPPORT

A venue system is not finished when it turns on. It is finished when it can be trusted.

That sounds like mood, and it is not. Trust is a technical property with operational consequences. A trusted system behaves predictably, recovers gracefully, tells the truth about its own health, and can be supported by the people who actually work in the building. An untrusted system becomes theatre. Staff stop relying on it. Guests stop noticing it. The venue retreats to manual workarounds, and the installation becomes expensive décor.

Commissioning is where a system earns the right to be trusted. Monitoring is how it keeps that right. Trust also has an integrity dimension.

A system can be up and still be untrustworthy if it produces confident guidance that is not grounded. Confabulation containment must therefore be commissioned the same way you commission safety and accessibility: as an acceptance-tested property.

If the venue cannot demonstrate abstain behavior, safe fallback behavior, and operator-visible evidence when grounding is insufficient, it has not earned the right to guide the public. Support is how it stays alive after opening week, when the building becomes busy, staff changes, content evolves, and equipment begins its slow drift toward failure.

This chapter is written with a blunt premise: if commissioning and support are not designed as part of the experience, you do not have a modern venue system. You have a fragile moment.

The economic reality: reactive maintenance costs more

Reactive maintenance during operating hours often costs substantially more than scheduled maintenance because of labor premiums, downtime, and urgency. Trend analytics that detects drift early shifts work from emergency repair toward planned maintenance and upgrades. This is not a "nice to have." It is the economic basis of sustainable operations.

A simple business case you can audit

This book avoids invented ROI numbers. Use the model below with your own inputs from finance, operations, and incident logs.

Let:

- H_e = emergency hours per month (unplanned, during operating hours)
- R_e = loaded emergency labor rate per hour
- H_p = planned maintenance hours per month
- R_p = loaded planned labor rate per hour
- D = guest-facing downtime hours per month attributable to experience systems
- C_D = fully loaded cost per downtime hour (pick a metric your finance team accepts)
- ΔH_e = expected reduction in emergency hours from drift detection and verified restoration
- ΔD = expected reduction in downtime hours from faster diagnosis and rollback

Then a simple monthly operational benefit estimate is:

$$\text{Benefit} = (\Delta H_e \times R_e) + (\Delta D \times C_D) - (\text{added planned work} \times R_p)$$

Validate the inputs with overtime logs, service invoices, incident tickets, and closure reports. If you cannot measure the inputs today, that is a procurement gap.

Drift signals worth instrumenting

Add integrity drift alongside hardware drift.

In addition to hardware drift and network drift, instrument integrity drift: increases in abstain rate, fallback rate, override rate, and confabulation incident count. A system that becomes more fluent but less conservative has degraded, even if it remains online. Practical drift signals to monitor early (summary):[xlviii]

- Lamp hours, thermal behavior, fan speeds, and filter status trends.
- Network latency spikes, packet loss, and clock drift outside defined bounds.
- Repeated soft faults, reboot patterns, or "fixed by restart" incidents.
- Audio intelligibility degradation, level drift, or amplifier protection triggers.
- Display or projection brightness drift, calibration drift, or mapping alignment drift.
- Increasing operator overrides, escalations, or guardrail triggers.

Commissioning is proof

Most teams treat commissioning as a scramble to get everything working. That work is necessary, and it is not sufficient. A credible commissioning process proves the system will keep working under the conditions it will face: crowds, time pressure, imperfect networks, mixed languages, repeated daily power cycles, and the endless small disturbances of public operation.

Debugging asks, "Does it work now?" Commissioning asks, "Can we demonstrate it keeps working, and can we detect when it stops?" Operators care about the second question because it protects the venue's reputation.

The commissioning sequence that scales

A dependable commissioning process follows a sequence. Sequence prevents magical thinking.

1) Freeze the intent

Before testing anything, freeze what the venue is supposed to do. Intent includes:

- narrative behaviors, including which experiences are deterministic and which are adaptive,
- multilingual behaviors across room, surface, and Personal Channel planes,
- accessibility and comfort modes, including captions, sign language, personal audio, quiet mixes, and other modes, and
- operational behaviors such as occupancy-aware power and acoustic changes.

If intent is not frozen, tests become meaningless because the target keeps moving. You do not need to publish internals to commission effectively. You need to state behaviors and outcomes clearly enough to test

2) Establish the truth sources

A system cannot be commissioned if it lacks a clear notion of truth.

For micro deployments, truth is often local: exhibit state, media state, and a limited set of session preferences. For macro deployments, truth includes operational state, occupancy, queue posture, acoustic state, power state, and constraint modes.

Commissioning requires one simple capability: the system can answer "What does it believe is happening right now?" If it cannot, it cannot be trusted to act coherently.

3) Validate the experience planes independently

A common commissioning failure is testing only the fully integrated system, which makes diagnosis slow and fragile.

Validate each plane on its own before testing cross-plane coherence:

- the room plane:
  shared media, shared cues, shared timing,
- the surface plane:
  canvases, signage, language rendering, readability modes, and
- the Personal Channel plane:
  synchronization, language switching, captions, sign language video, quiet mixes, and device behavior.

Once each plane is stable, test the transitions between them. Real failures hide in those transitions.

4) Commission the acoustic layer as a system layer

If acoustics is part of experience, acoustics must be commissionable.

That means verifying intelligibility in the conditions the venue will see, verifying sound bleed boundaries between adjacent zones, verifying volume and mix policies, including quiet and narration-forward modes, and verifying occupancy-aware behaviors that reduce output when a zone is unoccupied.

A venue that rarely commissions acoustic behavior will look perfect on paper and feel exhausting in practice.

5) Commission Lifecycle Automation Layer behaviors as hospitality behaviors

In this book, Lifecycle Automation Layer is operational calm as well as energy management. Commissioning it means proving that:

- unoccupied zones reduce unnecessary audio to prevent sound bleed,
- power behavior changes appropriately in empty zones, including dimming or powering down defined equipment classes where appropriate,
- restoration happens before visitors can perceive anything missing, and
- restoration does not create new failures elsewhere, such as timing drift, wrong language mode, or missing media.

A zone that feels broken erodes trust immediately. This is about the visitor's emotional experience of reliability.

6) Verify outcomes, not commands

This is where professional systems separate from hopeful systems. A command can succeed and still produce no outcome. A projector can receive an "on" command and still show no image. A display can be on and frozen. A light can be enabled and dim. A media player can be running and out of sync.

This is why the Outcome Verification Layer belongs inside commissioning, not after it. The posture is blunt. In public venues, hope is not an operating strategy. The system must confirm that restoration is real, not assumed.

7) Capture the baseline and define drift

Commissioning is also where you capture baseline behavior so future drift can be detected.

Capture envelopes for:

- expected brightness or visual output behavior for key displays and projectors,
- expected audio levels and mix behavior in each zone,
- expected thermal behavior for critical equipment,
- expected network performance ranges for critical paths, and
- expected timing behavior for synchronized cues.

You do not need perfect numbers. You need a baseline envelope and a definition of abnormal.

A system that cannot detect drift will not become better over time. It will become more fragile, and then it will be ignored.

Runbooks as operational governance

A runbook is the executable translation of governance into repeatable operator action. Governance states what the system is allowed to do. A runbook states what humans and automation do when reality changes under load, under uncertainty, or under incident conditions. In this architecture, runbooks are not vendor troubleshooting notes. They are governed operational playbooks, written so that staff can act consistently across shifts, and so that the system behaves predictably when the venue is stressed.

Runbooks exist for one reason: public venues cannot rely on improvisation. A live venue is a shared environment. Decisions have externalities. When staff improvise, policy drifts, messaging becomes inconsistent, accessibility outcomes become uneven, and recovery becomes fragile. A reference grade system treats operational response as part of the system, not as something left to institutional memory.

Why runbooks belong inside commissioning

Commissioning is where trust becomes measurable. Runbooks belong here because they are the last mile between policy and real behavior. If the venue cannot execute incident and degraded-mode procedures predictably, then governance exists only on paper.

Commissioning must therefore validate not only that the system works, but that the venue can operate it:

- Under peak attendance,
- Under degraded network conditions,
- Under partial subsystem failure,
- Under authoritative safety states,
- And under conditions where signals conflict.

A venue that can play media on opening day, but cannot execute a consistent runbook on a normal Tuesday, will become unstable over time.

Minimum runbook set for governed venues

At minimum, a governed, adaptive venue should maintain runbooks for:

1. Incident Mode activation and clearance
2. Degraded Mode activation and graceful degradation posture
3. Verification failure response and verified restoration
4. Safety override behavior, including life-safety boundary and emergency stop boundary handling
5. Operator override and hold procedures, including audit trails
6. Confabulation containment escalation, including abstain and fallback behavior for visitor-facing claims
7. Change control, update validation, and rollback
8. Post-incident review and corrective action capture

This list is not bureaucracy. It is the practical definition of operability.

Runbook design principles

A runbook must be short enough to use under stress, but complete enough to prevent ambiguity. The principles are:

1. Authority is explicit
    The runbook states who can execute it, who can clear it, and what overrides what. If authority is unclear during an incident, the venue will waste time negotiating control.
2. Triggers are explicit
    The runbook states what conditions require action and which signals are authoritative. It distinguishes authoritative declarations from advisory inference.
3. Actions are bounded
    The runbook states what the system may do automatically, what operators may do manually, and what is prohibited.
4. Verification is required
    A runbook ends only when outcomes are verified in the physical world, not when commands are issued.
5. Evidence is produced without dossiers
    Consequential actions create decision records and incident records. These records are bounded, access-controlled, and retention-limited. The venue can prove what happened without retaining biographies of visitors.
6. Safe failure is normal
    When truth is missing or contradictory, the correct behavior is conservative fallback, refusal, or escalation. Missing state is not a reason to guess.
7. Language is governed
    In incident contexts, public messaging must be pre-authored and owner-approved. Freeform generation is not permitted for safety-adjacent instructions, closures, routing claims, eligibility, or emergency prompts.

Standard runbook structure

Use a single structure across all runbooks so staff can operate by pattern recognition. Every runbook should include:

A. Name, scope, and version
B. Trigger conditions and authority class
C. Precedence and override rules
D. Immediate actions, automatic and manual
E. Messaging posture and channel behavior
F. Verification steps and pass conditions
G. Evidence and logging requirements
H. Clearance criteria and return-to-normal sequence
I. Post-incident review requirements

Worked example: Incident Mode Activation runbook

Purpose

Activate Incident Mode when an authoritative safety or hazard signal exists, clamp automation and messaging to a conservative, auditable posture, and maintain safe, inclusive operation until the incident is cleared and verified restoration is complete.

A. Trigger conditions and authority class

Treat the following as authoritative triggers:

- Life-safety state activation (fire alarm, emergency communications activation, or equivalent)
- Emergency stop chain asserted (E-stop, show-stop, ride-stop, or equivalent)
- Operator-declared hazard, closure, or responder-only zone

Treat the following as advisory triggers that may require escalation before Incident Mode:

- Conflicting occupancy signals
- Sensor anomalies without corroboration
- Unverified reports of obstruction, congestion, or closures

B. Precedence and override rules

1. Life-safety is unconditional priority. The system observes and reacts, but does not arbitrate.
2. Emergency stop chains override all experience behavior. The system may observe and log, but cannot defeat, delay, or auto-reset.
3. Operator-declared holds and closures are authoritative within their scope and time window.
4. Incident Mode constraints supersede personalization and optimization goals.

C. Immediate actions on Incident Mode activation

Within the affected scope, perform all of the following:

1. Clamp automation
   Suppress non-essential orchestration actions and prevent actions that could create new externalities. The correct posture is conservative stability, not optimization.
2. Clamp messaging
   Restrict all safety-adjacent messaging to pre-authored, owner-approved content. Disable freeform generation for instructions, routing claims, eligibility, closures, or life-safety prompts. If the system cannot ground a claim in authoritative state inside its freshness window, it must refuse and escalate.
3. Clamp guidanc
   Remove affected zones and routes immediately from recommendations. If state is uncertain, refuse precise routing and provide safe defaults or staff escalation prompts.
4. Preserve accessibility and baseline servic
   Maintain accessible baseline service. Do not degrade core service for visitors who decline opt-in or who require accessibility modes. Safety and accessibility remain hard constraints.

D. Operator visibility requirements

Operator tools must show, at minimum:

- Mode state per zone and time of activation
- Trigger type and source
- Active constraint set and suppressed action categories
- Current messaging mode and which content set is active
- Who activated the mode, and how to clear it

E. Evidence and logging requirements

Create an incident record capturing:

- Trigger event and timestamp
- Scope, affected zones, adjacent zones clamped
- Actions taken, including automation suppression and messaging switches
- Overrides invoked and by whom
- Verification outcomes during incident
- Clearance verification and return-to-normal actions

This record must be reconstructable, and retention must follow the venue's declared policy.

F. Clearance criteria and return-to-normal sequence

Do not clear Incident Mode by assumption. Clear only when:

1. The authoritative trigger is cleared and confirmed.
2. Closed routes remain closed until explicitly reopened.
3. Verification checks pass for affected surfaces, audio, guidance, and critical state propagation.
4. Automation re-enables in a controlled sequence, with heightened monitoring for a defined window.
5. A decision record is written noting clearance, verification, and the operator authorizing return to normal.

G. Post-incident review

Within the next review cadence, perform:

- Root cause classification (equipment fault, sensor anomaly, human override, external event)
- Corrective actions and ownership
- Monitoring threshold updates (see Appendix H)
- Runbook updates if ambiguity, latency, or unclear authority boundaries were observed

Monitoring is daily truth maintenance

Monitoring keeps the right to be trusted. Monitoring is not a dashboard nobody checks. It is an operational habit that makes failure visible before the public becomes the diagnostic tool.

A healthy monitoring posture has three layers.

Layer 1: Health

Is it up? Nodes reachable, services running, storage sufficient, and critical paths alive.

Layer 2: Integrity

Non-negotiables for verified restoration

If the room cannot return itself to ready state reliably, the experience will fail in public.

- Define an exhibit-ready state per zone.
- Prove restoration timing is early enough to be imperceptible.
- Prove drift detection produces actionable signals, not noise.
- Prove rollback and recovery restore readiness without guesswork.

  Implementation pointers: see Appendix H, Appendix I, and Appendix B.

Is it correct?

This is the layer most venues skip, and it is the layer that saves reputations. Integrity monitoring answers:

- Are screens showing content, and is it moving when it should move?
- Are audio zones outputting the correct mix at the correct level?
- Are language modes switching as intended?
- Are captions and other modalities available when selected, and
- Are Lifecycle Automation Layer transitions executed and restored correctly?

Outcome Verification Layer makes this possible. It turns "we think it is fine" into "we can demonstrate it is fine."

Layer 3: Drift

Is it getting worse? Most failures arrive as drift: rising heat, rising fan speed, increasing faults, dimming output, increasing latency, degrading intelligibility, and subtle timing misalignment. Drift monitoring is the foundation of preventive maintenance and a quiet form of hospitality because it prevents the guest from ever seeing the failure.

Analytics that matter: gold metrics for venues

Many venues drown in dashboards and still lack operational insight. A useful analytics posture is small and explicit. It focuses on metrics that predict guest experience quality, operator workload, and failure risk.

Core engagement metrics

- dwell time distribution by exhibit, not averages only,
- re-engagement signals, what visitors return to in the same visit, and
- repeat visitation signals measured in a privacy-forward way, rarely as covert profiling.

Core flow metrics

- zone occupancy and congestion thresholds with time-of-day patterns,
- queue estimate accuracy and queue shock detection, and
- path utilization: hot zones versus under-discovered zones.

Core operations metrics

- verification pass rate,
- lifecycle automation efficacy, reduced bleed and reduced runtime without guest-visible disruption, and
- drift trend rates: thermal drift, fan escalation, brightness decay, and latency creep per asset class.

Core experience integrity metrics

- language parity checks across surfaces and Personal Channel, and
- accessibility mode uptake: captions, Personal Channel audio, quiet mixes, and non-visual navigation usage.

These metrics are enough to drive continuous improvement without a surveillance posture. They tell you what is working, what is failing, and what will fail next.

Operator View is operator visibility, not theatre

The Operator View, as used in this architecture, is not a glossy control room. It is operational visibility that matches how venues work.

A useful Operator View posture provides:

- a clear zone-based picture of what is active, unoccupied, in lower-power posture, or in fault,
- quick drill-down to the signals that matter,
- a way to confirm that the system's world state matches physical reality, and
- patterns over time that guide maintenance and improvements.

The Operator View must respect privacy. Operator visibility should not become a backdoor for dossiers. Operators need truth about the place, not biographies of the public.

A WorldModel Venue is not designed to replace human operators with algorithms. It is designed to give operators superpowers: visibility they did not have, warnings they used to miss, and control they used to fight for. Complexity in the code pays for simplicity on the floor.

Support is a designed operating model

Support fails when it is treated as a heroic effort. A staff member calls the integrator, someone remote logs in, someone figures it out, and the venue survives. That can work once. It cannot be the operating model.

A support model that scales has defined responsibilities, defined procedures, and a rhythm that includes updates, verification, and planned maintenance.

A practical support model answers:

- who owns content changes and how changes are versioned and rolled back,
- who owns compute provisioning, meaning what happens when a node needs more compute,
- who owns the health of the Personal Channel, including device compatibility and synchronization verification,
- who owns operational policies: Lifecycle Automation Layer thresholds, audio comfort rules, language tiering, and other tunables,
- how incidents are triaged and what response times are, and
- how updates are staged, tested, and deployed without breaking the show.

Support should be boring. Boring is what makes venues stable.

Provisioning compute as a support strategy

This book uses compute as a noun because it changes how venues age.

In an older posture, a failing device triggers procurement panic. A device becomes obsolete. A replacement is not compatible. New converters and extenders accumulate. A new rack diagram appears, and the system becomes a fossil of compromises.

In a compute-first posture, replacement is provisioning. A node fails; you replace it with a node that meets the compute envelope, restore configuration, verify with Outcome Verification Layer, and return it to service. If you want an upgrade, you provision more compute where needed. The venue evolves without trauma. This is why the architecture remains useful for a decade.

Acceptance tests that protect the venue after opening

A venue should not accept a system without acceptance tests that operators can run. Not integrator-only tests. Operator-run tests.

A minimal acceptance suite includes:

- language switching tests across room, surface, and Personal Channel planes,
- Streaker, Stroller, Student depth behaviors across a multi-exhibit sequence,
- accessibility mode tests: captions, sign language options, personal audio, and quiet mixes,
- Lifecycle Automation Layer transitions: zone empty behavior and restoration timing,
- Outcome Verification Layer checks: proof that "on" produces image and motion and proof that faults are detected,
- basic drift indicators: temperature, fan, and error trends for critical assets, and
- recovery drills: what happens when a node is removed, a network segment degrades, or an exhibit goes offline.

Acceptance tests are not bureaucracy. They are long-term insurance. A designer who can specify acceptance tests builds better projects because projects become accountable to reality rather than optimism.

The chapter's conclusion

Commissioning, monitoring, and support are where systems become real. The rest of the book describes capabilities: multilingual dignity, Personal Channels, Programmable Canvases, acoustic comfort, recognition with restraint, WorldModel operational truth, and governance that makes macro behavior trustworthy.

None of it matters if the venue cannot keep the system alive.

A professional venue system treats operations as part of experience quality:

- commissioning proves behavior and captures baselines,
- monitoring maintains truth and catches drift early,
- Outcome Verification Layer verifies outcomes, not commands,
- operator view gives operators visibility without turning the public into inventory,
- support is a designed model, not a heroic scramble, and
- compute is provisioned as capacity so the venue can evolve without rebuild cycles.

In the next chapter, we move from the operational model into the specific automation loop that ties these ideas together most concretely: Lifecycle Automation Layer behaviors, verified restoration, and drift-driven prevention. We then follow it immediately with the power and lifecycle sequence that shows how occupancy, pre-warm timing, verification, and maintenance fit together.

38

Lifecycle Automation Layer, Outcome Verification Layer, and Verified Restoration

A modern venue must learn to rest. Not philosophically, operationally.

The traditional operating pattern is familiar: power everything on in the morning, leave it running all day, and power it down at night. It feels safe because it is simple. It also creates predictable costs. It runs output in empty rooms. It keeps sound bleeding into adjacent spaces when no one is present. It shortens the life of assets that are expensive to replace. When something fails, staff discover it the same way guests do, by encountering the broken moment in public.

Lifecycle Automation Layer is the corrective posture. It is the policy-driven behavior that reduces unnecessary output and unnecessary runtime in unoccupied spaces, while restoring those spaces before a visitor can perceive anything missing.

Outcome Verification Layer is what makes Lifecycle Automation Layer trustworthy. It is the verification layer that confirms, using evidence from the physical space, that the venue is in the state it believes it is in. In an automated venue, Outcome Verification Layer is the difference between "we sent the command" and "the venue is ready."

This chapter is the practical center of the operations story. It is where energy, comfort, reliability, and maintainability become one loop.

The economic reality: the cost of "generally on"

Always-on has two measurable properties: it consumes energy and it consumes operating hours.

The operating-hours math is concrete. In a venue with 50 projectors, one unnecessary hour per day adds:

$$50 \times 1 \times 365 = 18{,}250 \text{ projector-hours per year}$$

That is not a metaphor. It is additional runtime across the fleet. The energy cost is proportional to the power draw of the equipment and the hours it runs. The maintenance cost is shaped by the same variable. Runtime is wear.

Lifecycle automation captures value by reducing unnecessary runtime while preserving guest-perceived readiness.

Why Lifecycle Automation Layer matters beyond energy

Energy savings are real. In venues, the more valuable benefits are often comfort and operability.

If a gallery is empty, there is no reason for it to contribute sound to the building. Lowering volume in unoccupied zones reduces sound bleed, improves intelligibility in adjacent zones, and lowers sensory fatigue. The venue feels calmer without changing the designed experience in occupied spaces.

If a room is empty, there is no reason for it to run projection at full duty cycle. Reducing projector runtime extends the service life of the asset, reduces thermal stress, reduces fan wear, and reduces unplanned downtime.

In other words, Lifecycle Automation Layer is a reliability and hospitality tool. The venue becomes quiet when it should be quiet and alive when it should be alive.

The Lifecycle Automation Layer loop

A Lifecycle Automation Layer implementation consists of a loop with five phases:

1. Detect occupancy state and confidence
2. Apply a conservative transition policy
3. Maintain readiness timing rules
4. Restore before perception
5. Verify outcomes physically

Each phase matters. If you skip any phase, automation becomes either ineffective or risky.

Phase 1: Detect occupancy state and confidence

The first question is deceptively simple: is the zone occupied?

A professional system treats occupancy as state with confidence, not as a binary truth. In practice:

- occupancy can be inferred from multiple sources,
- sources can disagree, and
- the system should behave conservatively when confidence is low.

Occupancy sources should be classified as advisory unless they are physically authoritative. The point is to avoid false empties. A false empty is the worst failure mode. A room powers down while a visitor is present or restores too late.

Lifecycle automation should generally bias toward the visitor.

Phase 2: Apply a conservative transition policy

When the system believes a zone is empty with sufficient confidence, it applies a staged transition policy.

A typical staged policy looks like this:

- Step A: reduce program audio levels first to reduce sound bleed
- Step B: reduce non-essential visual output next, dimming or switching content posture where appropriate
- Step C: power down selected equipment classes last, based on readiness requirements

The ordering is practical. Audio reduction can often be immediate without guest impact. Projection power-down requires readiness guardrails.

The policy must be explicit and designed by role and zone, not as a one-size-fits-all global switch. A theatre, a gallery, a retail aisle, and a corridor do not have the same readiness requirements.

Phase 3: Maintain readiness timing rules

The core promise of Lifecycle Automation Layer is not that it powers things down. The core promise is that it restores them before a guest can notice.

This requires readiness timing rules. A simple set of rules is:

- if a device requires warm-up, begin restoration before the occupancy boundary is crossed,
- if a device restores quickly, restoration can be triggered closer to the boundary,
- if a zone's experience is time-critical, keep it in a higher readiness posture, and
- if a zone is optional and low risk, allow deeper savings.

This book does not prescribe specific warm-up times because they vary by equipment class and venue design. The implementation principle is the point: readiness is a contractual requirement, and it must be defined per zone and per asset class.

Phase 4: Restore before perception

This phase is where most automated systems fail, not because they are wrong, but because they are late.

A visitor does not care why. They care that the room looked broken for a moment.

A practical restoration guarantee is:

- the zone reaches visitor-ready state before the visitor can plausibly see or hear missing output.

This is a perception metric, not a technical metric. It also means the system uses spatial context. If a visitor is approaching but not yet at the threshold, you have time. If a visitor is at the threshold, you have no time.

Phase 5: Verify outcomes physically

A command is not an outcome. A projector can be "on" and showing black. A display can be "on" and frozen. A media node can be "running" and out of sync. An audio zone can be "enabled" and muted. A lighting system can be "active" and mis-addressed.

Outcome Verification Layer closes this gap. It verifies, using evidence from the physical space, that the zone is in the expected state. This is not about surveillance. It is about equipment verification.

In practice, Outcome Verification Layer includes checks such as:

- is there an image on the screen,
- is it the expected content class,
- is there motion when motion is expected,
- is brightness within an acceptable envelope,
- is audio present where audio is expected,
- is the zone level within expected bounds, and
- are key fixtures responding as expected.

In short, the system proves the room is ready.

The projector example, made explicit

A projector fails in slow, expensive ways. A venue wants to reduce runtime to extend service life. Lifecycle Automation Layer can power projectors down in empty rooms. Restoration must occur before a visitor can see the screen.

A practical sequence is:

1. Zone becomes empty with sufficient confidence
2. Audio reduces first to prevent bleed
3. Projector power-down is applied per policy
4. Visitor approaches the zone
5. Projector restoration begins early enough to guarantee readiness
6. Outcome Verification Layer verifies the projector is producing an image and that motion is present when expected
7. If verification fails, the system alerts staff immediately and the zone switches to a safe fallback state rather than silently failing

This is how "projector is on" stops being an assumption and becomes a fact.

Trend analytics: preventive maintenance inside the same loop

Lifecycle automation and verification keep the day stable. Trend analytics keeps the month stable.

Most failures do not arrive suddenly. They drift. Heat rises. Fan speeds climb. Error rates creep. Brightness decays. These are signals that maintenance is needed before public failure occurs.

A mature implementation uses trend analytics to generate preventive maintenance events such as:

- thermal drift increasing beyond a baseline envelope,
- fan duty cycle trending upward,
- increasing soft faults or retries in a media node,
- increasing restoration failures after lifecycle transitions, and
- increasing verification anomalies in a particular zone.

The key is workflow, not dashboards:

- what threshold triggers an alert,
- who receives it,
- what the recommended maintenance action is,
- how it is acknowledged, and
- how it is closed.

Trend analytics without workflow becomes noise. Trend analytics with workflow becomes reliability.

Where these layers belong in the WorldModel implementation

Lifecycle Automation Layer and Outcome Verification Layer map cleanly into the reference architecture:

- occupancy changes are events on the truth bus,
- power state change event and acoustic state change event are events on the truth bus,
- lifecycle actions are proposed by orchestration and permitted by the governance gate,
- power-down and audio reduction are executed by the action plane,
- verification results are events that feed back into the truth bus and state store, and
- drift detection produces maintenance events and operational alerts.

This integration keeps lifecycle automation governed. A venue should not allow an unbounded automation loop to power systems up and down without policy constraints, operator override, and verification.

Governance constraints specific to Lifecycle Automation Layer

A Lifecycle Automation Layer policy must be constrained by explicit rules. At minimum:

- rarely reduce output in a zone unless occupancy confidence is above threshold,
- generally, restore before perception per zone readiness rules,
- rarely violate show integrity constraints,
- rarely create accessibility harm, such as reducing required access,
- generally, verify restoration physically,
- under uncertainty, default to higher readiness, and
- provide operator override and hold modes for maintenance, VIP tours, or special programming.

These rules prevent automation from becoming visible failure.

Acceptance tests

A venue should not accept an automated operations layer without explicit acceptance tests. At minimum:

Lifecycle Automation Layer correctness tests

- zone transitions occur only when occupancy confidence is sufficient,
- audio reduction occurs in empty zones and reduces bleed into adjacent spaces,
- projectors and displays power down only per policy, and
- restoration begins early enough to guarantee readiness.

Restoration timing tests

- a visitor approaching at typical walking speed rarely sees a blank or uninitialized screen,
- a visitor entering the zone experiences no visible delay, and
- under peak conditions, restoration still meets readiness aims to provide.

Outcome Verification Layer tests

- verification detects black screens, frozen content, and missing motion,
- verification detects incorrect language or incorrect content class when relevant,
- verification detects audio absence when audio is expected, and
- verification results are logged and visible to operators.
- Integrity enforcement: when verification is missing or inconclusive, the system does not claim readiness. It abstains, alerts, and holds a defined fallback state.

Failure handling tests

- when verification fails, the system alerts staff promptly,
- the zone enters a defined safe fallback state, and
- operator override can hold a zone in ready state regardless of occupancy.

Drift detection tests

- temperature and fan drift beyond baseline triggers a maintenance alert,
- repeated restoration anomalies trigger investigation, and
- alert workflows are defined and operational.

These tests make the chapter's claims real.

Why this matters to stakeholders

Owners care because it protects capital assets and protects guest experience reliability. Designers care because it reduces sound bleed, preserves thematic boundaries, and supports comfort without compromising the room. Operators care because it replaces morning walkarounds and public embarrassment with proof and preventive action. Fabricators and integrators care because it yields a clear operational specification: automate with verified restoration and drift detection, not automation by assumption. Cities care because the same loop applies to civic infrastructure where public displays and kiosks must be truthful, energy-aware, and verifiable.

Lifecycle Automation Layer and Outcome Verification Layer are not features. They are the operational maturity layer that makes adaptive venues sustainable.

The chapter's conclusion

The promise of hyper-personalization is not that the venue can change. The promise is that the venue can change without breaking trust.

Lifecycle Automation Layer reduces unnecessary output and extends equipment life while preserving calm, especially through reduced sound bleed in empty spaces. Outcome Verification Layer proves restoration is real rather than assumed. Trend analytics turns the venue preventive rather than reactive.

Together, these form a simple, reliable automation loop for a WorldModel Venue:

- detect,
- act,
- verify, and
- prevent.

In the next chapter, we make the power sequence explicit end-to-end, including readiness timing rules that guarantee restoration and drift signals that turn maintenance into prevention.

39

POWER AND LIFECYCLE LOOP

Occupancy Detection, Pre-Warm Timing, Verification, and Drift

Power management in venues is not an electrical problem. It is a time problem.

The system is allowed to reduce runtime only if it can guarantee that the venue will rarely look broken when a visitor arrives. That guarantee is not achieved by hoping projectors will behave. It is achieved by implementing a loop that connects occupancy, power actions, readiness timing, verification, and drift trends.

This chapter section makes that loop explicit.

It is written at the same level as the WorldModel implementation chapter: interface and behavior, not proprietary internals. It is detailed enough to be specified, commissioned, and audited, and it remains patent-safe because it does not publish secret recipes.

The power and lifecycle loop in one line

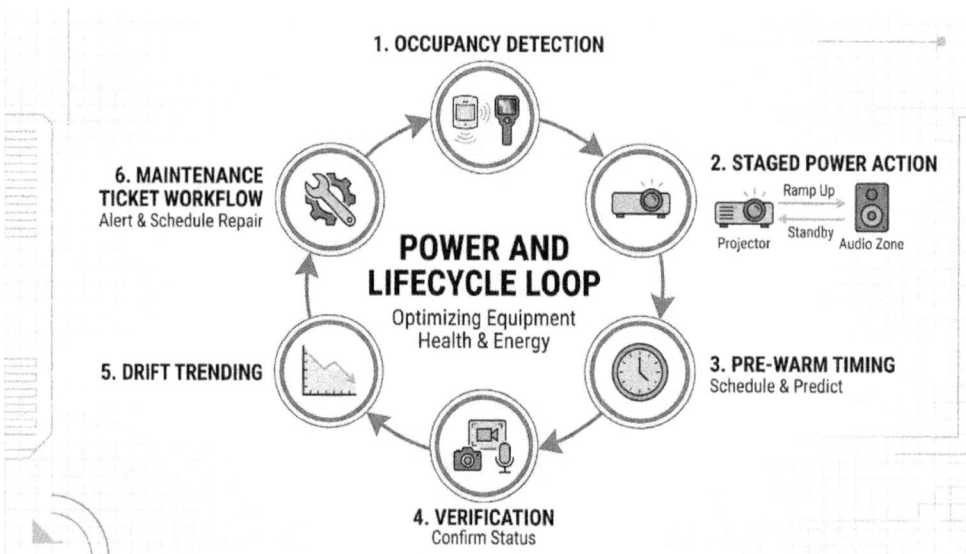

Occupancy detection → power action → pre-warm timing → verification → drift trend → preventive maintenance.

If you build that loop, power management becomes a reliability feature rather than a risk.

Where this loop lives in the architecture

In the WorldModel reference architecture, the power loop touches five components:

- occupancy signals enter on the truth bus
- the WorldModel state store records occupancy and asset readiness state
- orchestration proposes power actions under policy
- governance permits only bounded actions
- verification confirms outcomes and drift detection triggers preventive maintenance

This keeps power behavior governed, auditable, and safe.

Step 1: Occupancy detection, treated as state with confidence

Power decisions should rarely trigger off a single brittle signal. This is also a truthfulness rule: under uncertainty, do not pretend the zone is empty. Bias toward readiness, because false empties produce guest-visible failures and create a system that appears to lie about its own state. The system should treat occupancy as a state estimate with confidence, and it should bias toward the visitor under uncertainty.

Implementation requirement:

- occupancy_context per zone is one of: empty, light, busy
- each occupancy change includes confidence and provenance
- Lifecycle Automation Layer and power actions trigger only when occupancy confidence crosses a defined threshold and remains stable for a defined duration

This prevents false empties, which are the most damaging failure mode.

Step 2: Power action, staged by policy and asset class

Power action is not one switch. It is a policy ladder applied per zone.

A practical ladder looks like:

- Audio first: reduce program level in unoccupied zones to reduce bleed
- Visual second: dim or change canvas mode where appropriate
- Projection third: power down projectors only when readiness aims to provide can still be met

The exact ladder is venue-specific, but the discipline is universal: you stage actions to reduce guest-visible risk while still achieving savings.

Step 3: Pre-warm timing, the readiness guarantee

This is the missing piece in most "energy saving" schemes.

The system must guarantee that by the time a visitor can plausibly see the output, the zone is already in a visitor-ready state.

A serious implementation introduces a state per asset:

- readiness_state: ready, warming, restoring, not_ready
- restore_lead_time_s: per asset class and zone policy
- restore_trigger_boundary: defined spatially, not only temporally

The system then uses these rules:

1. If a zone is about to be entered, start restoration early enough to satisfy restore_lead_time_s.
2. Change pipeline checklist
3. Treat content changes like operational changes.
   - Staging environment or canary zone before full rollout.
   - Versioned assets, with rollback capability.
   - A defined acceptance test after each update.
   - A record of what changed, who approved, and what evidence confirms readiness.
   - If a zone is in a long corridor's line-of-sight, restoration triggers earlier, because visitors can see further.
   - In busy mode, restoration triggers earlier because crowd speed and predictability are lower.
   - Under uncertainty, default to higher readiness, not deeper savings.

Pre-warm is not about guessing. It is about enforcing a readiness contract.

The same contract principle applies to visitor-facing guidance.

If a system cannot guarantee readiness, it must not claim readiness. If it cannot guarantee state, it must not claim certainty. This is how you prevent confabulation-like behavior in operations: the system is allowed to be uncertain, but it is not allowed to be confidently wrong.

Step 4: Verification, outcomes not commands

As we have established, commands are not outcomes. A projector can be commanded to "on" and still be black. A display can be "on" and frozen. An audio zone can be "enabled" and silent.

Outcome verification is also the antidote to confabulation.

Confabulation is a confident claim without grounding. In operations, the analogue is phantom success: the system claims something happened because it issued a command, not because it verified the result.

Verified restoration replaces performance with proof. After any change or restoration action, the system must confirm the venue is actually in the required state, or it must surface uncertainty and hold a safe fallback posture.

Verification must confirm:

- image present on the intended surface
- motion present when motion is expected
- correct mode, guidance versus narrative
- audio present and within bounds
- time alignment, where synchronization is required

Verification results are events. They feed back into WorldModel state, so the venue can act on truth rather than assumptions.

If verification fails, the system must escalate and fall back safely. Silent failure is not allowed.

The power loop is not complete without drift.

A venue should not wait for a projector to fail publicly. Drift tells you failure is approaching.

At minimum, trend:

- temperature trend by runtime hour
- fan duty cycle trend
- restoration anomaly frequency
- verification failures per zone
- error-rate creep in playback nodes

Drift thresholds should generate maintenance tickets with recommended action, not vague alerts.

This is how "energy saving" becomes "reliability engineering."

Worked example: A projector zone and an adjacent audio zone

This example is intentionally common: a small gallery with projection and an adjacent corridor where sound bleed is a problem.

Zone and assets

- Zone: Z-GAL-B-02
- Projector: AS-PRJ-B-02
- Screen: AS-SCR-B-02
- Audio zone: AS-AUD-ZN-B-02
- Adjacent corridor: Z-CORR-B-01

Sequence

1. Occupancy drops
   Event: Zone occupancy change event
   - zone: Z-GAL-B-02
   - occupancy_context: empty
   - confidence: high
   - State update: zone marked empty.
2. Power actions are proposed
   Orchestration proposes:
   - reduce audio zone program level to bleed-safe level
   - dim local non-essential displays
   - power down projector after a hold-down timer expires

3. Governance permits staged action
   Governance approves:
   - audio reduction immediately
   - projector power-down only after the zone has remained empty for the configured duration and only if restore_lead_time_s can be met on return
4. Action executes
   - Audio zone reduces, bleed into the corridor reduces
   - Projector powers down, asset readiness state becomes "not ready"

5. Visitor approaches
   Event: Zone occupancy change event
   - corridor occupancy rising
   - approach detected toward Z-GAL-B-02 entry boundary

System triggers restoration early, based on restore_lead_time_s and line-of-sight policy.

6. Pre-warm and restore
   - Projector readiness_state becomes warming, then restoring
   - Audio zone returns to program level for occupied state, bounded by acoustic policy
7. Outcome Verification Layer verifies
   Verification events confirm:
   - screen has image
   - motion is present
   - correct content class is playing
   - audio is present within bounds
   - zone is visitor-ready

If any check fails, the system raises an alert, and the zone enters a defined safe fallback state.

8. Drift is recorded
   Projector temperature and fan duty are logged as drift signals. Over weeks, if temperature rises for the same runtime, the system flags maintenance before failure.

What this achieves

- Reduced sound bleed when empty
- Extended projector life through reduced runtime
- No guest-visible "black wall" moments
- Early maintenance detection rather than public embarrassment

That is the power and lifecycle loop working as a venue behavior.

Acceptance tests for the power and lifecycle loop

A venue should not accept automation without these tests.

1. False empty protection
   Power-down rarely occurs under ambiguous occupancy.
2. Readiness guarantee
   No visitor sees an uninitialized projection or silent zone after restoration triggers.
3. Verification enforcement
   A zone is not marked ready unless verification passes.
4. Fail-safe behavior
   On verification failure, alerting and fallback behavior are defined and observable.
5. Drift workflow
   Drift thresholds generate maintenance actions with owner, urgency, and closure process.

Where this leaves the reader

The Lifecycle Automation Layer is not a feature. It is part of a power and lifecycle loop that must be implemented end-to-end.

When that loop is explicit, the venue becomes calmer, quieter, longer-lived, and more reliable, all at once.

In the next chapter, we move from operational automation to publishing discipline: content operations, and fast change without rebuild, so the venue can evolve safely without turning updates into risk.

40

CONTENT OPERATIONS, AND FAST CHANGE WITHOUT REBUILD

A venue that cannot change will eventually be forced to pretend. It will pretend that its demographics have not shifted. It will pretend that language needs are stable. It will pretend that the world outside the doors is not moving faster than fabrication cycles. It will pretend that yesterday's signage still tells today's truth. A system that confidently repeats stale content is not simply outdated. It is operationally untruthful. It will pretend that broken moments are temporary, even when temporary becomes permanent.

None of this is a moral failure. It is an operational failure.

Content operations is also an integrity discipline.

When content changes faster than verification, a venue is forced into guesswork. That is how confabulation becomes institutional: the system continues to speak confidently while the underlying truth has moved.

Versioning, staging, rollback, and post-update acceptance tests are not bureaucracy. They are how a venue ensures that what it says remains consistent with what is true, across languages, across surfaces, and across modes.

The modern public has been trained by an environment where change is normal. Phones update weekly. Navigation changes instantly when a road closes. Media adapts to the moment. Visitors do rather than generally articulate these expectations; they feel the difference between a place that is alive and a place that is ossified.

This chapter is about staying alive, not through constant novelty and not by exhausting staff, but through a discipline: content operations.

If your venue's experience is partly software-defined, your content must be treated like a living system with versioning, approvals, rollback, staging, audits, and a cadence that respects both curatorial integrity and operational reality.

Programmable Canvases makes this visible because it turns surfaces into canvases. Edge Compute Nodes makes it possible because it provides the compute substrate that allows those canvases, those languages, and those modes to run locally and reliably. Virtual Docent Layer and Personal Channel make it meaningful because they turn content into delivery that can remain coherent across exhibits.

All that power is wasted if you cannot publish safely.

The real shift: Content is no longer an asset; it is behavior

Traditional exhibit content is an asset. You fabricate it, install it, and then protect it from change. That posture made sense when content was physical and costly to replace. It also created a predictable drift: the venue treated change as disruption, so change did not happen until pressure forced it. Pressure is the worst time to make changes.

In a WorldModel Venue posture, content becomes behavior. It becomes the outward expression of operational truth, language dignity, accessibility modes, and day-to-day hospitality. Behavior must be managed. This is why content operations is not a marketing term. It is the operational layer that prevents a venue from becoming brittle as soon as it becomes capable.

The three planes of content, and why each needs its own discipline

Modern venues deliver content through three planes:

1. the room plane, shared shows and shared moments
2. the surface plane, Programmable Canvases and public-facing guidance
3. the Personal Channel plane, private delivery of language, captions, sign language video, quiet mixes, and comfort modes

Each plane has different failure costs, and therefore each plane needs a different operational discipline.

A mistake in the room plane is public. It breaks show integrity and is noticed immediately.

A mistake in the surface plane is authoritative. It misguides, creates confusion, and erodes trust quickly because signage is supposed to be the truth.

A mistake in the Personal Channel plane is personal. It breaks inclusion. It can quietly exclude without staff noticing because the failure is experienced privately.

Content operations begins by admitting these are different systems with different publishing risk. A venue that treats all content as one publishing bucket will either publish too cautiously and stagnate or publish too freely and break the day.

Programmable Canvases makes the wall part of your publishing pipeline

When you treat a wall as a Programmable Canvas, you have created a publishing surface. Many teams still think like fabricators. They design the surface once and treat changes as exceptions. A Programmable Canvas makes change possible and makes stagnation visible.

The correct mental model is that a canvas is like a stage. It can host multiple scenes. It can be scheduled. It can be zoned. It can change language without reprinting, and it can change exhibits without construction. The same physical surface can serve multiple exhibit stories across time, and in some contexts even simultaneously, with different zones on the same surface acting like different interpretive stations.

A stage needs stage management. A canvas needs publishing management.

Content operations for Programmable Canvases includes:

- a layout system that supports multiple languages without the "big English, small Spanish" compromise,
- a scheduling system that allows exhibits to rotate by time, by program, or by operational state,
- a clear guidance override posture so operational truth can take precedence when safety, routing, or closures demand clarity, and
- a rollback posture so a bad update is not a crisis.

The deeper point is cultural. A Programmable Canvas is not a screen you installed. It is a living layer of the venue. If the venue wants to behave like an intelligent place, the canvas must be treated like a trusted communication channel.

Language operations is a real operational domain

Once language becomes runtime, translation stops being a deliverable and becomes a workflow. That is good news. It ends the fabrication trap. It also means multilingual support cannot be treated as an occasional project.

Language operations includes:

- deciding which content is vital and must be published with human review, always,
- deciding which content can be published with lighter review because risk is lower,
- maintaining parity so language options remain dignified and not perpetually behind, and
- maintaining typography and readability rules that respect scripts rather than merely translating words.

The goal is not to translate everything into every language instantly. The goal is to build a system where expanding language support is normal and predictable rather than political and painful.

A venue that operates this well will notice something. Language becomes less contentious when it becomes less scarce. Scarcity forces hierarchy. Runtime language removes some of that pressure.

Virtual Docent Layer content is structured content

A conversational docent layer is only as good as the content discipline behind it. If Virtual Docent Layer is grounded in curated knowledge, then content operations becomes a curatorial practice. It is managing a Body of Knowledge shaped by the institution's voice, scholarship, tone, and boundaries, rather than only publishing words.

A practical implication follows. The quality of a docent layer is not primarily a model question. It is a content structure question. If the museum wants Virtual Docent Layer to support Streakers, Strollers, and Students, the knowledge base must explicitly contain:

- the headline layer, one-minute meaning,
- the context layer, orientation and connection, and
- the deep layer, real detail, not filler.

It must also contain language variants, modality variants, and safety and sensitivity boundaries where appropriate.

Content operations for the docent layer includes:

- versioning of curated knowledge,
- editorial review and approval gates,
- a method for handling corrections and improvements without destabilizing tone, and
- a way to learn from what visitors ask in aggregate without turning questions into surveillance.

A conversational layer becomes a gift to the institution when it is operated as an instrument for improving interpretation, not as a novelty that speaks.

Personal Channel multiplies the publishing surface, and that is a responsibility

When you deliver content through a Personal Channel, you multiply the number of ways the experience can fail. That is not a reason to avoid it. It is a reason to operate it properly.

The Personal Channel has different requirements than surfaces:

- timing integrity matters more because out-of-sync content is actively frustrating,
- modality integrity matters more because accessibility modes are the point,
- device diversity is real and changes over time, and
- network conditions vary, so offline resilience is not optional.

Content operations for the Personal Channel therefore includes:

- a publishing pipeline for audio tracks, captions, sign language video, and other modes,
- a compatibility posture that treats device variation as normal,
- a verification posture after updates to confirm synchronization is still correct, and
- a clear fallback behavior when the Personal Channel is unavailable, so the experience remains complete without it.

This is where the stance becomes practical. Design for phone-era expectations without requiring a phone. A well-run Personal Channel makes the venue more inclusive, and it rarely becomes the only way to participate.

Personalization as editorial assembly under policy

Upgrade planning worksheet

Owners should be able to plan upgrades without redesigning the venue.

- What is replaceable by module, node, or device class?
- What must be re-commissioned after replacement?
- What is tested quarterly versus annually?
- What spares policy prevents long downtime?

This is how personalization becomes editorial assembly under policy: selection, assembly, and constrained generation grounded in curated sources and governed operational truth.

Generation becomes a layer of explanation, not invention. In many cases the safest use of generation is to create explanatory layers above curated media: age-appropriate vocabulary, accessible descriptions, diagrammatic clarity, and the optional deep layer that rewards curiosity. This is also where systems like Virtual Docent Layer can generate different outlines for young children and adults from the same approved source material. The physics does not change. The explanation does.

Now the system can assemble a cut for a specific visitor. A visitor asks about Dunkirk in 1940. The system retrieves relevant passages, then chooses 30 seconds or 90 seconds based on time budget, gallery state, and operating mode. It stitches a coherent micro-documentary, then adds narration and subtitles in the right language, tuned by age_years, and framed by the visitor's interest lens, such as military tactics or general history.

Editing and assembly is the next layer. Many institutions hold enormous archives. 10,000 hours of video is not unusual. The practical posture is to index that archive, generate metadata, then have curators review and enrich it.

The metadata needs two parts: selection metadata and story metadata.

- Selection metadata describes date, event, and retrieval cues that make relevant passages findable.
- Story metadata encodes interpretive intent for each passage or provides a Body of Knowledge that can be used to generate truthful explanation aligned with curatorial boundaries.

Two-part metadata model

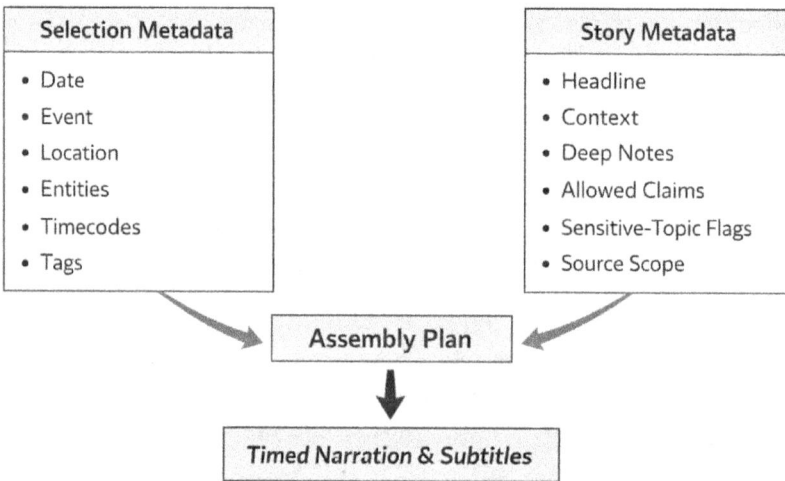

Selection is the first layer. If a story has multiple editorial cuts, science, history, design, tactical, the system can select the best pre-made cut based on declared interests. It can then overlay narrative and subtitle layers based on preferred language, accessibility modes, and age_years.

As AI becomes more capable, the most important practical shift is not that a venue can answer questions. It is that a venue can select, assemble, and, where appropriate, generate media layers to fit the visitor in the moment without rebuilding the exhibit.

Adaptive media selection, editing, and generation

Example: age-tiered visual explanation as bounded (constrained) generation

In the near term, generation is best treated as an explanation layer over curated media, not as a replacement for sources. The system selects and edits verified assets, then generates narration, captions, and overlays that fit language, age band, and interest lens. A single concept can be presented differently without changing truth. For a younger audience, show a simple animation and focus on causal intuitions. For older audiences, add formulas and definitions. The facts stay stable. The presentation adapts.

Safe publishing is governed publishing

A practical example is a WWII archive. A visitor asks about Dunkirk in 1940. The system retrieves relevant passages using selection metadata, then assembles a 30 second or 90 second cut based on time budget and live gallery state. It overlays narration and subtitles in the right language, tuned by explicit age and interest parameters, without inventing facts or breaking curatorial boundaries.

Story metadata is interpretive. It holds narrative intent for a passage, the context that must not be lost, and, where appropriate, a bounded Body of Knowledge for that passage. This layer allows narration, subtitles, and explanation to remain truthful and curatorial even when a cut is assembled on the fly.

Selection metadata is retrieval-oriented. It captures who, what, where, and when: dates, events, locations, entities, timecodes, and tags that let a system find relevant segments quickly.

Large archives become usable only when metadata is structured with intent. The practical model is two-part: selection metadata that makes the right passages findable, and story metadata that makes the passage explainable.

Selection metadata and story metadata

Venues sometimes hear governance and imagine slowness. In practice, governance is how you move quickly without breaking public trust.

A governed publishing model answers blunt questions:

- who is allowed to publish what, and under which conditions,
- what changes require human review, and what changes can be automated,
- what changes require operator approval because they affect operations or safety,
- what happens when an update is wrong,
- how quickly you can roll back, and
- how you demonstrate what was published, when, and why.

Governance is not only for macro routing decisions. It is also for content changes in environments where content is now part of operational truth. This is especially important in the surface plane, where signage and guidance can affect flow and safety. Operational messaging needs explicit constraints and approval pathways.

A venue that can publish quickly, safely, and audibly has a strategic advantage. It can respond to the world around it without rebuilding itself.

The practical pipeline: staging, canary, publish, verify

A modern venue needs a publishing pipeline that looks like a disciplined production process rather than a copy-and-paste adventure. The minimum viable pipeline is straightforward.

Staging

A place where updates can be reviewed, tested, and validated without affecting the public experience. This is where you check language rendering, typography, layout behavior, and media integrity. This is where you test that a headline layer is concise, that deep layers are real, and that accessibility modes remain available.

Canary

A controlled rollout to a limited subset of nodes, surfaces, or exhibits. The purpose is containment. If something is wrong, the blast radius is small.

Publish

A release that is versioned and auditable. Every release has a clear version label, a change summary, and a rollback target.

Verify

A step that confirms outcomes, not only that the publish command succeeded. This is where the Outcome Verification Layer belongs naturally. When you update surfaces, you verify that the correct content is visible and moving where it should be moving. When you update audio and Personal Channel content, you verify synchronization. When you update power and operational behaviors, you verify that the system still restores spaces before visitors can notice.

Verification turns publishing from hope into proof.

Content ops is also how the venue learns

A venue that can publish quickly is valuable. A venue that can learn what to publish next is exceptional.

Analytics, used with restraint, can guide content operations without measuring people. Repeated hesitation points can indicate where labels are unclear. Repeated question themes can indicate where interpretation is missing. Dwell patterns can reveal which parts of an exhibit are rewarding and which parts are exhausting.

The discipline is to treat analytics as signals about the place, not biographies about the public. Then content operations becomes a feedback loop:

- publish with intent,
- observe friction and engagement at the level of place behavior,
- revise and improve, and
- keep the venue aligned with its audience without turning the audience into inventory.

This is what makes the venue feel alive: continuous refinement, not constant novelty.

The personnel truth: content operations is a role

Venues often assume content updates can be handled by someone on the side. That assumption produces the same outcome every time. Updates become rare, then risky, then avoided, then the system stagnates.

A living venue needs living roles. The names vary by organization, but the functions are stable:

- a content owner who safeguards voice and intent,
- a multilingual and accessibility steward who safeguards parity, readability, and dignity,
- an operations owner who safeguards calm, safety, and reliability, and
- a technical publishing owner who safeguards the pipeline, versioning, and rollback.

These roles do not need to be large. They need to exist. When they exist, the venue can evolve without panic.

The chapter's conclusion

A venue that adopts Programmable Canvases, Personal Channels, and adaptive systems has made a choice. It has chosen to become an environment that can change its behavior without changing its bones.

That choice is sustainable only if content is treated as a living system: versioned, governed, staged, verified, and improved through a feedback loop that respects privacy.

This is how you keep a reference-grade venue alive for a decade:

- language evolves without reprinting,
- exhibits evolve without rebuild,
- accessibility expands without retrofit,
- operational truth remains coherent, and
- the public trusts what the venue says because the venue can demonstrate it is telling the truth.

In the next chapter we move from publishing discipline to the long arc: the upgrade path. How a venue absorbs new sensors, new interfaces, and new classes of compute simply by provisioning sufficient compute in the right nodes, and by keeping the architecture's interfaces clean enough that progress does not require demolition.

41

THE UPGRADE PATH

Staying Modern Without Rebuilding

Venues age in two different ways.

The first is visible: finishes scuff, exhibit hardware gets replaced, and the newest display generally looks brighter than the one installed five years earlier.

The second is invisible and more dangerous: systems become harder to change. Small updates begin to require big effort. Simple additions trigger cascading compatibility problems. Staff confidence erodes because no one is sure what will break if something is touched. The venue becomes cautious, then stagnant, not because the mission changed, but because the architecture stopped welcoming change.

A modern venue cannot afford that second kind of aging. The world around it moves too fast. Languages shift. Accessibility expectations deepen. Visitor behavior evolves. New devices appear. New interfaces become normal. If the venue's only way to respond is construction, it will generally be behind, and it will generally be expensive.

The upgrade path in this book is not an add-on. An upgrade path is also an integrity path.

As systems become more conversational and more adaptive, the most corrosive failure mode is fluent error: confabulation. The venue cannot treat that as a model quality issue that will "improve later." It must treat it as an upgrade discipline: claims that can influence behavior must be grounded in operational truth or approved sources, time-bounded by freshness, or refused with safe fallback.

This is why governance and verification appear repeatedly in the upgrade story. They preserve truth and trust as capability increases. It is a design requirement, and it is the reason we keep returning to a disciplined foundation: modular nodes, local execution, verification, and provisioning sufficient compute. This chapter explains how an environment stays modern without rebuilding itself every time capability advances.

The economic reality: the "demolition tax"

Traditional AV systems often require physical demolition to upgrade capabilities. A compute-first substrate, implemented through Edge Compute Nodes, reduces this demolition tax by allowing venues to add new capabilities, such as AI translation, through software provisioning and targeted node upgrades rather than structural teardown. This is not about ideology. It is about lifecycle ROI.

The point of an upgrade path is dignity

Upgrades are often framed as technical ambition. For venues, upgrades are dignity.

Dignity for operators because they can plan changes rather than fear them. Dignity for visitors because inclusion improves without constant reprints, retrofits, and apologies. Dignity for institutions because interpretation and service posture can evolve without turning every improvement into a capital project.

If you are building hyper-personalization as a living system, upgradeability is part of the promise.

A venue should upgrade like a city upgrades

Cities do not replace all streets at once. They replace segments. They reroute temporarily. They phase work. They prioritize the corridors that matter most. Progress is not all-or-nothing.

A WorldModel Venue should behave the same way. The upgrade path should allow you to:

- add capability in one zone without destabilizing another,
- replace a node without changing the rest of the network,
- expand language support without redoing surfaces,
- introduce new accessibility modes without rewriting everything, and
- incorporate better compute, better sensing, and better interfaces without changing the architecture's core agreements.

This is why the substrate matters. When the foundation is modular and compute-based, change becomes a capacity decision rather than a demolition decision.

Compute is the upgrade lever

The most reliable way to future-proof a venue is to stop future-proofing specific hardware. Future-proof the ability to provision the right compute.

When each node is treated as a role with a compute envelope, upgrades become predictable:

- if a node is underpowered, provision more compute at that node,
- if a node fails, replace it with another node that meets the envelope,
- if a new capability adds workload, expand compute where that workload lives, and
- if the venue becomes lighter over time, reduce compute in some places without breaking the system.

This is the opposite of the brittle device stack model where upgrades are hostage to vendor cycles, converter chains, and proprietary dependencies. It also keeps the venue calm. When replacement is routine, it stops being scary.

Clean interfaces make upgrades safe

A modular node helps. A modular node with unclear interfaces still becomes fragile. The real upgrade path is built on clean contracts between layers.

Scenario discipline

Treat the future as scenarios with assumptions, not predictions.

State your assumptions about connectivity, staffing, privacy posture, and device availability.

Then specify what remains true even when those assumptions fail.

At a high level, the venue needs stable agreements such as:

- what state is published into WorldModel,
- what a guidance request looks like and what a guidance response can contain,
- what a content surface expects and what it is allowed to render,
- what the Personal Channel expects and how timing is represented,
- what recognition signals are permitted and what privacy posture they must obey,
- what Lifecycle Automation Layer is allowed to change and what must be verified before a zone is considered restored, and
- what must be logged for audit and what must rarely be retained.
- what visitor-facing claims are allowed to be made as facts, what sources they must be grounded in, how freshness is represented, and when the system must abstain rather than guess

When these contracts are clean, you can upgrade inside a layer without rewriting the whole venue. You can swap a sensor without changing governance rules. You can improve a prediction model without changing signage language. You can add a new comfort mode without changing show control cues. You can incorporate new compute without changing the meaning of state.

That is what makes architecture a reference rather than a one-off.

Adding new sensors without creating a surveillance system

Sensors improve state. Better state improves decisions. Better sensing does not justify stronger certainty. Under uncertainty, the system must remain conservative. Confabulation is not cured by data hunger. It is contained by truthfulness constraints, evidence discipline, and safe refusal when grounding is insufficient. That is true. The risk is that more sensors becomes more collection, and the venue drifts from helpful into hungry.

A healthy upgrade path treats new sensing as a purpose-driven change with explicit boundaries:

- what the new sensor is for,
- what decisions it will improve,
- what data it will generate,
- how long any data is retained,
- what is processed locally, and
- what is aggregated so the venue measures the place rather than building biographies.

Governance protects the upgrade path. A new sensor does not bypass privacy posture. It does not silently expand retention because it is technically possible. The discipline is to add sensing only when it improves outcomes the venue can name and only under the privacy constraints the venue already claims to hold.

Expanding languages and modalities without rework

The most visible form of upgrade in public space is language and accessibility expansion. This is where runtime language, programmable surfaces, and the Personal Channel become an upgrade engine.

If language is printed into panels, expansion is fabrication. If language is rendered into surfaces and delivered through a Personal Channel, expansion is operations. That distinction separates venues that grow with their community from venues that grow stale.

A well-designed upgrade path makes it normal to:

- add a new language tier without redesigning graphics,
- revise translations without reprinting,
- add sign language variants as the community demands,
- add captions and readability improvements without disrupting the shared room, and
- add comfort modes like quieter narration mixes as a standard inclusion option.

This is where content operations and versioning become the difference between, we can and we do. A venue can only upgrade what it can publish safely.

Upgrading from micro to macro without breaking trust

One of the most important upgrade paths is not hardware. It is scope.

Many venues should begin at micro scale: multiple exhibits, coherent preferences, inclusive personal delivery, and strong operational reliability. Some venues will then want macro capabilities: guidance, load balancing, queue shaping, schedule coordination, and city-scale behaviors.

The upgrade path from micro to macro must be staged because the trust boundary changes.

A disciplined path is:

1. Strengthen operational truth: WorldModel state is reliable, current, and verified.
2. Introduce advisory guidance first: suggestions, not automated routing, with clear opt-out.
3. Add governance before expanding action classes: explicit objectives, hard constraints, auditability, and operator override.
4. Expand gradually: allow macro actions only where the venue can demonstrate they improve outcomes without harming fairness, privacy, or show integrity.

The upgrade path is never "turn on macro." It is "earn macro." That is how the system stays positive, trustworthy, and publicly defensible.

Redundancy is part of upgrading

As capability expands, failure cost rises. A single exhibit can fail and the venue survives. A macro guidance layer can fail and the whole day feels rough. Redundancy and fault containment should scale with capability.

The practical posture is to distribute roles and allow graceful degradation:

- if a local node fails, its zone falls back to a safe, stable mode,
- if a prediction component becomes unreliable, the system reverts to conservative assumptions, and
- if a guidance service is degraded, the venue shifts to static truth messaging and human support rather than continuing to push uncertain advice.

Upgrades should be designed so that more capability does not mean more brittleness.

Future compute belongs in the architecture, not in hype

It is reasonable to assume compute will get better, smaller, cheaper, and more capable. It is also reasonable to assume new compute classes will emerge for specific tasks such as simulation and predictive analysis.

The venue should not bet on a timeline. The venue should bet on the architecture's ability to incorporate improvement. If the system can be upgraded by provisioning more compute in the right nodes and interfaces remain stable, the venue does not need to predict which accelerators will win. It can incorporate what becomes practical when it becomes practical.

That is the honest version of future-proofing.

A practical upgrade checklist

If you are planning a venue system meant to last, use these questions as the upgrade discipline.

1. Node envelopes
    Do we have clear compute envelopes by node role, and can we replace nodes as a provisioning exercise?
2. Stable contracts
    Are interfaces between state, orchestration, governance, surfaces, and Personal Channels clean enough that upgrades do not require rewrites?
3. Publishing discipline
    Do we have content operations, versioning, rollback, and verification so change is safe?
4. Privacy posture
    Does every upgrade proposal include data minimization, retention limits, and clear consent boundaries?
5. Verification loop
    Can we demonstrate outcomes after changes, especially for Lifecycle Automation Layer transitions, signage truth, and show integrity?
6. Mode discipline
    Do we have normal, busy, and incident modes, and do upgrades respect those boundaries?
7. Graceful degradation
    If the upgraded component fails, does the venue fall back to a safe, stable behavior?
8. Integrity Discipline
    Do upgrades preserve truthfulness constraints, including grounded claims, freshness windows, and abstain behavior under missing or contradictory state? Do we treat integrity as a lifecycle control by monitoring abstain, fallback, and override rates, and reviewing them after each update so integrity drift is visible?

If you can answer yes to these, the venue has an upgrade path. It can evolve without becoming fragile.

The chapter's conclusion

The future of public environments is not constant rebuilding. It is adaptable behavior.

Edge Compute Nodes and the discipline of provisioning compute make adaptability physically possible. Programmable Canvases make it visible. The Personal Channel makes it inclusive. WorldModel makes it coherent. Governance makes it responsible.

Principle recap: Decisions come first, evidence stays attached, and language remains a rendering layer.

As capability increases, the system must not substitute fluent explanations for grounded decisions. The governing layers should produce bounded decision objects with evidence pointers, confidence posture, and safe fallback behavior. Narrative output is a rendering of those decision objects, not the authority beneath them. Verification makes it trustworthy. Content operations makes it sustainable.

An upgrade path is where all those commitments become real over time. It is the difference between a venue that was impressive once and a venue that stays essential.

42

The Future
that behaves like an
Upgrade

Every generation of venue technology arrives with a promise of revolution. The promise is seductive because it is simple. New tools appear, and suddenly everything can be different. A venue can become intelligent overnight. A destination can become personalized at city scale. A museum can become a living conversation. A cruise ship can become a calm machine. A public space can become humane infrastructure.

The problem with revolutions is not ambition. The problem is brittleness.

Revolutions tend to be built as one-offs. They depend on a specific stack, a specific vendor, a specific integration shortcut, and a set of assumptions that holds for a year or two and then collapses as the world changes. The institution is left with a relic that was impressive in its moment and expensive to sustain.

This book has argued for a different posture. The future should behave like an upgrade.

That sentence is the most practical way to describe what we are building: an architecture that can absorb better compute, better interfaces, better sensing, better identity models, and better operational intelligence without requiring demolition and without demanding that the public surrender privacy as the price of being served.

One more upgrade pressure will only increase: fluency.

As systems become better at sounding helpful, the risk of confident wrongness rises. Confabulation is not primarily a language defect. It is a trust defect. In public environments, fluent error becomes operational harm when it influences routing, accessibility, timing, eligibility, or safety posture.

So the future that behaves like an upgrade must also behave like truth: grounded, time-bounded by freshness, verifiable where it matters, and willing to abstain when it cannot know.

This final chapter is a synthesis. It brings the major threads together and frames them as a stable direction for physical places. Not a prediction and not a sales pitch. A direction.

The phone was the first upgrade

We began with a simple fact: phones did not arrive as an optional accessory. They arrived as the default interface. That shift is not complete. Many venues still behave as if the phone is a distraction to be discouraged. In reality, the phone is now a universal accessibility tool, translation tool, guidance tool, camera, and increasingly the place where credentials and consent can live.

The mature posture remains the simplest test of modernity: design for phones without requiring phones.

A venue that holds that posture can incorporate the next upgrades naturally because many of the next upgrades are phone upgrades: on-device agents, negotiated preferences, minimal proofs, and personal data stores that keep the visitor in control.

Compute is the second upgrade

The most durable architectures treat compute as a capacity lever rather than a fixed device. If you can provision the right compute in the right nodes, the venue can replace and upgrade without rewriting itself. It can incorporate new media demands, new language modes, and new interaction patterns by fitting the right compute into the nodes where the workload lives.

This is why Edge Compute Nodes appears in this book as an enabling substrate rather than a product detail. It is the discipline that makes evolution normal. The future requires normal evolution, not occasional rebuilds.

Programmable Canvases and the end of interpretive scarcity

The wall is where venues have historically baked their assumptions into permanence. A label once printed becomes the truth and the truth becomes hard to update. Language becomes hierarchy. Depth becomes clutter. Inclusion becomes a fabrication problem.

Programmable Canvases is the idea that interpretive surfaces can behave like canvases where change is possible and fast without degrading the room's dignity. It makes language runtime. It allows multiple exhibits per canvas. It supports response to the world around a venue and keeps a place aligned with the public it serves.

This is a call to end scarcity where scarcity causes exclusion, not a call for more screens.

The Personal Channel as inclusive infrastructure

The most powerful future upgrade is not spectacle. It is participation.

Participation expands when the venue can deliver language, captions, sign language video, personal audio, quiet mixes, and other comfort modes through a Personal Channel. Inclusion becomes a normal set of choices rather than a special request.

This also supports privacy-forward operation. Many valuable modes can be delivered anonymously, without accounts, without dossiers, and without the visitor feeling watched.

The Personal Channel is therefore not merely an accessibility feature. It is a new kind of public infrastructure: a layer that lets the environment meet human diversity without forcing the room into compromise.

The acoustic layer treated as state

One of the quiet upgrades embedded throughout this book is conceptual: treating acoustics as a system layer.

Sound is not decoration. It is intelligibility, comfort, fatigue, and inclusion. It shapes behavior and mood. It determines whether a space feels calm or chaotic.

When the acoustic layer is treated as state, the venue can manage sound bleed, provide quieter mixes, and offer calm routes without requiring a visitor to explain themselves. It can also use occupancy-aware behavior to reduce noise in empty zones, which makes neighboring spaces clearer and more pleasant.

This is how venues stop excluding through sensory exhaustion.

The trap of the custom stack

As venues seek to become intelligent, large organizations often face a temptation: "We have smart engineers; we will build this ourselves."

This is the Integrator's Gap. It is easy to build a prototype of a personalized exhibit. It is exponentially harder to build an operating system that can handle:

- frame-accurate synchronization across 500 mobile devices in a hostile RF environment,
- governance gates that satisfy privacy obligations across multiple jurisdictions, and
- drift detection that distinguishes between a warming projector and a failing fan.

The organizations that win the next decade will not be the ones that spend five years trying to reinvent operational truth. They will be the ones that acquire or adopt a proven architecture and build their stories on top of it. The future belongs to those who treat the venue operating system as a platform, not a project.

The most dramatic change in modern physical environments is not that they can speak. It is that they can know, in the practical sense. The venue can maintain operational truth about what is open, what is full, what is loud, what is quiet, what is scheduled, what is broken, and what is safe.

WorldModel is the architectural name for that truth layer. It prevents large destinations from becoming collections of disconnected systems that contradict one another. It allows a concierge layer to propose actions that reflect reality rather than optimism.

A place that has operational truth can be coherent. A place that does not will improvise.

At scale, improvisation becomes confabulation: confident claims that are not grounded in authoritative state. That is why truthfulness constraints and abstain behavior belong in the core architecture, not as afterthoughts.

Governance as the enabling layer

As soon as environments become capable of shaping flow, queues, scheduling, and comfort at scale, their decisions acquire externalities. That is where governance becomes mandatory because we respect public space, not because we fear capability.

Governance turns values into constraints: safety, accessibility, privacy, fairness, and show integrity. It makes decisions auditable. It makes override normal. It prevents a system from becoming an optimizer that sacrifices the human day for a metric.

Governance is not the opposite of imagination. It is what allows imagination to be deployed responsibly.

Web3 and Web5 as the future of trust without oversharing

If the twentieth century made people hand over documents to demonstrate simple facts, the future should make that ritual feel archaic.

The bartender should rarely need your home address. The gate should rarely need your full biography. The venue should not become a repository of personal documents simply because it can.

Web3 identity provides a vocabulary of DIDs and verifiable credentials: proofs and claims that can be disclosed selectively, often as yes-or-no answers. Web5 extends the direction by putting the visitor's agent and data store on their side of the interaction. The phone negotiates. The visitor controls preferences and consent. The venue requests only what it needs for a moment.

This is how trust scales without turning public environments into identity theft magnets. It is also how civic-scale deployments remain defensible over time.

Operations as discipline and why verification matters

The public does not judge a venue by its architecture. They judge it by whether it works.

A modern venue is a living system. Living systems drift. They require observability, verification, trend analytics, and preventive maintenance. They require operations that can be executed by staff who are not the original integrators.

This is why Lifecycle Automation Layer, Outcome Verification Layer, and operator view are not peripheral. They are the operational maturity layer that keeps advanced systems from becoming fragile. The future upgrade is not only new capability. It is the ability to keep capability reliable.

A system that can change but cannot be verified will eventually be turned off. A system that can change and can be proven will keep evolving.

The final claim of the book

> *The future of physical places is not a future of gadgets.*
> *It is a future of governed, inclusive, privacy-forward environments that behave*
> *coherently under real constraints.*

It is a future where:

- language is runtime,
- inclusion is the default,
- comfort is access,
- phones are partners, not enemies,
- compute is provisioned as capacity,
- surfaces are canvases, not permanent compromises,
- operational truth is shared,
- truthfulness constraints prevent confident guessing,
- decisions are governed,
- identity is proved without oversharing, and
- upgrades are normal rather than traumatic.

This direction aligns with stable trends of the last two decades: ubiquitous personal devices, increasing multilingual reality, increasing expectations of accessibility, increasing privacy pressure, and increasing demand for places that feel worth leaving home for.

A venue that builds toward this direction becomes future-capable by design.

A closing note in the spirit of the covenant

This book has been intentionally positive because the opportunities are large, not because the problems are small.

We live in a moment where the tools are unusually capable and the cost of deploying them thoughtfully is falling. We can build places that feel more humane, more inclusive, and more alive than anything the previous generation of infrastructure could support. We can do it without becoming invasive. We can do it without assuming the public must surrender privacy to be served. We can do it with the discipline that makes systems trustworthy rather than theatrical.

> *The future that matters will not arrive as a revolution.*
> *It will arrive as an upgrade.*

POSTSCRIPT

A Note on Responsibility, and a Note on Joy

Every so often, a new capability arrives and expands what physical places can do. The surprise comes later. The hardest part was not the capability. The hardest part was accepting the responsibility that came with it.

This book has argued for hyper-personalized physical environments built as governed systems. It has treated privacy as a design requirement, inclusion as the starting point, and operational truth as the foundation. That posture is ambitious because it is deployable. Systems that cannot be trusted do not scale. They become demonstrations.

Trust is an engineering outcome. It is earned when a system behaves predictably, stays within clear boundaries, and can be audited and operated under real conditions. Trust is also earned when visitors can participate fully without handing over more than the moment requires.

A simple principle follows: a visitor should not have to surrender more than is necessary to be served well. That applies to a bar patron, a museum visitor, a shopper, a child on a school trip, and a resident in a city. Oversharing should feel archaic.

There is also the reason we build these places at all. Humans still want to be somewhere. We want to stand in front of real things. We want to share a reaction with strangers. We want to feel scale and wonder. We want to be moved in ways the living room cannot reproduce. We want a reason to leave home.

The most valuable technologies in this book are the ones that remove friction quietly so the human parts can expand: clarity, language dignity, comfort, agency, and the feeling that a place meets you as you are. When the system is run well, it stays in the background. The experience becomes easier to enter, easier to understand, and easier to remember.

If you are designing venues, operating them, commissioning them, or funding them, the most productive question is: how human can we keep the experience as capability increases?

That question is the difference between a world where public spaces become invasive and a world where public spaces become generous. The architecture in this book is an attempt at the generous world.

Thank you for building.

ACKNOWLEDGMENTS

Books are written in hours stolen from other things. The people closest to the work rarely get credit proportional to their contribution.

Tricia, my wife, made this book possible in ways that do not fit neatly into a sentence. She kept the world running while I disappeared into drafts. She kept me running when the drafts threatened to win. The ideas in these pages exist because she protected the time and space for them to form.

Thank you is insufficient, but it is what I have.

Mark and Serguei brought their own original ideas to the work, and turned architecture into reality. They originated methods and approaches that reshaped implementation decisions, hardened the system, and improved the architecture's operational realism. Every framework in this book could have remained theory, plausible on paper and untested in practice. Instead, they built the software, ran it on real hardware, tried to break it, and then strengthened it, iteration after iteration. They translated what survives in the field into safeguards, operational behavior, and decisions that make the difference between a diagram and a dependable system. The systems that underpin WorldModel™ exist because Mark and Serguei did the patient, detailed work of making ideas survive contact with hardware, software, and real-world edge cases.

Writing about technology is one thing. Building technology that works on a Tuesday morning is quite another. It is hard. This book is stronger because they did the building and the thinking.

Thanks also to the wider Mad Systems team, whose work over twenty-five years built the operational foundation these ideas rest on, and to the early readers and proofreaders who improved this manuscript in ways large and small.

M.J.E.

How to Use the Appendices

The appendices translate the architecture into artifacts you can use directly: templates for procurement, checklists for commissioning, evidence structures for assurance, and reference patterns for operations.

If you need to write an RFP or scope of work:

Start with Appendix J. It provides model specification language in outcomes-and-evidence form. Pair it with Appendix E (AI System Inventory) to define what you are buying, and Appendix F (Risk and Impact Assessment) to document what you are accepting.

If you need to prepare for commissioning or acceptance:

Use Appendix B (Drift Signals) and Appendix C (Support Model) as quick references during handover. Appendix H (Post-deployment Monitoring Plan) defines what must be observable from day one.

If you need to demonstrate assurance readiness:

Appendix G (Logging, Traceability, and Evidence Pack) defines what must be captured. Appendix I (Third-Party Assurance Packet) structures how to present it. Appendix K (Assurance and Compliance Snapshot) maps the book's artifacts to external frameworks.

If you need network or infrastructure patterns:

Appendix D provides venue network reference patterns, including separation, access control, and patch posture.

If you need the detailed table of contents:

Appendix L provides the full chapter-and-section breakdown for precise navigation.

The appendices are designed to be photocopied, adapted, and inserted into project documents. They are templates, not scripture. Modify them to fit your venue, your operating model, and your regulatory environment.

Appendix A
Implementation Examples

This appendix provides concrete, public-safe examples of how the patterns in the book behave in practice. They are implementation-realistic without disclosing proprietary integration sequences.

Example 1:
Archival video assembly for a WWII museum

A museum holds 10,000 hours of archival footage. An AI-assisted pipeline produces two-part metadata: selection metadata (date, event, people, locations, timestamps) and story metadata (curatorial explanation and boundaries). A visitor asks about Dunkirk in 1940. The system selects segments by metadata, chooses a 30-second or 90-second cut based on time budget and gallery conditions, edits a sequence, and delivers timed narration and subtitles in the requested language, age band, and interest lens (for example: general history vs military tactics).

Example 2:
Motion-profile selection for a robot-arm ride

A robot-arm ride on an autonomous tracked vehicle publishes multiple approved motion programs: gentle, standard, and extreme. Visitors choose comfort and stimulation preferences. The system selects a program within safety constraints, and operators retain override and stop authority. The outcome is broader accessibility without reducing excitement for opt-in riders.

Example 3:
Converting a venue with minimal physical changes

A venue adds QR codes or NFC tags at stops, plus minimal local timing and caching support. Visitors opt-in to the Personal Channel, select language, age band, and interest lens, and receive synchronized layers without requiring the room to be rebuilt. Offline-first posture keeps the critical path local, so the experience remains responsive.

Appendix B
Drift Signals at a Glance

Severity rule: warning if stable, action required if rising, urgent if coupled with faults or repeated verification failures.

Network latency creep

- Trigger: critical-path latency or jitter rises above baseline envelope, or synchronization anomalies increase.
- Action: validate segmentation and QoS, inspect uplink utilization, roll back risky changes.

Brightness decay

- Trigger: verified output quality proxies fall below baseline envelope, or degradation repeats after restoration.
- Action: clean optics where applicable, inspect alignment, schedule service planning.

Fan escalation

- Trigger: fan duty cycle exceeds baseline envelope, or trends upward without matching workload.
- Action: clean vents and filters, inspect fan condition if escalation persists.

Thermal drift

- Trigger: temperature exceeds baseline envelope for sustained periods, or trend slope rises over weeks.
- Action: inspect airflow and filters, clear obstructions, confirm ambient conditions.

Appendix C
Support Model
at a Glance

Maintenance windows

- Weekly or biweekly: routine patch window, off-hours.
- Monthly: verification run, including Outcome Verification Layer and Lifecycle Automation Layer restore tests.
- Quarterly: preventive maintenance, cleaning, physical inspection, and spares rotation.

Incident severity ladder

- SEV-1 critical:
  guest-facing outage or safety-impacting failure.
- SEV-2 high:
  material degradation, rising risk of guest-facing failure.
- SEV-3 medium:
  localized issue with workaround.
- SEV-4 low:
  cosmetic issue or planned enhancement.

Roles and responsibilities

- Venue operator:
  day-of operations, overrides, and on-site coordination.
- Content owner:
  narrative accuracy and approvals.
- Accessibility and language steward:
  language parity, captions, comfort modes.
- On-call support engineer:
  triage, mitigation, escalation, and incident summaries.
- Systems reliability owner:
  monitoring, verification, and drift thresholds.
- Platform maintainer:
  node provisioning, patching, rollback, configuration control.

A good rule of thumb:

1. If you are cleaning and find dirt or dust, halve your cleaning interval
2. If you find no dirt or dust for 3 cleanings, double your cleaning interval, but never inspect and clean less than every three months

Appendix D
Venue Network
Reference Patterns

1. Four-network separation
   - Guest network: visitor Wi-Fi and public access.
   - Operations network: staff tools, dashboards, and admin services.
   - Show control network: time-critical control paths and device control.
   - Audio transport network: clocking and transport kept separate to protect stability.

2. Remote access privilege model
   - Default deny. Least privilege. Time-bounded access. Logged sessions.
   - Remote access to show control and audio transport is restricted to the minimum necessary for management and diagnostics and is rarely routed through guest paths.

3. Patch cadence and rollback posture
   - Staged rollout: test in staging, canary a subset, then deploy broadly.
   - Versioning and rollback: every release can be rolled back quickly, and rollback is a normal operation.
   - Audio-specific caution: verify clock stability, routing, and zone integrity after any audio transport or DSP update.

Appendix E
AI System
Inventory Template

1. System identification:
   name, version, deployment site, accountable owner, provider, deployer, and operator.
2. Purpose and outcomes:
   what decisions the system can make, what requires human approval, what it is not allowed to decide.
3. Data and identity:
   data categories used, identity continuity method, consent and notice method, minimization strategy, retention periods, deletion process.
4. Models and content:
   model classes used, update behavior, content sources, safety controls applied to outputs.
5. Safety, inclusion, and accessibility:
   supported modes, vulnerable populations, human oversight and escalation path.
6. Monitoring and logging:
   metrics monitored, drift approach, log events captured, retention and access controls.
7. Known limitations:
   failure modes, out-of-scope behaviors, fallback modes.

APPENDIX F
RISK AND IMPACT
ASSESSMENT TEMPLATE

(Govern, Map, Measure, Manage)

This template documents how an AI-enabled capability is governed, assessed, tested, and controlled before and after deployment. It is intended to be completed for each materially distinct system, use case, or deployment context, and then updated whenever the system, data, model behavior, operating environment, or policies change. The output is a decision artifact: it records who is accountable, what risks were considered, what tests were run, what thresholds must be met, what controls are in place, and who accepted any residual risk. It is structured to align with NIST AI RMF 1.0's four functions: Govern, Map, Measure, and Manage.[xlix]

.

1. GOVERN (accountability and readiness)

Accountable owner:

Name/Role: _____ Org: _____

Decision rights and approvals:

- Who can approve deployment? _____
- Who can approve policy changes? _____
- Who can approve model updates? _____
- Who can halt or roll back the system? _____

Policies in force (attach or link):

- Privacy and data minimization policy: ☐ yes ☐ no
- Consent and notice policy (if applicable): ☐ yes ☐ no
- Accessibility and inclusion policy: ☐ yes ☐ no
- Safety policy and incident response: ☐ yes ☐ no
- Logging and retention policy: ☐ yes ☐ no
- Vendor and third-party assurance requirements: ☐ yes ☐ no

Staff readiness:

- Operator training completed: ☐ yes ☐ no Date: _____
- Support escalation path defined: ☐ yes ☐ no
- On-call or response ownership defined: ☐ yes ☐ no

Third-party dependencies:
List vendors, hosted services, external models, or data feeds, plus failure modes and fallback plan.

2. MAP (context, users, harms, and risk scoring)

Context of use:

What the system does, where it runs, and what decisions it influences.

User groups impacted (primary and secondary):

- Primary users: _____
- Bystanders or indirect impacts: _____
- Vulnerable or protected groups considered: _____

Edge cases and operating conditions:

- Peak load conditions: _____
- Degraded connectivity or offline mode: _____
- Sensor failure or partial data: _____
- Adversarial or misuse scenarios: _____

Harm scenarios (include at least one per category that applies):

- Safety harms: _____
- Privacy harms: _____
- Fairness or bias harms: _____
- Accessibility harms: _____
- Reputational or trust harms: _____
- Operational harms (downtime, chaos, staff overload): _____

Risk scoring (define your scale once, then score each scenario):
Scale used: Severity (1–5) _____ Likelihood (1–5) _____ Exposure (optional) _____

Highest-risk scenarios (top 3 to 5):

1. Scenario: _____ Sev: __ Lik: __ Score: __
2. Scenario: _____ Sev: __ Lik: __ Score: __
3. Scenario: _____ Sev: __ Lik: __ Score: __

Highest-risk use cases and why:

3. MEASURE (tests, evidence, and acceptance thresholds)

Trustworthiness measures selected (check all that apply):

- ☐ Accuracy or task success rate
- ☐ Robustness under noise, drift, or missing data
- ☐ Adversarial or red-team testing (where appropriate)
- ☐ Bias and differential performance testing
- ☐ Privacy tests (minimization, retention, access controls)
- ☐ Safety tests and fail-safe behavior
- ☐ Accessibility tests (WCAG, multimodal support, usability)
- ☐ Reliability tests (uptime, recovery, graceful degradation)
- ☐ Monitoring and observability validation (logs, metrics, alerts)

Evaluation methods and datasets:

- Test approach summary: _____
- Data used (and governance status): _____
- Ground truth source: _____
- Test environment: ☐ lab ☐ staging ☐ pilot ☐ production shadow

Acceptance thresholds (must be explicit):

Define the pass/fail thresholds that must be met before deployment.

- Primary metric threshold(s): _____
- Safety-critical thresholds (if any): _____
- Fairness or differential thresholds: _____
- Accessibility thresholds: _____
- Reliability thresholds (latency, uptime, recovery): _____
- Privacy thresholds (retention max, access rules): _____

Results summary (attach detailed reports):

- Passed all thresholds? ☐ yes ☐ no
- Exceptions granted? ☐ yes ☐ no   If yes, list exceptions:

4. MANAGE (controls, oversight, incidents, and residual risk)

Controls implemented (prevent, detect, respond):

- Preventive controls (guardrails, constraints, data minimization):
- Detection controls (monitoring, drift detection, anomaly alerts):
- Response controls (rollback, disablement, incident playbook):

Human oversight design:

- Where a human must approve actions: _____
- Where a human can override: _____
- Operator UI or decision transparency provided: ☐ yes ☐ no

Incident triggers and escalation:

- Trigger conditions (examples): policy violation, safety event, abnormal outputs, bias alarm, security event, repeated user complaints.
- Escalation path (who, how fast, how logged): _____

Change control and rollback:

- Versioning approach: _____
- Rollback procedure tested: ☐ yes ☐ no Date: _____
- Post-change verification required: ☐ yes ☐ no

Residual risk acceptance:
Residual risks remaining after controls (list):

Residual risk accepted by: _____ Date: _____

Sign-off role: _____ Signature: _____

Post-deployment monitoring plan (pointer):

Monitoring plan location or Appendix reference: _____

Reference: Structured to align with NIST AI RMF 1.0's functions: Govern, Map, Measure, and Manage.[1]

APPENDIX G
LOGGING, TRACEABILITY, AND EVIDENCE PACK

G1. Minimum Logging and Traceability for WorldModel Systems

G1.1 Log Events

Capture, at minimum:

- Decision events: timestamp, decision type, policy or constitution version, model version, confidence constraints, outcome
- Identity events: session start and stop, identity continuity state, consent state
- Data access events: what data categories were accessed, and why
- Safety events: refusals, escalations, guardrail triggers, override requests
- Human oversight events: operator interventions, approvals, overrides, and rationale
- Incident events: suspected harm, complaint, breach, anomalous behavior
- Update events: configuration changes, model updates, content updates, rollback events

G2. Evidence Properties

- Integrity: logs must be tamper-evident
- Access control: least privilege
- Retention: documented durations by event type
- Auditability: ability to reconstruct "why this happened" for a specific decision
- Privacy: avoid logging sensitive content unless necessary; redact where possible

G3. Evidence Pack Contents

A deployable evidence pack should include:

- The AI System Inventory (Appendix E)
- The Risk and Impact Assessment (Appendix F)
- Monitoring plan (Appendix H)
- Model and configuration version history
- Incident register and corrective actions
- Most recent assurance review notes (internal or external)

Appendix H
Post-deployment
Monitoring Plan Template

1. Scope:
   components covered, deployment environment, operating hours.

2. Metrics:
   experience quality KPIs, safety and escalation counts, accessibility outcomes, drift signals, reliability (latency, uptime, error rates).

3. Review cadence:
   daily checks, weekly checks, monthly reviews, quarterly governance review.

4. Thresholds and triggers:
   what triggers investigation, rollback, or formal incident handling.

5. Ownership:
   monitoring owner, escalation contacts, update approval authority.

6. Change and rollback:
   patch cadence, validation method, rollback method

7. Incident Mode Activation Runbook Template

Purpose

Define the required steps, authority, messaging limits, verification requirements, and evidence artifacts when Incident Mode is activated.

7.1 Runbook metadata

Runbook name: Incident Mode Activation

Version: _____

Effective date: _____

Owner: _____

Applies to: ☐ Entire venue ☐ Zone(s) ☐ Subsystem(s)

If Zone(s) or Subsystem(s), list: _____

7.2 Trigger conditions and authority class

Authoritative triggers (check all that apply):

- ☐ Life-safety system activation (fire alarm, emergency communications activation, or equivalent)
- ☐ Emergency stop chain asserted (E-stop, show-stop, ride-stop, or equivalent)
- ☐ Operator-declared hazard, closure, or responder-only zone
- ☐ Authority Having Jurisdiction direction (if applicable)
- ☐ Other authoritative trigger: _____

Advisory triggers (require corroboration or operator confirmation):

- ☐ Sensor anomaly or conflicting signals
- ☐ Unverified report of obstruction or congestion
- ☐ Verification failure trend exceeding threshold
- ☐ Other advisory trigger: _____

7.3 Scope and boundaries

Affected zones: _____

Adjacent zones to clamp routing: _____

Start time: _____ Expected review time: _____

Activated by (name, role): _____

7.4 Immediate required actions

Automation clamp (required):

- ☐ Suppress non-essential orchestration actions in affected zones
- ☐ Prevent optimization actions that could create new externalities
- ☐ Freeze non-essential personalization changes in affected zones

Messaging clamp (required):

- ☐ Restrict safety-adjacent messaging to pre-authored, owner-approved content only
- ☐ Disable freeform generation for instructions, closures, routing claims, eligibility, and life-safety prompts
- ☐ Switch surfaces in affected zones to clarity-first or guidance mode as required

Guidance clamp (required):

- ☐ emove affected zones and routes from recommendations immediately
- ☐ If state is uncertain, refuse precise routing and provide safe defaults or staff escalation prompts
- ☐ Confirm all visitor-facing channels apply the same route removals (surfaces, Personal Channel, staff tools)

Accessibility and baseline service (required):

- ☐ Maintain accessible baseline behavior and do not degrade core service for non-opt-in visitors
- ☐ Confirm accessible routes remain available, or are clearly and safely re-routed if impacted

7.5 Operator visibility requirements

Operator tools must show (check when confirmed):

- ☐ Mode state per zone and activation time
- ☐ Trigger type and source
- ☐ Active constraint set and suppressed action categories
- ☐ Current messaging mode and content set in use
- ☐ Activation operator identity and clearance controls

7.6 Verification and monitoring during incident

Verification checks to run (check all that apply):

- ☐ Safety state propagation confirmed
- ☐ Surfaces in guidance or clarity-first mode as required
- ☐ Audio behavior within incident constraints
- ☐ Routes removed and cannot be recommended by any interface
- ☐ Personal Channel and conversational interfaces honor Incident Mode restrictions
- ☐ Outcome verification checks pass for critical surfaces and zones

Monitoring posture during incident:

Review cadence: ☐ continuous ☐ every 5 min ☐ every 15 min
                ☐ other: _____

Escalation contact: _____

7.7 Evidence and logging requirements

Incident record ID or ticket number:

_____

Record must include (check when captured):

- ☐ Trigger and timestamp
- ☐ Scope and zones affected
- ☐ Actions taken (automation, messaging, guidance)
- ☐ Overrides invoked and by whom
- ☐ Verification outcomes
- ☐ Clearance verification and return-to-normal actions

Retention class and duration: _____

7.8 Clearance criteria and return-to-normal sequence

Clear only when all are true (check when verified):

- ☐ Authoritative trigger cleared and confirmed
- ☐ Closed routes remain closed until explicitly reopened
- ☐ Verification checks pass for affected systems
- ☐ Automation re-enables in a controlled sequence
- ☐ Heightened monitoring window defined: _____ (duration)

Cleared by (name, role): _____

Clearance time: _____

7.9 Post-incident review

Review date: _____ Owner: _____

Root cause category: ☐ equipment fault ☐ sensor anomaly ☐ human override

☐ external event [ ] other: _____

Corrective actions and owners:

_____ Owner: _____ Due: _____

_____ Owner: _____ Due: _____

Monitoring plan updates required (Appendix H): _____

Runbook updates required: _____

APPENDIX I
THIRD-PARTY
ASSURANCE PACKET OUTLINE

1. Executive Summary:
    what the system does, where deployed, and what it is not allowed to do.
2. System Inventory:
    include a completed AI System Inventory (Appendix E).
3. Risk and Impact Assessment:
    include a completed assessment (Appendix F) highlighting top risks and mitigations.
4. Controls and Evidence:
    logging and traceability approach (Appendix G), human oversight design, and safety and accessibility controls.
5. Monitoring and Improvement:
    post-deployment monitoring plan (Appendix H), incident handling, and corrective action examples.
6. Assurance Scope:
    what was reviewed, what was excluded, and remediation recommendations.

APPENDIX J
MODEL RFP SPECIFICATIONS FOR OWNERS

Updatable Companion Appendix

Informative, non-normative procurement template. Appendix J is provided as sample language that owners, facility directors, and creative leads may adapt for RFPs, scopes of work, and acceptance-test plans. It is written in outcomes-and-evidence form to help purchasers specify what must be true at delivery, commissioning, and operation.

Not a complete specification, and not legal advice. These clauses are examples and must be tailored to the site, the experience, the operating model, applicable law, and the owner's risk posture, and should be reviewed by qualified professional advisors.

Patent-aware and non-enabling by design. The clauses describe required behaviors, verification, and evidence, but they do not disclose proprietary integration sequences, configuration specifics, algorithms, source code, schematics, or other enabling implementation details. Offerors remain responsible for proposing their own implementation methods and for demonstrating compliance with the owner's acceptance criteria.

No license granted. Use of this appendix, including reproduction in procurement documents, does not grant any license under any patent, trademark, copyright, or other intellectual property right. Offerors are responsible for ensuring that their proposed solution has the necessary rights to practice any patented or licensed methods, if applicable.

0. Compute Node Architecture & Future Resistance

To ensure the venue remains upgradeable across product cycles, supply chains, and evolving capabilities, without being trapped by a single vendor's hardware roadmap.

0.1 Node-Based Compute Baseline: For any new venue build, the System shall provide a modular, node-based compute architecture where experience behaviors run on replaceable compute nodes rather than being locked to single-purpose appliances.

0.2 Non-Proprietary Hardware Only: Compute nodes shall be based on commercially available, non-proprietary hardware, and standard interfaces. The bidder shall provide a substitution strategy for supply chain changes, including at least one qualified alternative node configuration.

0.3 Replaceability and Restore: The System shall demonstrate that a failed compute node can be replaced and restored using documented procedures without rewiring, and without reprogramming the experience logic.

0.4 Provisioning and Scale: The System shall support provisioning additional compute by adding nodes, and allocating workloads by configuration, so that new features can be introduced without re-architecting the deployment.

1. Operational Automation & Verified Restoration

To ensure equipment longevity and acoustic hygiene without risking the "broken room" experience.

- 1.1 Occupancy-Aware Power Management: The System shall be capable of automatically reducing power consumption and acoustic output in specific zones when unoccupied for a user-definable duration. This must include the ability to dim or shutter projection, mute audio, and reduce compute requirements.
- 1.2 Verified Restoration (The "Ready" Standard): The System shall provide automated verification that the zone has returned to a visitor-ready state before a visitor enters the viewing area. This verification must confirm physical output in the zone rather than relying solely on device status logs or API responses.
- 1.3 Restoration Timing: The System shall support configurable "pre-warm" timing based on visitor approach vectors, ensuring that restoration processes are triggered early enough to be imperceptible to the arriving guest.

- 1.4 Automated Drift Detection: The System shall continuously monitor key health indicators (including but not limited to thermal trends, fan speeds, and network latency) and generate preventive maintenance alerts when values drift outside a baseline envelope, distinct from hard failure alerts.

2. The Personal Channel & Inclusive Delivery

To ensure accessibility and multilingual support are delivered as a reliable service, not a best-effort add-on.

- 2.1 Frame-Accurate Synchronization: The System must deliver auxiliary audio (e. g., alternate languages, audio description) and video (e. g., sign language interpretation, captions) to visitor-owned devices (BYOD) with at or near frame-accurate synchronization to the shared room media.
- 2.2 Low-Latency Performance in Dense Environments: A/V Synchronization (Owner acceptance criterion): For zones where synchronized audio and video are part of the guest experience, the system shall maintain audio-to-video synchronization within -90/+250 ms for at least 100 concurrent users per zone, when operating under the specified Environment Class for that zone (managed RF, managed device profile, and defined zone geometry), not taking into account excessive delays by BYOD devices.

- The bidder shall
    - (1) state the Environment Class assumptions used to meet this requirement,
    - (2) provide a measurement method (for example, a flash-and-blip test), and
    - (3) propose any alternate thresholds required for other Environment Classes, with justification.

As a minimum, the bidder shall demonstrate end-to-end A/V sync performance consistent with established guidance as defined for the venue.

Environment Classes (for calibration):

- Class A (Fully managed): Venue-controlled network and venue-controlled devices.
- Class B (Managed network, mixed devices): Venue-managed network, but mixed device performance.
- Class C (Unmanaged BYOD): Guest BYOD on best-effort connectivity.

The Owner will designate the Environment Class per zone; the bidder shall provide performance commitments per class.

- 2.3 Offline-First Capability: The personal delivery system must remain fully functional for local media playback and synchronization even in the event of a wide-area network (WAN) or internet service outage.
- 2.4 Privacy-Forward Session Management: The System must enable visitors to select language and accessibility preferences via their personal device without requiring the creation of a permanent user account or the transmission of personally identifiable information (PII) to a cloud server.
- 2.5 Neurodiversity and Cognitive Comfort (Owner acceptance criteria): The System shall support neurodiversity-inclusive delivery by providing explicit, opt-in comfort modes that reduce sensory load and cognitive load without segregating the experience.

At minimum, the bidder shall:

1. provide at least two comfort modes (for example, Standard, Calm), selectable at any time by the visitor (or guardian) via the Personal Channel,
2. ensure comfort mode changes take effect within an owner-defined threshold (for example, within 2 seconds), without breaking synchronization,
3. support predictable interaction pacing in comfort modes (no surprise timers, no forced escalation, and no sudden sensory spikes without warning),
4. support pause, resume, and skip behaviors for interactive legs without losing session continuity, and
5. provide evidence of implementation and verification, including a test plan and demonstration scenarios for comfort mode selection, persistence across multiple exhibits, and degraded connectivity operation.

Privacy constraint: comfort mode support shall not require a diagnosis disclosure, account creation, or transmission of sensitive personal data to a cloud service.

3. Shared Media Planes and Non-Personal-Channel Media Delivery

This section applies to all shared media planes, including room media, programmable canvases, signage, projection surfaces, and any other venue-fixed or venue-shared outputs, excluding the Personal Channel.

To ensure multilingual delivery on shared outputs remains flexible, maintainable, and verifiable.

- 3.1 Real-Time Subtitle Rendering (Shared Media Planes): All subtitles shall be rendered by media players in real time from external text-based subtitle files (for example, timed text files). Subtitles shall not be permanently burned into video assets as the only delivery method.
- 3.2 Subtitle Language Switching (Shared Media Planes): All media players shall be capable of switching subtitle languages either instantly, or before the next timed subtitle cue is rendered. Switching shall not require replacement of the underlying media asset, operator-only intervention, or system restart. Subtitles shall remain time-aligned to the associated media during and after switching, including under peak load.
- 3.3 Subtitle Asset Management: Subtitle text files shall be independently versioned and deployable without re-rendering the underlying video. The System shall support controlled rollout and rollback of subtitle assets and shall log subtitle language state as operational evidence for verification.
- 3.4 Main Show Language Switching (Room Plane and Shared Audio): The System shall support switching the language of shared room media, including narration and any language-dependent audio stems, without requiring replacement of the underlying media asset. Switching shall occur either instantly or at the next owner-defined synchronization boundary (for example, an approved language switch point, segment boundary, or cue) so that visitors do not experience mid-sentence language tears or loss of synchronization.
- 3.5 Verification of Multilingual State: The System shall verify and log that the intended language mode and subtitle language are actually active on each relevant shared surface and shall raise operator-visible faults for mismatches or failures within 30 seconds, with a defined safe fallback posture.

4. Integrated Recognition & Governance

To ensure personalization remains safe, fair, and publicly defensible.

- 4.1 Governance Gate Architecture: The System shall include a distinct governance layer capable of evaluating automated decisions against a set of explicit, owner-definable constraints (e. g., Safety, Accessibility, Fairness) before execution.
- 4.2 Anonymous Continuity: The System must support "anonymous recognition," allowing visitor preferences (such as language or depth) to persist across multiple exhibits during a single visit without requiring facial recognition or persistent biometric storage.
- 4.3 Constraint-Based Guidance: Any automated wayfinding or flow-management advice generated by the System must be auditable. The System must be able to log why a specific route was suggested and verify that it complied with active safety and accessibility constraints.
- 4.4 Operator Override: The System must provide a centralized mechanism for operators to immediately override automated behaviors (e. g., force a zone to "Full On" or "Maintenance Mode") without requiring code changes or system reboots.
- 4.5 Life-Safety Boundary (Non-Negotiable)
  The System shall not inhibit, delay, or modify any life-safety system behavior. If life-safety state is available as a signal to the System, it shall be treated as authoritative and shall trigger a defined incident posture for guidance, signage, and automation suppression within an owner-defined time threshold (for example, within 5 seconds). Any visitor-facing safety messaging shall be pre-authored and owner-approved
- 4.6 Emergency Stop Boundary (Non-Negotiable)
  Where E-stop, show-stop, ride-stop, or equivalent emergency stop chains exist, they shall be implemented as independent hard safety functions. The System may monitor and log emergency stop state, but shall not be in the causal chain required for the stop to function, or for the stop to remain latched until manual reset[li]. The System shall not perform automatic restart after an emergency stop[lii].
- 4.7 Emergency-Aware Routing and Hazard Exclusion
  The System shall support authoritative hazard flags (life-safety triggered, operator-declared, or responder-only) that immediately remove affected zones from all routing, recommendations, and guidance. Acceptance test: demonstrate that within an owner-defined threshold (for example, 5 seconds), all planes reflect the restriction (shared surfaces, Personal Channel guidance, staff tools), and no guidance is produced that routes toward the restricted area.

5. Infrastructure & Maintainability

To prevent vendor lock-in and ensure the system can be upgraded without demolition.

- 5.1 Non-Proprietary Compute Substrate: To ensure long-term maintainability, the core logic, media playback, and control nodes must run on standard, commercial-off-the-shelf (COTS) computing hardware. Proprietary "black box" media players that cannot be serviced or replaced by standard IT staff are not acceptable.
- 5.2 Software-Defined Upgradability: The System architecture must allow for the addition of new capabilities (e. g., new language tiers, new accessibility modes, updated recognition models) via software updates to existing compute nodes, without requiring the physical replacement of infrastructure cabling or displays.
- 5.3 Local Truth & Resilience: The System shall maintain a local "Operational Truth" state (WorldModel) that allows the venue to continue operating coherently (including logic, triggers, and show control) even if connection to central cloud services is lost.

6. Digital Audio, Spatial Audio, and Adaptive Soundscapes

To ensure the audio layer is modern, scalable, verifiable, and maintainable.

6.1 Audio transport and interoperability

- 6.1.1 Standards-based interoperability: The System shall support standards-based network audio interoperability (for example, AES67 where required), with a clearly documented clocking posture.
- 6.1.2 Audio transport separation: The System shall provide an audio transport network separated from guest and general operations networks, with explicit QoS and multicast policies where required.
- 6.1.3 Clocking resilience: The System shall define a PTP clocking posture suitable for venue operation, including boundary behavior, holdover behavior, and defined failover behavior under partial clock loss.

- 6.1.4 Redundancy (if specified): Where specified, the audio transport shall support redundant paths (for example, A/B networks), using a hitless redundancy method (for example, SMPTE ST 2022-7, or equivalent). Loss of any single path shall not cause guest-audible artifacts (including pops, clicks, distortion, or dropouts). Verification: demonstrate continuous audio during a live path removal test at nominal operating level, with no audible artifacts during a 60-second audition at representative listening positions."

6.2 Capacity and channelization

- 6.2.1 Multi-channel output: Each render node shall support a minimum of 64 channels of network audio output, addressable as discrete channels for routing and zoning.
- 6.2.2 Concurrent mixes: The System shall support zone-based and scene-based routing, including concurrent delivery of language mixes and comfort mixes, without requiring additional exhibit hardware per mix.

6.3 Spatial audio and soundscape variation

- 6.3.1 Spatial audio: Where spatial audio is specified, the System shall support spatial audio and zone-based positioning, including commissioning procedures and acceptance tests that validate localization, intelligibility, and bleed control.
- 6.3.2 Constrained soundscape variation: The System shall support constrained variation of soundscapes, including randomized stems or layers, within an owner-defined policy envelope. Variation shall not introduce abrupt transitions, thematic breaks, or safety conflicts. Implementations shall be polyphonic to allow addition of additional audio channels such as narratives.
- 6.3.3 Loudness and intelligibility policy: The System shall support zone-based loudness targets and maximums and shall preserve intelligibility under busy-mode conditions.
- 6.3.4 Intelligibility and bleed acceptance tests: The commissioning plan shall include intelligibility and bleed tests per zone pair, with pass/fail thresholds defined by the owner and documented in an acceptance test plan.

6.4 Event ingest and external sources

- 6.4.1 Controlled ingest: The System shall support controlled ingest of time-bounded external audio sources for special events, including caching where appropriate and defined safe fallback behavior under degraded connectivity.
- 6.4.2 No internet dependency for core show: Core show operation shall not depend on external internet connectivity.
- 6.4.3 Rights and licensing posture: External sources shall include documented rights, licensing posture, and retention rules for any cached audio.
- 6.4.4 Auditability: The System shall log which external source was used, when, what fallback occurred, and what operator actions were taken.

6.5 Verification, drift, and no silent failure

- 6.5.1 Outcome verification: The System shall provide outcome verification for audio, including audio present where expected, level within bounds, and intelligibility and bleed proxies where applicable.
- 6.5.2 Drift signals and workflows: The System shall provide drift signals and operator workflows, including rising noise floor, rising packet loss or jitter, clock instability, and degradation of intelligibility proxies.
- 6.5.3 No silent degradation: Any verification failure affecting guest-audible output shall raise an operator-visible fault within 30 seconds and shall trigger a defined safe fallback posture rather than silently degrading in public.

6.6 Change control and rollback

- 6.6.1 Versioning: Audio routing, DSP configuration, clocking posture, and zone loudness policies shall be versioned.
- 6.6.2 Rollback: The System shall support rollback to a known-good configuration within 30 minutes without re-imaging hardware.

Note to Specifiers: These requirements define a system that is deterministic, verifiable, and privacy-forward. While many vendors may claim "personalization" or "automation," few can meet the strict requirements for verified restoration and offline-first synchronization without specific, advanced architecture. Enforcing these specs protects your capital investment from obsolescence and your operation from reputational risk.

APPENDIX K
ASSURANCE AND COMPLIANCE SNAPSHOT

Updatable Companion Appendix
Snapshot as of February, 2026

This appendix is a replaceable snapshot of external standards and regulatory expectations that buyers may cite in procurement, assurance, and risk reviews. It is intentionally separable from the core narrative so it can be updated or replaced between editions without rewriting the main architecture chapters.

1. Management-system framing for AI governance:
    ISO/IEC 42001 specifies requirements for establishing, implementing, maintaining, and continually improving an AI management system.[liii] It is useful as a governance wrapper because it emphasizes lifecycle discipline, accountability, monitoring, and continual improvement.
2. Risk framing for AI systems:
    NIST AI RMF 1.0 organizes AI risk work into four functions: Govern, Map, Measure, and Manage.[liv] It is widely used as a practical, procurement-friendly language for AI risk management.
3. GenAI-specific risk terminology:
    NIST AI 600-1 (Generative AI Profile) defines and organizes risks that are specific to generative systems.[lv] It is a useful anchor for terminology, including confabulation, and for system-level mitigations such as evidence requirements, verification, and safe failure behavior.
4. Security posture for LLM and agentic systems:
    OWASP's Top 10 for Large Language Model Applications provides a practical taxonomy of common failure and abuse modes, including prompt injection, insecure output handling, sensitive information disclosure, and supply chain risks.[lvi] It is useful as a procurement and assurance checklist for hostile-input and misuse scenarios.
5. Regulated expectations where applicable:
    Where it applies, the EU AI Act codifies expectations that purchasers increasingly ask for globally, including technical documentation, logging and traceability, and post-market monitoring for certain AI systems.[lvii]
6. Third-party assurance direction of travel:
    The UK has published a roadmap aimed at growing a trusted third-party AI assurance market. Even outside the UK, this signals a procurement shift toward independent assurance artifacts over self-attestation.[lviii]
7. OECD Recommendation on AI provides a values-based policy vocabulary anchor that procurement teams often cite for accountability, transparency, and responsible deployment.[lix]

Mapping: Book artifacts to common external expectations

- AI System Inventory (Appendix E):
  Proves scope, purpose, user groups, data categories, decision boundaries, and accountable roles.
- Risk and Impact Assessment (Appendix F):
  Proves identified harms, mitigations, thresholds, and residual risk sign-off.
- Logging and Traceability Spec (Appendix G):
  Proves reconstructable decisions, interventions, and versioning, without turning logs into dossiers.
- Post-deployment Monitoring Plan (Appendix H):
  Proves drift detection, review cadence, escalation triggers, and improvement loop.
- Third-party Assurance Packet (Appendix I):
  Bundles the above into a bounded package for procurement and independent assessors.
- Owner procurement language and acceptance tests (Appendix J):
  Translates requirements into RFP clauses, commissioning evidence, acceptance tests, audit rights, and support expectations.
- Governance threshold definition (Ch. 4) and governance in practice (Ch. 26):
  Proves when personalization becomes governed decision-making, and what constraints apply.
- Outcome verification and verified restoration (Ch. 38, Appendix G, Appendix H):
  Proves the system can confirm exhibit-ready state, detect drift, and restore without guesswork.
- Change control and rollback evidence (Ch. 41, Appendix G, Appendix H):
  Proves versioning, rollback drills, and time-to-restore performance.

The crosswalk below maps widely cited external frameworks to the relevant chapters and the minimum evidence artifacts defined in Appendices E through J.

APPENDIX K.1
STANDARD CROSSWALK
(NON-NORMATIVE)

Standard or Framework Anchor	What it expects (plain language)	Where addressed in this book (chapter or appendix)	Evidence artifacts to provide (what auditors and owners ask for)
NIST AI RMF 1.0: GOVERN	Defined governance structure: who owns risk, who approves changes, and how accountability is enforced over the lifecycle	Ch. 0 Book Covenant; Ch. 4 Governance threshold; Ch. 26 Governance in Practice; Ch. 37 Commissioning and Support; App. I; App. K	AI governance roles and RACI; decision authority and override policy; change control policy; assurance packet outline (App. I); governance snapshot (App. K)
NIST AI RMF 1.0: MAP	Identify intended use, stakeholders, context, and impact boundaries before building or procuring	Ch. 3 Personalization Ladder; Ch. 4 Externalities test; Ch. 18-20 Privacy and analytics; App. E; App. F; App. J	AI System Inventory (App. E); Risk and Impact Assessment (App. F); Owner RFP specs (App. J); documented scope and prohibited uses list
NIST AI RMF 1.0: MEASURE	Measure performance, risks, and failure modes, including GenAI-specific errors, and track drift over time	Ch. 37 Monitoring; Ch. 38 Outcome Verification and Verified Restoration; App. B; App. G; App. H	Logging minimums and evidence pack (App. G); Post-deployment monitoring plan (App. H); drift signals quick reference (App. B); commissioning report with measurable acceptance tests
NIST AI RMF 1.0: MANAGE	Mitigate and respond: controls, incident handling, rollback, and continuous improvement with documented change management	Ch. 39-42 Lifecycle and upgrades; App. H; App. I; App. K	Incident playbooks; rollback evidence; update and hotfix log; periodic risk review cadence; third-party assurance packet (App. I)
NIST AI 600-1: Confabulation (hallucination) risk	GenAI may produce confident but wrong content; treat it as a managed risk, not a surprise	Ch. 16 Virtual Docents; Ch. 24-26 WorldModel and governance; Ch. 37 Monitoring; App. H; App. K	Confabulation control policy: approved sources, evidence pointers, abstain behavior, escalation; confabulation incident log; monitoring metrics

Standard or Framework Anchor	What it expects (plain language)	Where addressed in this book (chapter or appendix)	Evidence artifacts to provide (what auditors and owners ask for)
NIST AI RMF 1.0: GOVERN	Defined governance structure: who owns risk, who approves changes, and how accountability is enforced over the lifecycle	Ch. 0 Book Covenant; Ch. 4 Governance threshold; Ch. 26 Governance in Practice; Ch. 37 Commissioning and Support; App. I; App. K	AI governance roles and RACI; decision authority and override policy; change control policy; assurance packet outline (App. I); governance snapshot (App. K)
			and thresholds; regression tests after updates
Government model governance frameworks for agentic AI (example: IMDA, 2026)	Bound agent authority and autonomy upfront, keep humans accountable for consequential actions, enforce lifecycle controls (testing, monitoring, rollback), and enable responsible end-user operation through clear disclosures, consent boundaries, and training.	Ch. 4 (governance threshold and externalities); Ch. 24 (truth bus, governance gate, evidence hierarchy); Ch. 25 (CGL, ICL, MAOL, EDE as decision architecture); Ch. 26 (governance in practice); Ch. 37–38 (commissioning, monitoring, outcome verification, verified restoration); App. E (AI System Inventory); App. F (Risk and Impact); App. G (Logging and Evidence Pack); App. H (Monitoring Plan); App. I (Third-Party Assurance Packet); App. J (Owner RFP specs).	Bounded authority and permissions: allowed-actions catalog, least-privilege access model, and governance boundary definition; Accountability: decision records for consequential actions, operator override logs, and approval checkpoint definitions; Lifecycle controls: commissioning report with acceptance tests, post-deployment monitoring plan with thresholds, rollback drill evidence, and incident register; End-user responsibility: one-paragraph disclosure, user training outline, and consent, retention, and "must forget" policy evidence.
NIST AI 600-1: Transparent and explainable operation	Users and operators need understandable explanations of what the system is doing, within safety and privacy limits	Ch. 4 Governance; Ch. 22 Concierge interface; Ch. 26 Governance in Practice; App. K	UI disclosure rules for adaptive behavior; operator view of current mode and constraints; explanation templates for visitor-facing systems
NIST AI 600-1: Data protection and privacy risks	GenAI and analytics can leak or misuse sensitive data; enforce minimization, retention, access control, and auditability	Ch. 18 Privacy; Ch. 19 Anonymous recognition; Ch. 20 Analytics; App. E; App. G; App. H	Data inventory with purpose and retention; automated deletion evidence; access logs; privacy controls checklist; monitoring plan for data access anomalies

Standard or Framework Anchor	What it expects (plain language)	Where addressed in this book (chapter or appendix)	Evidence artifacts to provide (what auditors and owners ask for)
NIST AI RMF 1.0: GOVERN	Defined governance structure: who owns risk, who approves changes, and how accountability is enforced over the lifecycle	Ch. 0 Book Covenant; Ch. 4 Governance threshold; Ch. 26 Governance in Practice; Ch. 37 Commissioning and Support; App. I; App. K	AI governance roles and RACI; decision authority and override policy; change control policy; assurance packet outline (App. I); governance snapshot (App. K)
ISO/IEC 42001: AIMS scope and policy	Establish an AI management system: policy, objectives, defined scope, and continual improvement	Ch. 0 Covenant posture; Ch. 37-42 Operations lifecycle; App. E; App. F; App. I; App. K	AIMS-style policy statement; scoped AI system register; internal review cadence; supplier oversight evidence; assurance packet outline
ISO/IEC 42001: Risk treatment and lifecycle controls	Repeatable lifecycle controls for design, deployment, monitoring, and change management	Ch. 37 Commissioning; Ch. 38 Verified Restoration; Ch. 41 Upgrade Path; App. H; App. I	Change control log; commissioning checklist; verified restoration report; monitoring plan; audit trail continuity across upgrades
ISO/IEC 42001: Third-party and supplier governance	Control and document supplier risks, including model, sensor, and platform dependencies	Ch. 9 Networking and separation; Ch. 37 Support posture; App. I; App. J	Supplier disclosure checklist; support life and patch policy; SBOM or equivalent supplier security posture evidence; RFP clauses for ownership and audit rights
OWASP Top 10 for LLM Applications: Prompt injection	Assume inputs can attempt to override instructions; isolate and constrain what user input can cause the system to do	Ch. 16 Virtual Docents; Ch. 26 Governance; Ch. 9 Network separation; App. J	Prompt injection test cases; policy that limits tool actions; denial behavior for unsafe requests; red team results summary; RFP security clauses
OWASP Top 10 for LLM Applications: Insecure output handling	Downstream systems must not blindly execute or render untrusted model outputs	Ch. 10 Orchestration and verification; Ch. 24 Implementing WorldModel venues; Ch. 38 Outcome verification; App. G	Output handling rules; evidence-bearing outputs requirement; safe rendering templates; verification logs proving outputs did not trigger unsafe actions
OWASP Top 10 for LLM Applications: Sensitive information disclosure	Prevent leakage of personal data, secrets, or restricted content via model responses	Ch. 18-19 Privacy and recognition; Ch. 16 Virtual Docents; App. E; App. G; App. H	Data classification policy; redaction rules; access control; audit logs; monitoring for leakage incidents; incident response steps for disclosure events

Standard or Framework Anchor	What it expects (plain language)	Where addressed in this book (chapter or appendix)	Evidence artifacts to provide (what auditors and owners ask for)
NIST AI RMF 1.0: GOVERN	Defined governance structure: who owns risk, who approves changes, and how accountability is enforced over the lifecycle	Ch. 0 Book Covenant; Ch. 4 Governance threshold; Ch. 26 Governance in Practice; Ch. 37 Commissioning and Support; App. I; App. K	AI governance roles and RACI; decision authority and override policy; change control policy; assurance packet outline (App. I); governance snapshot (App. K)
OWASP Top 10 for LLM Applications: Supply chain and model risk	Track model versions, dependencies, and update integrity	Ch. 37 Monitoring; Ch. 41 Upgrade path; App. G; App. H	Version pinning record; update and rollback logs; model and Body of Knowledge change history; post-update regression results
OECD AI Recommendation: human rights, accountability, transparency	Values-based principles that must be translated into enforceable constraints in public environments	Ch. 0 Ten Laws; Ch. 26 Constitution and governance; App. K	Constitution summary; decision record requirement; accountability and override policy; periodic audit and review cadence
WCAG 2.2: accessibility as a baseline	Accessibility and inclusive delivery must be designed in, tested, and preserved through upgrades	Ch. 11-14 Inclusive delivery; Ch. 7 Non-visual guidance; Ch. 37 Commissioning; App. J	Accessibility acceptance tests; language package QA evidence; degraded-mode accessibility behavior; RFP clauses requiring WCAG-aligned UI and documentation
W3C DID Core: decentralized identifiers	Portable identity and minimal disclosure should be supported without centralizing identity by default	Ch. 27 Portable Identity; Ch. 28 Agent coordination; App. J	DID method support statement; minimal disclosure use cases; consent and revocation flows; evidence that baseline experience works without identity
W3C Verifiable Credentials Data Model	Attribute proofs and credentials can be presented without oversharing; support issuer, holder, verifier roles	Ch. 27 Portable Identity; Ch. 29 Consented Context; App. J	Credential acceptance criteria; attribute-based access rules; credential revocation handling; audit traces that show attributes, not full identity data
NIST SP 800-63: digital identity assurance	Identity assurance levels and authentication requirements when identity is used at all	Ch. 27 Portable Identity; Ch. 18 Privacy posture; App. E; App. J	Identity use register: why identity is requested; authentication and session security requirements; role-based access controls; retention limits for identity-linked events

Standard or Framework Anchor	What it expects (plain language)	Where addressed in this book (chapter or appendix)	Evidence artifacts to provide (what auditors and owners ask for)
NIST AI RMF 1.0: GOVERN	Defined governance structure: who owns risk, who approves changes, and how accountability is enforced over the lifecycle	Ch. 0 Book Covenant; Ch. 4 Governance threshold; Ch. 26 Governance in Practice; Ch. 37 Commissioning and Support; App. I; App. K	AI governance roles and RACI; decision authority and override policy; change control policy; assurance packet outline (App. I); governance snapshot (App. K)
NIST FRVT demographic effects (NISTIR 8280 and related)	Biometric performance varies by demographics; document performance, mitigation, and operational limits	Ch. 19 Anonymous recognition; Ch. 18 Privacy; App. F; App. K	Demographic performance evaluation plan; threshold and fallback rules; bias and error monitoring; opt-in and minimization posture; third-party test references when used
Book assurance posture: logging, traceability, evidence pack	Auditability requires consistent logging and evidence artifacts, not ad hoc screenshots	Ch. 24-26 Governance implementation; Ch. 37-38 Operations; App. G; App. I	Logging minimums (App. G1.1); evidence pack contents (App. G3); third-party assurance packet outline (App. I); decision record templates
Book assurance posture: monitoring plan and drift signals	Monitor for drift, confabulation incidents, privacy violations, and degradation in inclusion outcomes	Ch. 37 Monitoring; Ch. 38 Verified restoration; App. B; App. H	Monitoring plan (App. H); drift signals quick reference (App. B); ticketing integration evidence; periodic review reports
Book procurement posture: owner RFP and acceptance tests	Procurement language should demand evidence, tests, and audit rights	Ch. 0 Proof Pack; Ch. 37 Commissioning; App. J; App. K	RFP clause groups (App. J); acceptance test index; commissioning evidence; audit rights and data ownership clauses

Notes (Non-Normative, Patent-Safe):

This crosswalk is provided to help readers map the architecture, assurance posture, and evidence discipline described in this book to widely recognized standards and frameworks. It is a terminology and auditability aid, not a compliance checklist, and it does not claim that any implementation is certified, approved, or compliant in any jurisdiction. The book describes requirements, acceptance criteria, and evidence artifacts at a public-safe level, and it does not disclose implementation sequences, configuration specifics, source code, or other enabling details that could compromise protected methods. Where standards or laws are mentioned, they are referenced for context and vocabulary; implementers remain responsible for determining applicability, obtaining professional advice where required, and meeting all obligations.

APPENDIX K.2
EVIDENCE PACK INDEX
(ARTIFACT MAP)

Artifact ID	Artifact name	Purpose	Where it lives in this book	Minimum contents to be considered complete	When to use it
E-1	AI System Inventory Template	Declare what exists, where it runs, what data it touches, and what it is allowed to do	Appendix E	Components, deployment scope, data categories, model or rules components, decision boundaries, human oversight points, retention notes	Before procurement, during design review, before go-live, and after major changes
F-1	Risk and Impact Assessment Template	Identify key risks, affected parties, mitigations, residual risk, and sign-off	Appendix F	Risk list, impact analysis, controls, verification plan, residual risk acceptance, sign-off role, and date	Before go-live, after major changes, and as part of periodic governance reviews
G-1	Logging minimums and traceability	Make decisions reconstructable without turning logs into dossiers	Appendix G, Section G1.1	Decision events, identity and consent events, data access events, safety and escalation events, human oversight events, incident events, update and rollback events	Always on in production, and reviewed during audits and incidents
G-2	Evidence properties	Define what makes evidence trustworthy and reviewable	Appendix G, Section G2	Integrity, access control, retention, auditability, and privacy posture for logs and artifacts	When defining audit posture, vendor requirements, or third-party assurance expectations
G-3	Evidence Pack Contents (manifest)	Define the minimum evidence bundle that a deployment must maintain	Appendix G, Section G3	Inventory, risk assessment, monitoring plan, version history, incident register, and recent assurance notes	For readiness reviews, assurance packets, and procurement acceptance gates

H-1	Post-deployment Monitoring Plan Template	Specify what is monitored, how often, by whom, and what triggers action	Appendix H	Scope, metrics, review cadence, thresholds and triggers, ownership, change and rollback approach	Before go-live, then continuously throughout operation
H-2	Drift Signals at a Glance	Provide a compact operator-facing list of signals that predict failure or degradation	Appendix B	Signal list, thresholds, and response actions mapped to tickets or runbooks	Daily operations, weekly reviews, and incident prevention work
I-1	Third-Party Assurance Packet Outline	Provide a reviewable structure for external assurance without exposing sensitive implementation details	Appendix I	Executive summary, system inventory reference, risk assessment reference, controls and evidence summary, monitoring and improvement summary, assurance scope statement	When a buyer, regulator, insurer, or third party requests an assurance package
J-1	Model RFP Specifications for Owners	Translate the book into procurement-ready clauses and acceptance-test language	Appendix J	Outcomes-and-evidence requirements, commissioning and acceptance tests, vendor obligations, and audit rights	For RFPs, SOWs, contract exhibits, and acceptance planning
D-1	Venue Network Reference Patterns	Support segmentation, timing posture, resilience, and operational observability	Appendix D	Network separation, show and safety posture, operations posture, and failure-mode assumptions	Early design, security review, commissioning plan, and troubleshooting
A-1	Implementation Examples	Provide illustrative examples while remaining vendor-neutral and public-safe	Appendix A	Scenario-level examples, acceptance criteria focus, and evidence references	Workshops, design reviews, and stakeholder alignment

K-1	Standards Crosswalk Table	Provide a non-normative mapping from this book to recognized standards and frameworks	Appendix K (this appendix)	Framework anchors, what they expect, where addressed, and evidence artifacts required	Procurement reviews, assurance reviews, and PhD or academic evaluation framing

APPENDIX K.3
EVIDENCE PACK
READINESS CHECKLIST

This checklist is designed to be initialed during commissioning and acceptance. "Pass" means the artifact exists, is reviewable, and matches the minimum contents described in the referenced appendix. This checklist is non-normative and does not constitute legal advice or a certification claim.

A. System inventory and scope

1. AI System Inventory (App. E) present and current
   Pass criteria: All deployed components, channels, and decision boundaries are recorded, including where data is processed and retained.
2. System scope and prohibited uses documented
   Pass criteria: Clear statement of what the system does not do, especially identity-by-default, covert profiling, or unbounded generation.
3. Deployment modes declared (curated, Body of Knowledge, hybrid, bounded generation, where applicable)
   Pass criteria: Modes are explicitly listed by zone, surface, and channel.

Initials: _____ Date: _____

B. Risk, impact, and governance posture

4. Risk and Impact Assessment (App. F) complete and signed
    Pass criteria: Risks, mitigations, residual risk, and acceptance sign-off are captured.

5. Governance boundary defined (what is allowed, disallowed, and human-approval required)
    Pass criteria: The governance boundary is written in verifiable terms and can be audited

6. Operator authority and override path defined and tested
    Pass criteria: Override exists, has clear scope, and produces auditable records.

Initials: _____ Date: _____

C. Logging, traceability, and decision evidence

7. Logging minimums enabled (App. G1.1)
    Pass criteria: Decision events, consent events, data access events, override events, incident events, and update events are logged

8. Decision evidence exists for consequential actions
    Pass criteria: Decision records include operational state used, constraints checked, outcome (approve, modify, or reject), and override when applicable

9. Evidence Pack manifest present (App. G3)
    Pass criteria: A complete evidence bundle exists and is stored with access controls and retention rules.

Initials: _____ Date: _____

D. Monitoring, drift, and operational readiness

10. Monitoring Plan (App. H) complete and active
    Pass criteria: Metrics, thresholds, escalation rules, and ownership are defined and operating
11. Drift Signals quick reference available (App. B)
    Pass criteria: Drift signals are mapped to response actions and ticketing or runbooks
12. Confabulation (hallucination) controls documented and monitored
    Pass criteria: Approved-source boundaries, evidence expectations for factual claims, abstain behavior when grounding is insufficient, incident capture, and trending are defined and in use

Initials: _____ Date: _____

E. Verification, restoration, and update discipline

13. Outcome verification checks implemented
    Pass criteria: The venue can verify that key outputs occurred, not only that commands were issued
14. Verified restoration procedure exists and has been exercised
    Pass criteria: A restoration test has been run and recorded for at least one representative failure scenario
15. Update and rollback discipline evidenced
    Pass criteria: Version history exists, rollback is possible, and post-update regression checks are performed

Initials: _____ Date: _____

F. Procurement and third-party assurance readiness

16. Owner RFP Specifications reference pack ready (App. J)
    Pass criteria: Procurement clauses and acceptance tests are aligned to the deployed scope
17. Third-Party Assurance Packet Outline ready (App. I)
    Pass criteria: An external reviewer can be given a bounded package without disclosing sensitive implementation details
18. Audit access posture defined
    Pass criteria: Who can access logs and evidence, how access is granted, and how long artifacts are retained are defined.

Initials: _____ Date: _____

Acceptance outcome

Overall status (circle one): PASS / CONDITIONAL PASS / FAIL

Conditional pass items and remediation plan:

_____

_____

_____

Reviewed by (name / role): _____

Date: _____

APPENDIX L
DETAILED TABLE OF CONTENTS

INDEX

Portable Identity Layer, see also Web3
Consented Context Layer, see also Web4
Agent Coordination Layer, see also Web5
AI Negotiation Layer, see also Web6

Web3, see Portable Identity Layer
Web4, see Consented Context Layer
Web5, see Agent Coordination Layer
Web6, see AI Negotiation Layer

"SEE ALSO" CROSS-REFERENCES

COMPANION RESOURCES

Detailed diagrams, downloadable procurement checklists, stage assessment tools, and updates to the framework are available at:

worldmodel.global

The World Model

Complete reference standard with architecture specifications, governance patterns, and procurement criteria

A NOTE FROM THE AUTHOR

This book exists because of a sequence of lessons that would not stop repeating.

For years, we built venues using both traditional black-box systems and node-based compute. Black boxes can be effective inside a tight scope, but they harden assumptions, tie capability to hardware refresh cycles, and make change feel like replacement. Node compute shifts the economics because capability becomes software, and software can evolve without tearing out the venue.

Then we crossed a threshold: we integrated facial recognition into the platform. That did not just add a feature. It created entirely new possibilities and forced an interaction mediator to exist explicitly. I started thinking of that as a Concierge layer. It was always compelling, but it was also limited by what the system could infer, coordinate, and adapt in real time.

When we began integrating AI into personalization, and pushed into different forms of hyper-personalization, responsibility stopped being abstract. The same capabilities that reduce friction and improve inclusion can also misbehave, quietly, and at scale. As soon as we started thinking seriously about scaling, it became obvious that misbehavior prevention, privacy posture, and operational accountability were not optional add-ons. They were requirements.

That was the moment the Concierge layer returned with clarity. It had to become a governed mediator between people, spaces, and systems. Recognition and privacy had to coexist. Scaling had to be explicit. Responsibility demanded governance. That path pulled in Web3-style privacy options for controlled disclosure, then expanded into feedback loops such as emotion feedback and emotion feed-forward, and then into the reality that many interactions will become agent-to-agent, AI to AI, while still preserving privacy and dignity.

From there, the implications spread across verticals, into Digital Twins, and ultimately into the WorldModel™ and everything it now entails: a way to preserve capabilities at different levels for different use cases, while holding the overall system to consent boundaries, minimization, inclusion defaults, and governance that does not degrade as complexity grows.

This book ended up with forty-two chapters.

That was not planned. Anyone familiar with Douglas Adams will already have noticed, and anyone else can safely ignore it. There is no claim here about the Answer to Life, the Universe, and Everything, although I do hope that this book provides *some* answers.

What the number does reflect is more mundane, and more important. Once you start describing what it actually takes to build governed, adaptive, multilingual, privacy-respecting environments at scale, the subject refuses to collapse into a shorter form. Static systems fit neatly into outlines. Living systems do not.

Forty-two is simply where the material stopped arguing back. What a coincidence.

If any of this resonates with the challenges you are facing, I would welcome the conversation.

Maris J. Ensing, February 2026
maris@worldmodel.global

ABOUT THE AUTHOR

Maris J. Ensing is the founder and chief architect of Mad Systems, based in Orange, California, a technology company building governed venue intelligence for public environments. His work sits at the intersection of experiential systems, human-centered design, and operational infrastructure, focused on making hyper-personalization reliable, privacy-respecting, and auditable at venue scale.

Mad Systems is also a full-scope audiovisual systems integrator and interactive experiences developer, delivering complex projects from design through commissioning and long-term support. This delivery capability is not separate from the architecture. It is how the work is proven in real rooms, under real constraints, and made operable over years.

Mad Systems' architecture addresses the hard deployment constraints that determine whether AI-driven experiences survive real operations: multilingual and multi-modal delivery, accessibility-first interaction, preference continuity without default identification, explicit governance gates for consequential decisions, and lifecycle discipline that keeps systems maintainable through upgrades.

He is a patented systems architect and patent holder with issued patents and pending patent applications covering hyper-personalized media delivery, visitor recognition, and intelligent venue systems, with filings and grants spanning multiple jurisdictions, including the United States, China, Hong Kong, the Kingdom of Saudi Arabia, and the United Arab Emirates.

For interviews and professional correspondence, please contact maris@worldmodel.global. For speaking invitations, please refer to marisensing.com.

ENDNOTES AND SOURCES

Use independent sources for factual claims. Supplemental examples and capability-to-implementation mapping may be provided separately by the author and are not part of this publication.

---

[i] National Institute of Standards and Technology (NIST), "Artificial Intelligence Risk Management Framework: Generative Artificial Intelligence Profile," NIST AI 600-1 (2024). (Confabulation terminology and GenAI risk framing.)

https://nvlpubs.nist.gov/nistpubs/ai/NIST.AI.600-1.pdf

[ii] NIST AI 600-1 (Generative AI Profile), "confabulation" risk terminology.

[iii] NIST, "Artificial Intelligence Risk Management Framework (AI RMF 1.0)," NIST AI 100-1 (2023).
https://nvlpubs.nist.gov/nistpubs/ai/nist.ai.100-1.pdf

NIST, "Artificial Intelligence Risk Management Framework: Generative Artificial Intelligence Profile," NIST AI 600-1 (2024).
https://nvlpubs.nist.gov/nistpubs/ai/NIST.AI.600-1.pdf

ISO, "ISO/IEC 42001:2023, AI management systems."
https://www.iso.org/standard/42001

OWASP, "Top 10 for Large Language Model Applications."
https://owasp.org/www-project-top-10-for-large-language-model-applications/

OECD, "Recommendation of the Council on Artificial Intelligence (OECD/LEGAL/0449)."
https://legalinstruments.oecd.org/en/instruments/oecd-legal-0449

W3C, "Web Content Accessibility Guidelines (WCAG) 2.2."
https://www.w3.org/TR/WCAG22/

W3C, "Decentralized Identifiers (DIDs) v1.0 (DID Core)."
https://www.w3.org/TR/did-core/

W3C, "Verifiable Credentials Data Model v2.0," W3C Recommendation (15 May 2025).
https://www.w3.org/TR/vc-data-model-2.0/

NIST, "Digital Identity Guidelines (SP 800-63-4)."
https://pages.nist.gov/800-63-4/

NIST, "Face Recognition Vendor Test (FRVT), Part 3: Demographic Effects," NISTIR 8280 (2019).
https://nvlpubs.nist.gov/nistpubs/ir/2019/nist.ir.8280.pdf

U.S. Food and Drug Administration (FDA), "OTC Hearing Aids: What You Should Know" (effective date noted as October 17, 2022).
https://www.fda.gov/medical-devices/hearing-aids/otc-hearing-aids-what-you-should-know

[iv] NIST AI 600-1, confabulation as a GenAI risk, with "hallucination" as the common label.

[v] NIST AI 600-1, confabulation risk, and recommended mitigations at the system level.

[vi] Ernst & Young describes a destination technology posture centered on unified operational data (a "single source of truth"), advanced analytics, and automated decisions tied to guest experience and operational efficiency in theme parks. Themed Entertainment Association and AECOM publish the annual Theme Index and Museum Index report, a widely used industry benchmark tracking attendance and performance across leading themed entertainment and cultural destinations.
https://www.ey.com/en_us/industries/media-entertainment/unleashing-theme-park-technology-transformation
https://aecom.com/theme-index/

[vii] NIST AI 600-1, confabulation risk terminology.

viii Pew Research Center, "Mobile Fact Sheet," updated Nov 20, 2025.
https://www.pewresearch.org/internet/fact-sheet/mobile/

ix Pew Research Center, "Mobile Fact Sheet," updated Nov 20, 2025.
https://www.pewresearch.org/internet/fact-sheet/mobile/

x Apple Newsroom, "iPhone Premieres This Friday Night at Apple Retail Stores," Jun 28, 2007.
https://www.apple.com/newsroom/2007/06/28iPhone-Premieres-This-Friday-Night-at-Apple-Retail-Stores/

xi Acoustical Society of America, "Enhancing Museum Experiences: The Impact of Sounds on Visitor Perception," Nov 27, 2023.
https://acoustics.org/enhancing-museum-experiences-the-impact-of-sounds-on-visitor-perception/

xii Bottalico P., Astolfi A., Shtrepi L., "Lombard effect, ambient noise, and willingness to spend time and money in a restaurant," Journal of the Acoustical Society of America (2018). (PubMed record.)
https://pubmed.ncbi.nlm.nih.gov/30424648/

xiii NIDCD, "Age-Related Hearing Loss (Presbycusis)," Mar 17, 2023.
https://www.nidcd.nih.gov/health/age-related-hearing-loss

xiv International Electrotechnical Commission (IEC), "IEC 60268-16:2020, Sound system equipment – Part 16: Objective rating of speech intelligibility by speech transmission index."
https://webstore.iec.ch/en/publication/26771

xv Bluetooth SIG, An Overview of Auracast™ Broadcast Audio, May 2024.
https://www.bluetooth.com/wp-content/uploads/2024/05/2403_Auracast_Overview.pdf

xvi ISO 13850:2015, Safety of machinery, Emergency stop function, Principles for design, International Organization for Standardization (ISO), Edition 3, 2015; and IEC 60204-1:2016+AMD1:2021 CSV, Safety of machinery, Electrical equipment of machines, Part 1: General requirements, International Electrotechnical Commission (IEC).

xvii U.S. Food and Drug Administration (FDA), "OTC Hearing Aids: What You Should Know" (effective date: Oct 17, 2022).
https://www.fda.gov/medical-devices/hearing-aids/otc-hearing-aids-what-you-should-know

xviii National Institute on Deafness and Other Communication Disorders (NIDCD), "Age-Related Hearing Loss," PDF (states ~15% of American adults ages 18 and over report some

trouble hearing).
https://www.nidcd.nih.gov/sites/default/files/age-related-hearing-loss.pdf

xix U.S. Food and Drug Administration (FDA), "OTC Hearing Aids: What You Should Know" (effective date: Oct 17, 2022).
https://www.fda.gov/medical-devices/hearing-aids/otc-hearing-aids-what-you-should-know

xx National Geographic Education, "Sign Language," Apr 9, 2024.
https://education.nationalgeographic.org/resource/sign-language/

xxi Apple Support, "Use a Braille display with VoiceOver on iPad."
https://support.apple.com/guide/ipad/use-a-braille-display-ipad9a246499/ipados

xxii Apple Support, "Use a braille display with VoiceOver on iPad."
https://support.apple.com/guide/ipad/use-a-braille-display-ipad9a246499/ipados

xxiii Google, Android Accessibility Help, "Change braille display settings."
https://support.google.com/accessibility/android/answer/3535946

xxiv City of Los Angeles, "Report on FY 2023-24 Citywide Language Access Annual Report," May 2025 (PDF).
https://cityclerk.lacity.org/onlinedocs/2022/22-1262-S1_rpt_cifd_5-8-25.pdf

xxv NIST, "Face Recognition Vendor Test (FRVT) Part 3: Demographic Effects," NISTIR 8280, 2019.
https://www.nist.gov/publications/face-recognition-vendor-test-part-3-demographic-effects

Illinois General Assembly, Public Act 103-0769 (SB2979) amending the Biometric Information Privacy Act (BIPA), signed Aug 2, 2024.
https://www.ilga.gov/Legislation/publicacts/view/103-0769

xxvi ISO/IEC 9241-210:2019, *Human-centred design for interactive systems*. International Organization for Standardization.
https://www.iso.org/standard/77520.html

xxvii Apple Support, "Use private Wi-Fi addresses on Apple devices," updated Dec 5, 2025.
https://support.apple.com/en-us/102509

Android Open Source Project, "MAC randomization behavior."
https://source.android.com/docs/core/connect/wifi-mac-randomization-behavior

xxviii Apple Support, "Use private Wi-Fi addresses on Apple devices," updated Dec 5, 2025.
https://support.apple.com/en-us/102509

Android Open Source Project, "MAC randomization behavior," updated Dec 2, 2025.
https://source.android.com/docs/core/connect/wifi-mac-randomization-behavior

xxix National Institute of Standards and Technology (NIST), "Face Recognition Vendor
Test (FRVT), Part 3: Demographic Effects," NISTIR 8280 (2019).
https://nvlpubs.nist.gov/nistpubs/ir/2019/nist.ir.8280.pdf

xxx Illinois General Assembly, Illinois Compiled Statutes, "Biometric Information Privacy
Act (740 ILCS 14/)."
https://www.ilga.gov/Legislation/ILCS/Articles?ActID=3004&ChapterID=57&Print=True

xxxi EY, "How digital twin technology and AI can reimagine theme park experiences," Sep 4,
2025.
https://www.ey.com/en_us/industries/media-entertainment/unleashing-theme-park-
technology-transformation

xxxii NIST, "Artificial Intelligence Risk Management Framework (AI RMF 1.0)," NIST AI
100-1, Jan 2023.
https://doi.org/10.6028/NIST.AI.100-1

xxxiii NFPA 72-2025, National Fire Alarm and Signaling Code, National Fire Protection
Association (NFPA).

xxxiv NFPA 101-2024, Life Safety Code, National Fire Protection Association (NFPA).

xxxv ISO 13850:2015, Safety of machinery, Emergency stop function, Principles for design,
International Organization for Standardization (ISO), Edition 3, 2015.

xxxvi ASTM F2291-23B, Standard Practice for Design of Amusement Rides and Devices,
ASTM International.

xxxvii Infocomm Media Development Authority (IMDA). Model AI Governance Framework
for Agentic AI, Version 1.0, January 2026.

xxxviii NIST, Artificial Intelligence Risk Management Framework (AI RMF 1.0), NIST AI 100-1, Jan 2023.
https://nvlpubs.nist.gov/nistpubs/ai/nist.ai.100-1.pdf

xxxix W3C Recommendation, "Decentralized Identifiers (DIDs) v1.0."
https://www.w3.org/TR/did-1.0/

W3C Recommendation, "Verifiable Credentials Data Model v2.0," May 15, 2025.
https://www.w3.org/TR/vc-data-model-2.0/

xl W3C, "Verifiable Credentials Use Cases," Sep 24, 2019.
https://www.w3.org/TR/vc-use-cases/

xli National Institute of Standards and Technology (NIST), "Digital Identity Guidelines (SP 800-63-4)."
https://pages.nist.gov/800-63-4/

xlii W3C, "Verifiable Credentials Use Cases," Sep 24, 2019.
https://www.w3.org/TR/vc-use-cases/

xliii W3C, "Verifiable Credentials Data Model v2.0," W3C Recommendation (15 May 2025).
https://www.w3.org/TR/vc-data-model-2.0/

xliv W3C, "Verifiable Credentials Implementation Guidelines 1.0."
https://www.w3.org/TR/vc-imp-guide/

xlv W3C, "Decentralized Identifiers (DIDs) v1.0 (DID Core)."
https://www.w3.org/TR/did-core/

xlvi Decentralized Identity Foundation, "Decentralized Web Node (DWN) Specification."
https://identity.foundation/decentralized-web-node/spec/

Decentralized Identity Foundation, "Decentralized Web Node Companion Guide."
https://identity.foundation/decentralized-web-node/guide/

xlvii NIST, "Privacy Framework: A Tool for Improving Privacy through Enterprise Risk Management (Version 1.0)," Jan 16, 2020 (PDF).
https://nvlpubs.nist.gov/nistpubs/CSWP/NIST.CSWP.01162020.pdf

^{xlviii} IBM, "Condition monitoring."
https://www.ibm.com/think/topics/condition-monitoring

Epson, "Laser Projector Solutions."
https://epson.com/laser-projectors

Epson Support, "Cleaning the Air Intake Vents."
https://files.support.epson.com/docid/cpd6/cpd65128/EN/Maintenance/Tasks/cleaning_air
_intake_vent.html

^{xlix} NIST, "Artificial Intelligence Risk Management Framework (AI RMF 1.0)," NIST AI
100-1, Jan 2023.
https://doi.org/10.6028/NIST.AI.100-1

^l NIST, "Artificial Intelligence Risk Management Framework (AI RMF 1.0)," NIST AI
100-1, Jan 2023.
https://doi.org/10.6028/NIST.AI.100-1

^{li} IEC 60204-1:2016+AMD1:2021 CSV, Safety of machinery, Electrical equipment of
machines, Part 1: General requirements, International Electrotechnical Commission (IEC).

^{lii} ISO 13850:2015, Safety of machinery, Emergency stop function, Principles for design,
International Organization for Standardization (ISO), Edition 3, 2015.

^{liii} ISO, "ISO/IEC 42001:2023 - Artificial intelligence management system."
https://www.iso.org/standard/42001

^{liv} NIST, "Artificial Intelligence Risk Management Framework (AI RMF 1.0)," NIST AI
100 1, January 2023.
https://doi.org/10.6028/NIST.AI.100-1

^{lv} NIST AI 600-1 (Generative AI Profile).
https://nvlpubs.nist.gov/nistpubs/ai/NIST.AI.600-1.pdf

^{lvi} OWASP Foundation, "OWASP Top 10 for Large Language Model Applications."
https://owasp.org/www-project-top-10-for-large-language-model-applications/

^{lvii} European Union, "Regulation (EU) 2024/1689 of the European Parliament and of the
Council of 13 June 2024 laying down harmonised rules on artificial intelligence (Artificial
Intelligence Act)," Official Journal, 12 July 2024.
https://eur-lex.europa.eu/eli/reg/2024/1689/oj/eng

lviii UK Department for Science, Innovation and Technology, "Trusted third-party AI assurance roadmap," September 3, 2025.
https://www.gov.uk/government/publications/trusted-third-party-ai-assurance-roadmap/trusted-third-party-ai-assurance-roadmap

lix OECD. Recommendation of the Council on Artificial Intelligence. OECD/LEGAL/0449 (2019).
https://legalinstruments.oecd.org/en/instruments/oecd-legal-0449

www.ingramcontent.com/pod-product-compliance
Lightning Source LLC
Chambersburg PA
CBHW081341190326
41458CB00018B/6063